Characterisation of Bulk Solids

Characterisation of Bulk Solids

Edited by

Don McGlinchey
Centre for Industrial Bulk Solids Handling
Glasgow Caledonian University, UK

Blackwell
Publishing

CRC Press

Editorial offices:
Blackwell Publishing Ltd, 9600 Garsington Road, Oxford OX4 2DQ, UK
Tel: +44 (0) 1865 776868
Blackwell Publishing Asia Pty Ltd, 550 Swanston Street,
Carlton, Victoria 3053, Australia
Tel: +61 (0) 3 8359 1011

ISBN-10 1-4051-1624-2
ISBN-13 978-14051-1624-4

Published in the USA and Canada (only) by CRC Press LLC, 2000 Corporate Blvd., N.W., Boca Raton, FL 33431, USA
Orders from the USA and Canada (only) to CRC Press LLC

USA and Canada only:
ISBN 0-8493-2437-8

First published 2005

Library of Congress Cataloging-in-Publication Data:
A catalog record for this title is available from the Library of Congress

British Library Cataloguing-in-Publication Data:
A catalogue record for this title is available from the British Library

Set in 10/12 pt Times
by TechBooks
Printed and bound in India
by Gopsons Papers Ltd, Noida

The publisher's policy is to use permanent paper from mills that operate a sustainable forestry policy, and which has been manufactured from pulp processed using acid-free and elementary chlorine-free practices. Furthermore, the publisher ensures that the text paper and cover board used have met acceptable environmental accreditation standards.

For further information on Blackwell Publishing, visit our website:
www.blackwellpublishing.com

Contents

Contributors

L. Bates Ajax Equipment Ltd, Milton Works, Mule Street, Bolton, BL2 2AR, UK

D. Geldart Powder Research Ltd, 7 Westminster Gate, Harrogate, HG3 1LU, UK

M. Jones Division of Mechanical Engineering, The University of Newcastle, Callaghan Campus, Callaghan, NSW 2308, Australia

G. Lunn BASEEFA (2001) Ltd, Health and Safety Laboratory Site, Harpur Hill, Buxton, SK17 9JN, Derbyshire, UK

D. McGlinchey School of Engineering, Science and Design, Glasgow Caledonian University, Cowcaddens Road, Glasgow, G4 0BA, UK

A. Roberts Division of Mechanical Engineering, The University of Newcastle, Callaghan Campus, Callaghan, NSW 2308, Australia

N. Stanley-Wood SSW Partnership, St Peter Port, Guernsey, GY1 1RB, UK

T. Taylor Taymark Engineering, 19 Chantry Avenue, Hartford, Northwich, Cheshire, CW8 1LZ, UK

Preface

Characterisation of Bulk Solids aims to provide the reader with the breadth of knowledge to give a good general understanding of particulate materials characterisation, its importance and application. This should place the reader in a better position to be able to diagnose solids handling and processing problems in industry, and to deal with experts and equipment suppliers from an informed standpoint. It would also serve well as a primer in any of the specialist areas to inform further study.

The book covers the fundamental characteristics of individual particles and bulk particulate materials, including particle size, shape and density, bulk density, fluidisation behaviour, flow properties, and the characterisation of powders for explosion potential. Measurement techniques are described, and the use of material characteristics in design and industrial practice is discussed. The way characterisation data are currently used in industry is also considered.

The readership is expected to be post-graduate engineers, scientists and technologists, most likely chemists, chemical or mechanical engineers, or physicists. They may be young engineers or 'role changers' with little under-graduate education in particle technology or bulk solids handling who have responsibility for processing or production involving particulate materials. Alternatively, scientists or engineers embarking on post-graduate academic studies will find the book a valuable resource of referenced articles by leading authors in their respective fields.

The material contained in the book is designed to be equally applicable to engineers and scientists working in a broad spectrum of industries, such as agriculture, agrochemicals, cement, construction, food, bulk and fine chemicals industries, minerals and metals, petrochemicals, pharmaceuticals, plastics, pigments, power generation and waste handling.

Acknowledgements

I wish to express my thanks to all the authors who took the time from their very busy schedules to make such significant contributions to this book. It has been a great pleasure, opportunity and responsibility to be part of this enterprise alongside some of the best known and respected experts in the field of bulk solids handling. I would like to thank Paul Sayer and all at Blackwell Publishing for their patience and skill in dealing with a novice editor. I would also like to thank colleagues at Glasgow Caledonian University, Andrew Cowell, Elizabeth Knight, John Pugh and Ray Ansell, who all helped in the formation of this book.

1 Particle characterisation in bulk powders

NAYLAND STANLEY-WOOD

1.1 Introduction

Surveys have shown that the production of powdered material had an annual value in excess of $10 billion (Waldie 1980), whilst others have shown that an estimated 25–30% of the total output from the chemical industries was in the physical form of either powdered or agglomerated material (Oliver & Ford 1987; Stanley-Wood *et al.* 1993; Waldie 1989). The characterisation of the particulate material produced, together with the handling and behaviour of powdered material in bulk quantities, is therefore of paramount importance in the optimisation of many multiphasic and multiparticulate processes.

The emergence of the modern-day discipline of powder technology, and thus the measurement of irregularly shaped particles, occurred at approximately the same time in America, Europe and the UK during the late 1940s and early 1950s. However, according to a plenary lecture on 'The origins and development of particle size analysis' given by Harold Heywood at a conference on particle-size analysis (PSA 70) (Groves & Wyatt-Sargent 1970), it is believed that powder handling and classification of natural occurring sands and mineral ores can be traced to Egyptian times (Singer *et al.* 1954–1956). The art of sieving, in more 'modern' times, together with the settling of mined ores and the technical development of mineral dressings, has been vividly illustrated by Donald, in his history of Elizabethan copper (Donald 1989). This traces the formation of one of the first English companies to be granted legality to manufacture an article – copper – as opposed to a company whose sole reason for existence was to trade abroad.

One of the first publications dedicated solely to outlining the various mechanico-physical techniques for particle measurement was written by DallaValle (1943). This was closely followed by Herdan (1952), who concentrated on the conceptualisation of the statistics of particles by the combination of both number statistics and physical statistics, as seen in colloidal systems, to give rise to the new discipline of 'small particle statistics'.

Rumpf (1990), however, was concerned with process technology in terms of the transformation of matter by different mechanisms. In chemical process engineering, transformation of matter is deemed to occur by chemical reaction, while thermal process engineering is regarded as the transformation of thermodynamically defined phases. In both cases, the system parameters of pressure, temperature and composition follow the laws of thermodynamics. Rumpf presupposed that there was a third engineering technology which could characterise changes in the state of disperse systems and of mixtures consisting of particulate matter and continuous fluid phases. Rumpf created an area of academic and industrial interest, termed 'mechanical process technology', which was a forerunner of the now known discipline of *powder technology*. In reality, although a mechanico-physical unit processing operation may precede, and at times terminate, a chemical and thermodynamic unit processing operation, all three technologies are inter-correlated and essential to the transformation of matter into a desirable end product.

The disciplines of powder characterisation and powder technology, which initially emanated from the UK and continental Europe, have now become global in their ramifications and acceptability. The forms of particulate matter generated, fabricated, produced and processed nowadays have caused a dramatic advancement in the type of instruments available for particle-size measurement.

1.1.1 *Instruments for particle characterisation*

The wide range of different physical, optical, electromagnetic and, at times, electrochemical measurement systems currently used is partly due to the influx of computer technology. In recent years, the amount of, and the speed at which, measurement data can be collected by new and novel measurement systems, connected to plant items, has opened up new frontiers to the powder technologist (Trachtler & Mesch 1998). Computer-assisted particle-measurement instrumentation not only aids the collection of data but also places in the hands of the powder technologist the possibility of advanced statistical analysis of data gained from real-time measurement systems. The integration of computers to control and automate processing plants, be it chemical, mechanical or powder bulk handling, is extending the original role of off-line laboratory particle measurements to one in which measurement of particle sizes may be made on-line, in-line and in real time within a chemical processing plant. The ultimate measurement system, which may not be dependent solely on the measurement of particle size, would be an *in situ* array of sensors, within a reactor or processing unit. Such a development could detect and characterise one or more particle parameters to control a dynamic particulate process and produce a specifically desired end product. Current research, being now multi-disciplinary as well as multi-authored, is adopting a wider selection of physical criteria than that initially used for off-line laboratory analysis, and through the use of ultrasonic, optical and tomographic sensing techniques, advances have been made to implement in-line particle characterisation into the armoury of the particle/powder engineer (Scott *et al.* 1998).

1.1.2 *Functionality and particle characterisation*

The choice of a particle-size measurement technique is dependent upon the accuracy of the measurement which is required and the circumstances to which the measurements are to be applied. This rationale was forwarded and emphasised by Heywood (1937, 1954), who was regarded as a founding father of powder technology in the UK, with the words (Heywood 1966):

> *'However, it must be realised that particle-size analysis is not an objective in itself but is a means to an end, the end being the correlation of powder properties with some process of manufacture, usage or preparation'.*

This pronouncement has also been given by Herdan (1952), in *Small Particle Statistics*, who stated about 50 years ago, that:

> *'Knowledge of the size characteristics of particulate matter is generally of little value in itself; but particle-size measurements are often made to control the quality of the final product, because certain sizes may be correlated with certain desirable properties of the product'.*

It should be noted, however, that particle-size measurement by itself has little or no meaning unless the reason for the particle measurement is also specified.

1.1.3 *Selection of instrumentation*

One of the most important factors which must be borne in mind when approaching any particle-size analysis is that, although there are many ingenious, sensitive, reliable and repeatable techniques, no one method can be regarded as perfect. A compromise usually has to be made, in terms of cost and time available, to achieve the best particle characteristic or measurement of particle size. To these ends, a number of techniques have been chosen to illustrate the diversity that exists and can now be used by the powder technologist.

A particle primarily refers to an object which has a physical boundary regardless of the finite size. Because a particle is a discrete portion of matter which is usually regarded as being small in relation to the space in which it is encompassed, the upper size limit of an individual particle is therefore arbitrary. With this definition, planets, rocks, boulders, aggregates, powdered particles, colloids, molecules and sub-molecular and atomic entities may all be regarded as being technologically important to a powder technologist.

For practical purposes, however, an arbitrary definition of the size of powder particles has been chosen and, as stated in a British Standard (British Standards 1993), is defined as: '*A powder shall consist of discrete particles of any material with a maximum dimension of less than 1000 μm (= 1 mm)*'.

The lower limit these days, because of the introduction of ultracentrifugation, ultrasonics, light scattering and X-ray scattering, is commonly taken to be in the nanometre to colloidal dimensional range. Since a particle can be defined and described as a discrete entity of matter, it is possible to obtain size distributions of populations of solid, liquid or gaseous particles in solid, liquid and gaseous phases by using an appropriate measurement technique.

Although particle size is an important factor in the determination of how a particle will behave, the measurement of that irregularly shaped particle characteristic, dependent on the final objective of the measurement, may be achieved by a number of physical techniques. It is unusual, however, to solely measure one particle; thus, particle sizing and powder characterisation generally measure many particles, and the results are described in terms of a particle-size distribution. Thus, a particle-size distribution may be described in terms of the size and frequency of particles in a population or as the functional relationship between a quantity of particles and a dimensional measurement of the irregularly shaped particles.

1.1.4 *Particle averages and equivalent spherical diameter*

When the term 'particle size' is used, ambiguities can arise because comparison of the magnitude of the particle size dimension of an irregularly shaped particle obtained by one measurement technique with the magnitude of the particle size obtained by a different physical measurement technique may not be identical; there is no single definition of particle size. Thus, there is no theoretical justification why any irregularly shaped particle measured by two different techniques should be comparable. Only spherical particles would be comparable. This non-equivalence with irregular particles can be advantageous only if used to describe the shape of particles but has no merit when trying to achieve a consensus on the dimension or size of an irregularly shaped particle.

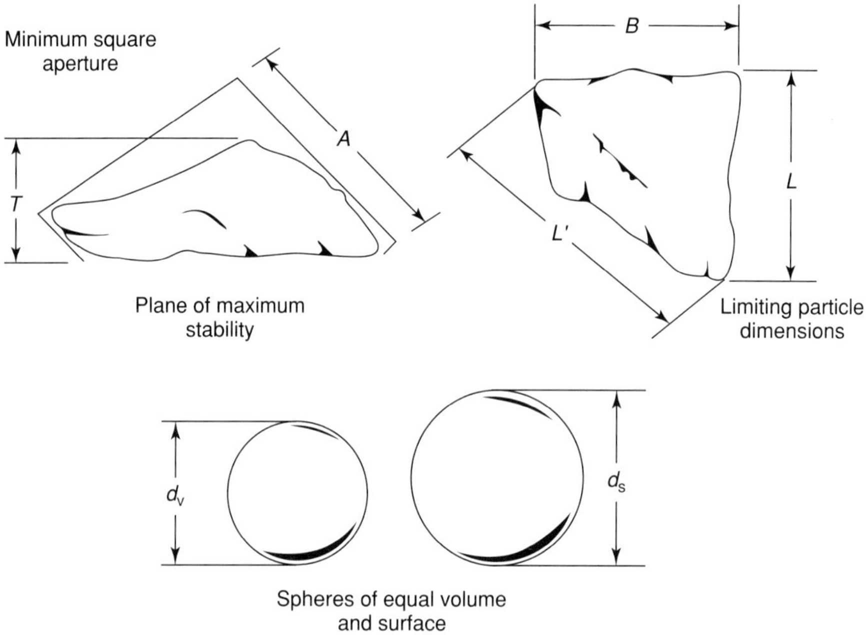

Figure 1.1 Equivalent particle diameters (after Heywood).

To overcome this potential confusion when 'particle size' is not fully notated, an Expert Group, under the auspices of the Bureau of Certified Reference (1980), undertook a certification campaign to measure five powdered reference materials in the size range 0.1–1000 μm. Consensus was reached that the classical treatment of size should be based, not solely on physical criteria, but on the concept of the diameter of an equivalent sphere and to use the traceable fundamental units of length, mass and time.

The size of the irregularly shaped BCR reference material particles was subsequently defined as the diameter of a sphere, which is equivalent to the particle's volume (volume diameter) or to its settling rate in a viscous fluid (Stokes diameter). An irregularly shaped particle has no unique dimension, and its size should be expressed in terms of the diameter of a circle or a sphere which is equivalent to the irregularly shaped particle in some stated particle property, *the equivalent spherical diameter* (Figure 1.1).

Although there are other definitions for particle size such as those based on the openings in a sieve or a statistical diameter measured by image analysis (Feret diameter), it should be mandatory that whenever the size of a particle is stated, an indication is given of how that particle size has been defined and measured.

If the method used to size particles is a mechanical sieving technique, the particle dimension is a sieve size, while if a sedimentation technique is used, the particle dimension obtained can be calculated and described in a number of ways; a free fall diameter, a drag diameter or a Stokes diameter. BS ISO Standard 9276-1 (1995) *Representation of results of particle size analysis – Part 1: Graphical representation* (ISO 1995a) has recommended that the symbol 'x' be used for particle size because the term 'd', for diameter, infers a spherical object.

1.1.5 *Particle-size distributions*

In BS ISO 9276-2 (2001) *Calculation of average particle sizes/diameter and moments from particle size distributions* (ISO 1998), a unique definition of average size, derived from the moments of a size distribution, is given as:

$$\text{Average particle size/diameter} = x_{k,r}, \tag{1.1}$$

where subscript r is the type or quantity of the distribution, and subscript k is the power of the moment of size x or the kth moment of the density distribution. A number of combinations can therefore be obtained, which shows the wealth of averages that can be calculated (Table 1.1). This elegant nomenclature is most useful as it illustrates that a distribution can have a number of 'average' sizes. Not all the theoretically possible averages have, however, a practical use in powder technology or bulk powder processing and handling. These single averaged or weighted parameters can thus convey, by a single number, the order of the particle size involved within a distribution, but such a single parameter gives no indication of the spread of the distribution. Powders can, however, have more than one modal value and particle-size distributions can be mono-, bi-, tri- and even multi-modal.

The distribution is usually determined either as a number distribution or as a distribution by volume or weight, and the information may be displayed graphically either as a histogram or as a continuous distribution curve. A cumulative distribution curve is the most common

Table 1.1 Calculation of averages and weighted averages (from BS ISO 9276-2 2001).

Average particle size/diameter ($x_{k,r}$)
Depending on the chosen number of the subscript ($k = -3, -2, -1, 0, 1, 2, 3$ and $r = 0, 1, 2, 3$) different average particle diameters can exist. The arithmetic mean particle diameters are, however, always based on a number density distribution thus $r = 0$.

Arithmetic average particle diameter ($x_{k,0}$)
The recommended arithmetic average particle diameters are:

$x_{1,0}$	$x_{2,0}$	$x_{3,0}$
Arithmetic average length	Arithmetic average surface	Arithmetic average volume

Weighted average particle diameter ($x_{1,r}$)
k is the power of the dimension used to characterise or measure the particle size, x, which can be a volume (x^3), a surface (x^2), a length (x^1), a number (x^0), a reciprocal length (x^{-1}), a reciprocal surface (x^{-2}) or a reciprocal volume (x^{-3}).
The recommended weighted average particle diameters for a number weighted average can be:

$x_{1,0}$	$x_{1,1}$	$x_{1,2}$	$x_{1,3}$
Number density	Length density	Surface density Sauter diameter	Volume density

Thus, for a volume weighted average:

$x_{3,0}$	$x_{3,1}$	$x_{3,2}$	$x_{3,3}$
Number–volume	Length–volume	Surface–volume	Volume–volume (aka mass average)

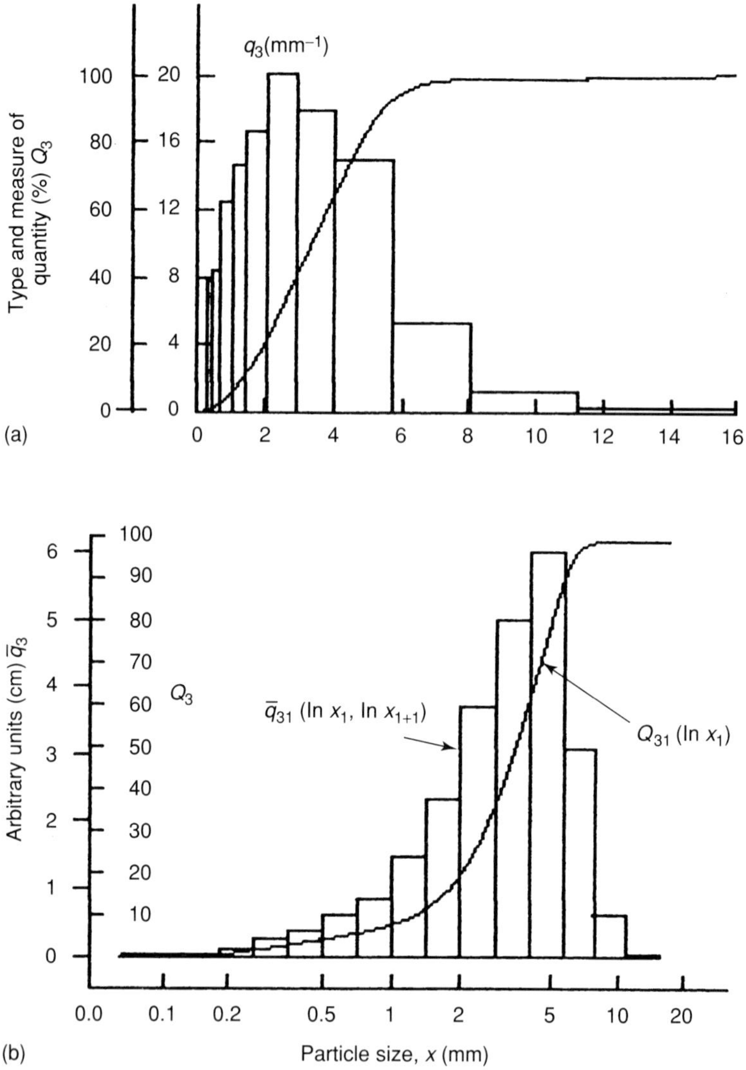

Figure 1.2 Graphical representation of particle size from ISO 9276-1.

way because the curves can be more readily interpolated and normalised. ISO 9276-1 recommends that measurement of a distribution be graphically displayed as either a discrete density distribution, q, or as a cumulative distribution, Q. The word density, in ISO 9276-1, denotes a statistical parameter akin to the frequency of an event in number statistics and should not be confused with the physical property of mass per unit volume (Figure 1.2).

1.2 Metrology of particle characterisation

In considering particle-measuring techniques, the first major distinction that should be made is between direct and indirect methods of measurement. If the particles can be directly observed by a static microscope or an electron microscope, then there is an indefinite amount of information which may be obtained from the image of the particles. However, there is a limitation on the number of particles which can be analysed in a reasonable time, and so image analysis techniques are usually used for more sophisticated experimental and research analyses.

Currently, there is a diversity of equipment commercially available which is based on relatively few physical principles. However, with the advent of nanotechnology and concern over the emission of particles into the environment, new physical principles, such as differential mobility analysers, small-angle X-ray scattering and ultrafine condensation nucleus counters, may have to be considered for the future measurement of nanoparticles as well as the modification of accepted metrology to discriminate between measured and monitored micrometre and nanometre particle sizes.

The methods that are commonly used these days for characterising industrial powders, off-line and in some cases on-line, are: sieving; sedimentation; electrical sensing zone; light and laser scattering; photon correlation spectroscopy; light and ultrasonic attenuation. At present, the first three are established off-line techniques (Table 1.2 and Figure 1.3) which are well tried and developed and from which reproducible and replicated results can be obtained. They are, however, less amenable to automation and on-line application. The last three methods are, at present, generally off-line measurement techniques but are amenable to further technological modifications for use in automated powder processing plants (Stanley-Wood 2000).

Table 1.2 Measurement methods in common use.

1. Direct dimensional measurement
Sieving and microscopy
2. Transport measurements
Sedimentation: Gravity, centrifugal and ultracentrifugation
3. Rapid physical response measurement
Electrical sensing zone
Light scattering (optical and laser)
Light diffraction
Photon correlation spectroscopy
Light blockage
Ultrasonics
4. Surface area and porosity
Permeametry
Gas adsorption (N_2, Kr, Ar, CO_2 and water)

1.3 Selected particle characterisation instruments

1.3.1 *Direct dimensional measurements: General principles*

This is typified by the microscopy and sieving techniques which are generally applied to particulate systems in the size range 10–25 000 μm. Particles undergoing size analysis by sieving have to be robust to withstand the sieve action which can cause attrition when a ‘dry’ sieving technique is used. For fragile particles, there is a ‘wet’ technique in which water or organic solvents may be used to support delicate materials, whenever wire and electroformed sieves are used. The particles must, however, be insoluble in the suspending fluid.

Physical criterion	Dimension measured	Size range
Direct dimensional measurement.		
Static image 	Projected area	Optical: 0.8 - 800 μm Electron: 1.0 - 15000 nm
Dynamic image	Size and Shape	30 μm - 30 mm
Mechanical 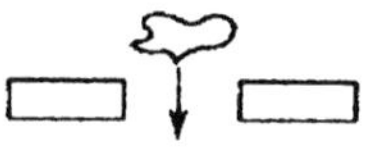	Sieve diameter	Dry : 40 - 1000+ μm Wet : 1.0 - 40 μm
Transport measurements.		
Dynamic 	Free fall Drag diameter Stokes diameter	Gravity : 10 - 100 μm Centrifugal : 0.05 - 50 μm Ultracentrifugal : < 50 nm
Photozone (Light blockage) 	Projected area Stokes diameter	Light : 10 - 100 μm X-ray : 0.05 - 100 μm Centrifugal : 0.01+ μm
Rapid response measurements.		
Electrical sensing zone 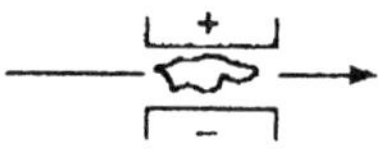	Volume diameter	0.6 - 2000 μm
Light diffraction 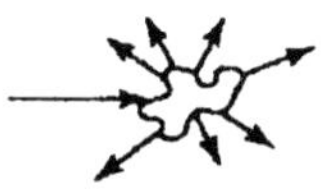	Surface volume	0.2 - 2000 μm
Light scattering (optical & laser)	Surface volume diameter	0.02 - 2500 μm
Photon correlation spectoscopy (DLS)	Dynamic/surface volume	<1.0 nm -1.0 μm
Topography (surface area and porosity) measurements.		
Permeametry 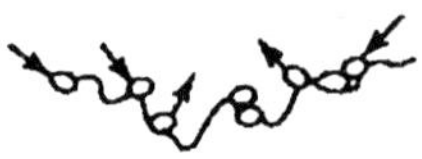	Surface diameter	10 - 1000 cm^2/g (20 - 2500 μm mean)
Gas adsorption [N_2, Kr, CO_2 & water]	Surface volume diameter	0.1 - 2000 m^2/g (0.001 - 20 μm mean)

Figure 1.3 Measurement methods in common usage.

1.3.1.1 *Sieving* Sieving, being one of the oldest techniques for powder classification, is applicable to free-flowing dry powders or carefully prepared slurries (wet sieving). The method involves shaking a powdered sample through a series of usually woven wire faces of known mesh or aperture size arranged in decreasing aperture size (Table 1.3).

Table 1.3 Mesh number and aperture size (μm) for woven wire sieves.

Mesh number	Nominal aperture size (μm)	Mesh number	Nominal aperture size (μm)
3	5600	36	425
3.5	4750	44	355
4	4000	52	300
5	3350	60	250
6	2800	72	212
7	2360	85	180
8	2000	100	150
10	1700	120	125
12	1400	150	106
14	1180	170	90
16	1000	200	75
18	850	240	63
22	710	300	53
25	600	350	45
30	500	400	38

Sieves can be stacked, with the largest aperture at the top and the smallest at the bottom of a stack. The size distribution is reported as the mass of retained particles on any given sieve size. The technical requirements of test sieves have been standardised by a number of ISO standards (ISO 1990a, b, 1999a, b). Since the passage of particles through the sieve mesh occurs in two phases or at two rates, there is no definitive end-point to the sieving process, so the procedure of sieving and the definition of the end-point of sieving has been standardised by the publication of three standards (ISO 2591-1: 1988 Test Sieving. Part 1: *Methods using test sieves of woven wire cloth and perforated metal plate*; BS 1796:1976 *Method of test sieving* and BS 410-2: 2000). The point at which the rates change is historically known as the Whitby point (Whitby 1959). Table 1.4 summarizes the salient points of sieving together with its advantages and disadvantages.

Table 1.4 Salient points of sieving.

Principle	Dry or wet passage through sieves of woven wire mesh, photoformed electroformed sheets or perforated metal plates
Size range	125 mm to 20 μm (wet: 10 μm)
Sample size	Free flowing powders g to kg
Calibration	Microscope or reference materials
Analysis time	Minutes for larger sizes. No end-point
Training	Low. Few hours
Strengths	Broad size range, simple method. Low costs
	One size level per sieve. Mass distribution
Limitations	Agglomeration with fine powders
	Fragile particles protected by wet sieving. Particles need to be insoluble in chosen solvent
ISO Standard	ISO 3310: -1&3 1990, 3310-2: 1999
Makers	Endecotts, Retsch, Rotex Inc.

1.3.1.2 *Static microscopy* Microscope-based techniques (optical and electron) require direct observation (visual and/or electronic) of irregularly shaped particles. Subsequent determination of the size of the particles is based on specifically defined diameter measurements, some of which (Feret, Martin and Projected area diameters) are shown in Figure 1.4.

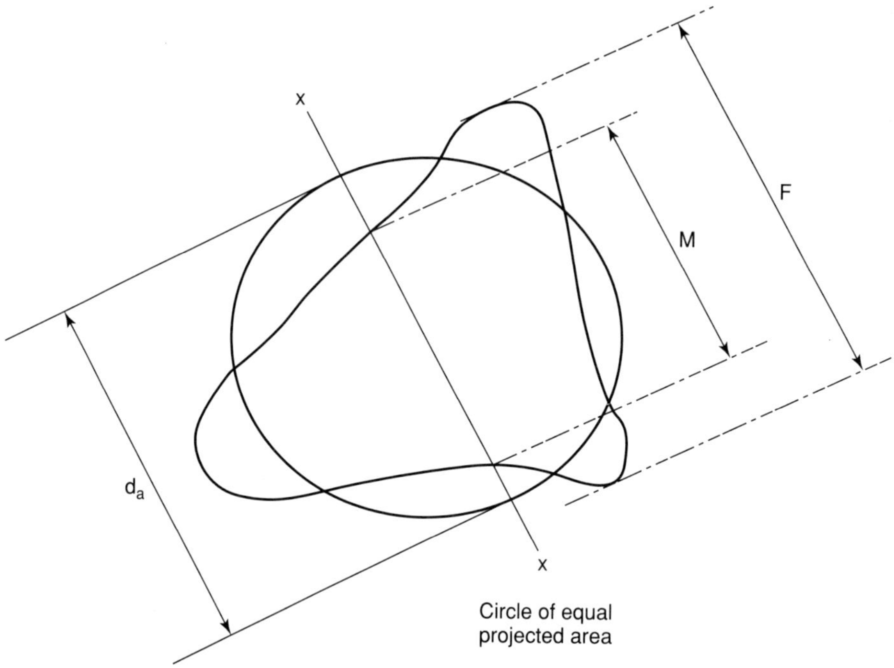

Figure 1.4 Image diameters (F, Feret; M, Martin; d_a, Projected area).

Although particle measurement by image analysis is regarded as relatively simple, great care has to be exercised to ensure (because of the small number of particles observed and measured) that any sample analysed is representative of the bulk powder under investigation. The size distribution of the observed particles can be expressed in a number of different ways (length, area, volume or mass) if limitations to the various mathematical manipulations are recognised.

1.3.1.3 *Dynamic microscopy* With dynamic image analysis, it is necessary to capture a still image from the moving particles. This can be achieved by the illumination of the particles for a very short duration; the captured image then yields a static picture of the moving particles. The period of illumination is determined by the magnification of the image desired from the image capture instrumentation and the resolution required. Moving particles can be observed by three systems:

(1) Sheath flow, in which the particle position is controlled by a sheath of liquid. The sheath directs the dispersed particle precisely to the focal point of the image capture instrument. A still image is then captured by short illumination periods of the flowing particles.

(2) The electrical sensing zone (ESZ) method, in which the image-capture equipment is focused on the orifice in the ESZ tube. As one particle passes through the ESZ orifice, an electrical pulse is generated which triggers a signal for the illumination of that particle, by a strobe flash light, at the precise instant the particle passes the focal point of the image capture instrument.

(3) The free-falling method, in which particles falling from a vibrating feeder are illuminated, and a still image is captured at the focal point.

Image analysers then digitise the image into an array of pixels. The primary measurements are thus the projected area of each particle in pixels (a), the perimeter of the particle (p), the longest dimension, the shortest dimension, and the maximum and minimum Feret diameter. From these measurements, the area equivalent diameter (*aed*) can be calculated from the relationship, square root of ($4a/\pi$). Other shape factors are the aspect ratio (the ratio of the maximum and minimum Feret diameter), the circularity (the ratio of the maximum Feret to *aed*) or sometimes *p*/*aed* (Figure 1.5 and Table 1.5, see also Section 1.5).

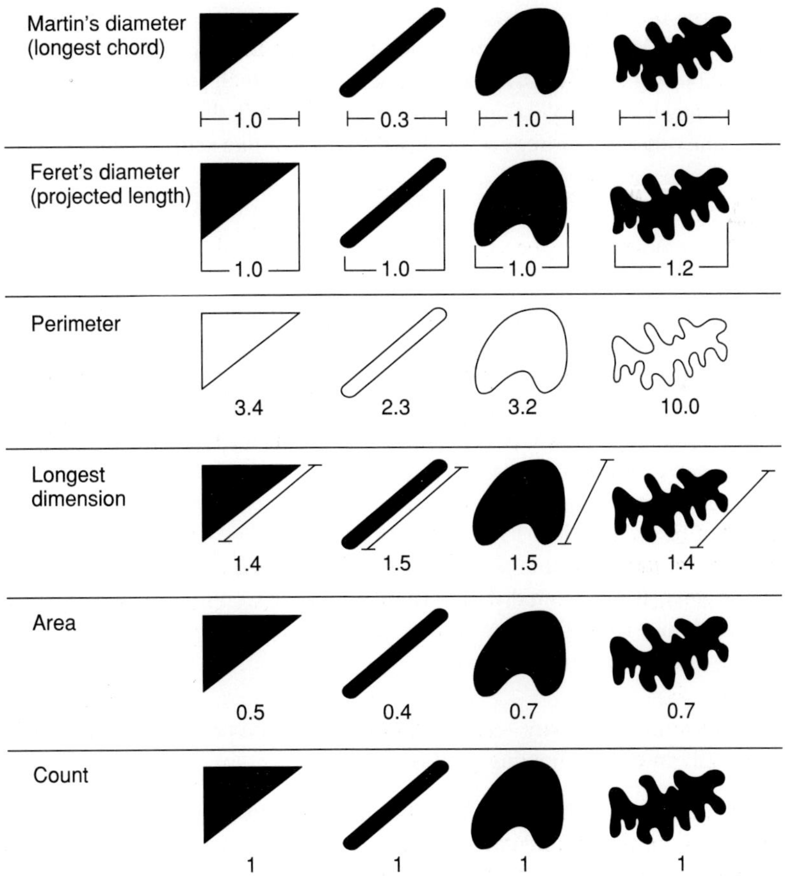

Figure 1.5 Size dimensions and shape profiles from microscopy.

Table 1.5 Salient points of microscopy.

Principle	Magnified image of particles
Size range	Optical: (white light) lower limit 1 μm Wavelengths < 630 nm limited to 0.2 μm Electron: SEM 10 nm; TEM 5 nm.
Sample size	Dispersed powders mg
Calibration	Certified graticules
Analysis time	If manual – slow. Automatic – faster
Training	Medium. Few (2) days with automatic instruments High. Months with manual optical methods
Strengths	Direct, absolute, simple method Shape identity and analysis of mixtures possible
Limitations	Slow and tiring. Experience needed
ISO Standard	Static: ISO/DIS 13322-1, Dynamic: ISO/WD 13322-2
Makers	Galai/Roth Scientific, Retsch/Horiba

1.3.2 *Transport measurements: General principles*

Metrology based on the transport phenomena of particles, as seen with Stokesian sedimentation and elutriation, covers a particle-size range of 10–1000 μm when the force causing particles to settle is gravity. Instruments using gravity to classify particles measure the terminal velocity or height of fall (h) in a given time (t), at which particles of density ρ_s settle in the fluid of known viscosity (μ) and density (ρ_L), in which they are dispersed, regardless of buoyancy and other drag forces acting upon the streamline settling of the particles.

$$\text{Stokes' Equation: } x_{St} = [18\mu h/(\rho_s - \rho_L)t]^{1/2}. \tag{1.2}$$

The particle sizes measured by centrifugal separation are smaller and generally in the range 0.05–25 μm.

Recently, it has been argued that the lower particle-size limits for gravitational sedimentation can be theoretically specified, in terms of height and time of fall, rather than the vague statement that the Reynold Number should be less than 0.25. The lower particle-size limits for gravity sedimentation were found to be inversely proportional to the cube root of the measurement height and inversely proportional to the fifth root of the measurement time (ISO 2001). Convection currents will always exist, however, within a particulate gravitational suspension system. Thus, protracted measurement time, to achieve a supposed measurement of particles below 1 μm, tends to be suspect because of mass flow and possible temperature variations within settling suspensions which cannot be measured with any degree of certainty. It is essential therefore that an increase in the degree of temperature control be mandatory when the sizes of particles being analysed, by gravity sedimentation, decrease below the 50 μm size range.

1.3.2.1 *Sedimentation: Gravity* The methods used in gravitational sedimentation can be classified broadly into the following groups:

(1) Cumulative, in which the weight of the sediment collecting on a scale pan placed at the bottom of the sedimentation column is recorded as a function of time.
(2) Incremental methods in which the concentration of settling particles at a measured depth and known time gives, from Stokes' equation, the size range of the particle

identified. The pipette method (Andreason Pipette) extracts a known volume of suspension (usually in water) which, on solvent evaporation, measures the weight of material sediment. Since the concentration of particles settling varies throughout the sedimenting fluid, a measurement of the system density at known levels using either 'divers' or a hydrometer can be used to determine a size distribution of the system. Alternatively, the concentration of particles can be measured by the attenuation of a narrow beam of light projected through the suspension onto a photelectric cell at a known depth (the Photosedimentation method). The attenuation of the beam is dependent upon the Lambert–Beer Law and gives a projected area diameter. To identify smaller particles than those that can be seen with visible light (wavelength of approximately 0.630 μm), the radiation used may come from an X-ray source. Attenuation of the X-ray intensity at various depths gives both the size and the mass distribution.

1.3.2.2 *Sedimentation: Centrifugal and ultracentrifugation* The use of centrifugal force reduces the time involved in the sedimentation of particles. There are two approaches to the solution of the equations which are used to determine particle size and the distribution of sizes by centrifugation.

When particles are injected into the centrifuge to form a thin layer on the surface of the spinning sedimentation liquid (the two-layer technique), particle diameter can be measured from Stokes' Law. The amount of material settling can be determined either by extraction of a known volume and thus mass of the sediment or by attenuation of a light beam, using a photosedimentation method to give a cumulative undersize distribution by mass. The photosedimentation technique measures particle size as a function of the cross-sectional area of the settling particles. Alternatively, particles can be classified into size classes by the attenuation of an X-ray beam, which is directly proportional to the mass of the settled sample.

If, however, the particles are uniformly dispersed throughout the sedimentation liquid (the homogeneous technique), there is no exact solution to equations which describe the movement of different-sized particles from different positions or the same-sized particles settling radially from different positions within the homogeneous suspension. Kamack (1951) simplified the mathematics used in centrifugation by the application of a radial dilution correction factor to Stokes' equation. Radial centrifugal settling of particles in a fixed depth centrifugal pipette operating with an initially homogeneous suspension, with the Kamack theory, was commercially marketed as the Simcar Centrifuge (Table 1.6).

1.3.3 *Rapid physical response measurements: General principles*

The size range measured with this group of instruments tends to be in the range 0.1–1000 μm. Rapid-response particle-size analysis can be achieved by electrozone sensing, Fraunhofer diffraction and nephelogenic (light scattering) techniques (Figure 1.6). In diffraction and light-scattering instruments, the intensity of light scattered by the particles, usually in suspension, is recorded as a function of the scattered angle. The light-scattering models used to determine particle size from rapid-response instruments depend on Fraunhofer diffraction, Mie theory or ray optics (Table 1.7).

Colour absorbency, coupled with radiation absorbency, in photozone sensing may cause some inaccuracies. Light-scattering instrumentation is dependent upon the computational

Table 1.6 Salient points of sedimentation.

Principle	Particle dispersed in non-aqueous and aqueous media. Gravity and centrifugal
Size range	Gravity: 300–10 μm: Centrifugal: 10 μm < 1 nm
Sample size	Dispersed powders g to μg
Calibration	Stokes' Law and streamline flow
Analysis time	Slow with gravity, hours < 10 μm. Automatic – faster
Training	Low. Few hours
Strengths	Andreasen, fundamental, inexpensive, semi-skilled
Limitations	Slow, not for mixed densities, dispersion and convection anomalies
ISO Standard	Gravity: BS ISO 13317-1. 2001; BS ISO 13317-2.2001 Pipette method and BS ISO 13317-3. 2001. X-ray Cent: BS ISO 13318-1 2001; BS ISO 13318-2. Photocentrifuge BS ISO 13318-3 X-ray.
Makers	Gravity-cumulative: Bostock sedimentation balance, Shmada sedimentation balance and Sartorious sedimentation balance Gravity-incremental: Andreason pipette, Bound Brook, Eel and wide angle photosedimentometers (Fritsch GmB, Ladal Ltd and Microscal Ltd). Micromeritics Sedigraph, Quantachrome Microscan Centrifugal and ultracentrifugation: Joyce Loebl disc centrifuge, Whitby centrifuge, Coulter/Kaye disc centrifuge, Brookhaven photocentrifuge and Dupont/Brookhaven scanning X-ray centrifuge, Simcar centrifuge, Horiba

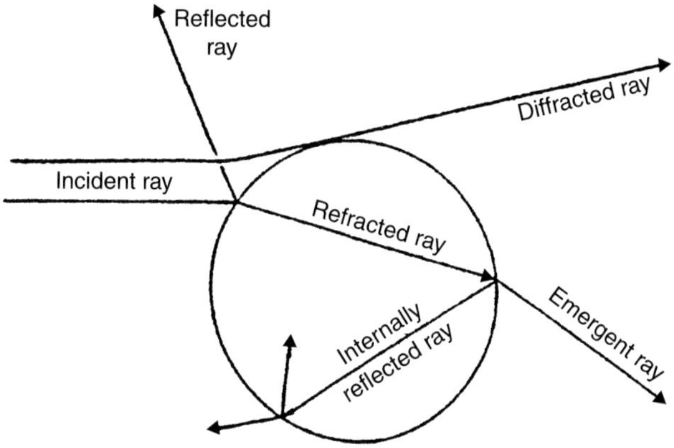

Figure 1.6 Light pathways through a spherical particle.

models which deconvolute the measured scattered light intensities falling onto photodetectors. It is always assumed that the light scattered emanates from spherical particles. Thus, either under- or overestimation of the finer particle sizes in a distribution can occur due to either the computer model or the insertion of incorrect particle and suspending fluid physical data into the computer program. The anomalies, seen in laser light-scattering

Table 1.7 Factors determining Mie scattering theory.

1. Ratio of diameter of the sphere (d) to the wavelength of light, usually 632.8 nm (λ), with m the refractive index of the dispersion media ($\alpha = \pi m d/\lambda$)
2. Scattering angle, θ, usually 90°
3. Refractive index, $m = m_1/m_2$ where m_2 is the refractive index of the dispersion media
4. State of polarisation of incident radiation $I(\theta) = I(\alpha, \theta, m, \pi)$

instruments may, at times, be due to instrument misalignments, but the main source of error is due to the choice and application of the incorrect computational model used for different types of particles, together with the incorrect refractive index and density of the particle.

Historically, the laser-diffraction technique began with the collection of scattered light at angles smaller than 14°, but now instrumental technology has advanced and laser light-scattering techniques can be divided into two categories:

(1) Low-angle laser light-scatter instruments (LALLS). These limited the minimum particle size which could be measured to 1 μm with these angles, and by the use of Fraunhofer diffraction the maximum size measured was approximately 1000 μm. The laser-scatter technique has now been broadened to capture scattered light of different wavelengths and polarisation over a much wider angular range. By placing lasers at various angles to the light-detection system, these geometrical arrangements have enabled scattered light from small particles to be collected at increasingly greater angles, thus enabling increasingly smaller particles to be measured (Figures 1.7 and 1.8).
(2) Multi-angled and multi-lasers.

Many recent instruments now have angles up to, and at times greater than, 150°, so that particle sizes smaller than 0.1 μm can be readily analysed with converging beams and more than one focusing lens. Instrumentation is now commercially available, with more than one laser at more than one wavelength to collect back-scattered light, and therefore particles smaller than 5 nm can measured using Mie theory (Figure 1.8).

Since all the light scattering theories used are based on the scattering of light from spherical bodies, the size of irregularly shaped particles can be theoretically extended to the nanometre size range. The measured/computed size for non-spherical particles is the

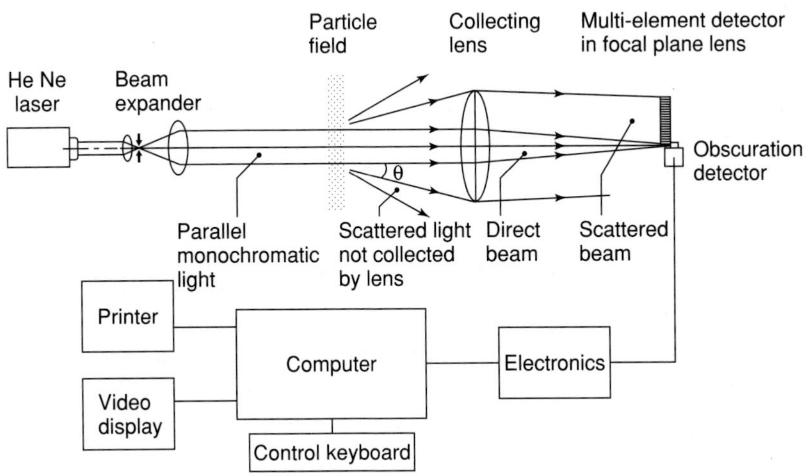

Figure 1.7 Schematic diagram of laser-light-scattering instrumentation.

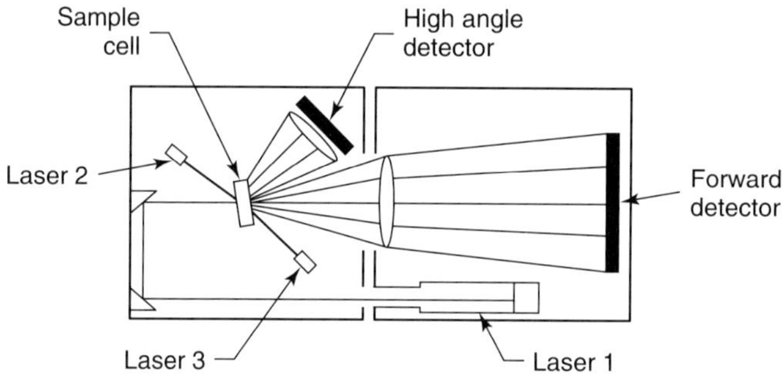

Figure 1.8 Multi-laser configuration for light-scattering instrument.

equivalent spherical size which is an equivalent cross-sectional diameter, generally termed the Sauter diameter or, in ISO terminology, the $x_{1,2}$ (average surface length) or an average volume-surface diameter, $x_{2,3}$.

To study the properties of suspensions and solutions of colloids, photon correlation spectroscopy (initially termed quasi-elastic light scattering (QUELS) and sometimes referred to, these days, as dynamic light scattering (DLS) or intensity fluctuation spectroscopy (IFS)) can be used. The intensity of light scattered from particles undergoing Brownian motion can be measured and then used to compute, from the Stokes–Einstein equation, particle sizes within the sub-micrometre and nanometre particle size range.

1.3.3.1 *Electrical sensing zone (ESZ)* The ESZ technique is based on the passage of an irregular particle through an orifice (of known size) submerged in an electrically conductive liquid. The conductivity changes when a particle passes through this orifice. The change in conductivity or electrical pulse generated by the passage of the particle through the orifice is proportional to the volume of both a spherical and an irregularly shaped particle. The magnitude of signal generated is evaluated from a calibrated instrument to size the particles; the number of particle(s) passing through the orifice is also counted, thus yielding a size, number and volume, distribution. The ESZ technique can provide size measurement distortions, because of incorrect volume displacement, with high conductivity, very porous materials or the passage of too many particles through the measurement orifice simultaneously (Figure 1.9 and Table 1.8).

1.3.3.2 *Laser light scattering* Initially, the laser light-scattering technique measured scattered light at small angles; thus, laser diffraction has been known as Fraunhofer diffraction, near-forward diffraction and LALLS. A powdered sample dispersed in either a gas or a liquid at adequate concentrations is passed through a beam of monochromatic light, usually a laser. The light scattered at various angles by the particles is then measured by a multi-element detector, and the magnitude of the scattered light intensity recorded. These light intensity values are then transformed mathematically to yield the proportion of particles in a discrete volume of a known class size and, ultimately, to compute a volumetric particle-size distribution (Table 1.9).

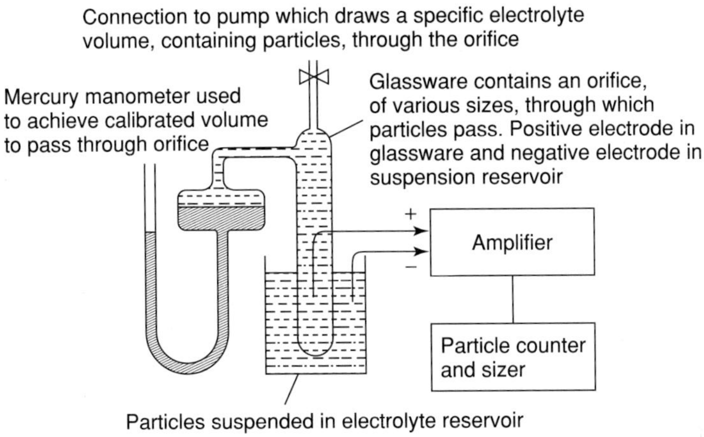

Figure 1.9 Electrical sensing zone instrumentation.

Table 1.8 Salient points of the ESZ method.

Principle	Particle passage through an orifice immersed in an electrolyte Multi-tube technique for potential on-line measurement
Size range	2–60% orifice size 2000 to < 1 μm Under favourable conditions and a small orifice, nanometre-sized particles may be measured
Sample size	Wetted, insoluble in the electrolyte; mg to μg
Calibration	Fundamental volume measurement
Analysis time	Minutes
Training	Medium. Few (2) days
Strengths	Calibration with known latex sizes Particle size traceable
Limitations	Porous particle Need for dilute suspensions Difficulty in finding inorganic electrolytes in which to suspend particles
ISO Standard	BS ISO 13319: 2000
Makers	Beckmann Coulter, Elzone

Table 1.9 Salient points of Fraunhofer diffraction.

Principle	Fraunhofer approximation from Mie
Size range	2–8000 μm. Mie for 0.1–2 μm
Sample size	Dispersed powders mg to μg
Calibration	Fundamental. Certified reference material also used
Analysis time	Seconds/minutes
Training	Medium. Few (2) days
Strengths	Quality control, on- and off-line. 4000:1 size range Dry and wet powders
Limitations	Resolution of size by inversion equation Optical properties needed
ISO Standard	BS ISO 13320-1: 1999
Makers	Cilas, Honeywell (formally Leeds & Northrup), Malvern, Beckmann Coulter, Sympatec, Fritsch, Seishin, Horiba, Shimadzu, Lasentec

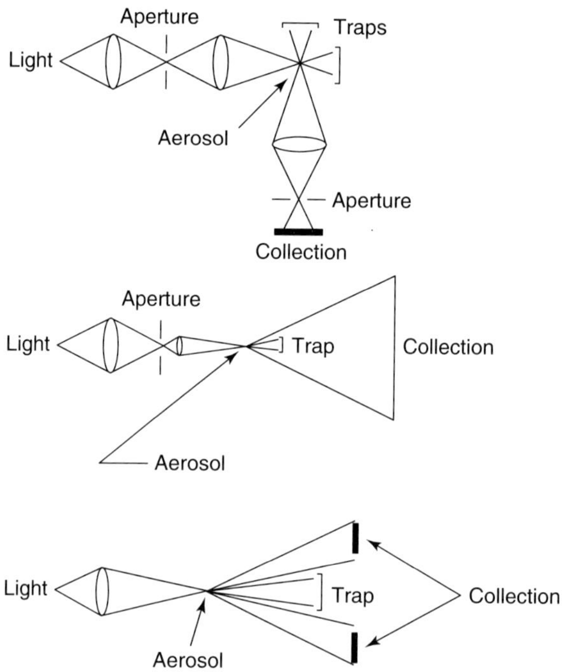

Figure 1.10 Schematic diagram of optical particle counter configurations.

1.3.3.3 *Single-particle optical counters (OPC)* There are two types of OPCs which are designed to view individual particles in suspension as particles flow between a light source (white light for older instruments, but now lasers are used with wavelengths of 633–780 nm for modern instruments) and a detector system which collects either an attenuated beam of light or the scattered intensity of light at various angles (Figures 1.10 and 1.11).

The light extinction type of OPC measures the amount of light removed from an initial beam passing through an illuminated area (the sensitive zone), through which each particle must pass. This type of OPC is mainly used with particles in liquid suspensions. When the size of particles approaches the wavelength of light used in attenuated light OPCs, a correction, the extinction coefficient, K, has to be applied due to the breakdown of geometric optics when small particles block the light beam. K is unity for large particles and rises to $K = 2$ as particles enter the Mie scattering region (Figure 1.11).

With light scattering OPCs, the amount of light scattered from a single particle over a specific solid angle, as the particle passes through the sensitive zone, is collected at various angles on a detector. This type of OPC is mainly used for the characterisation of particles in gases. Only the amount of light emitted from the OPC sensitive volume, generated by the presence of a single particle, is observed and measured. No image of the particle is taken, although within the collected signal, information on the shape of particles is innate. The change in emitted light intensity (scattered or attenuated) is then correlated with particle size (Table 1.10).

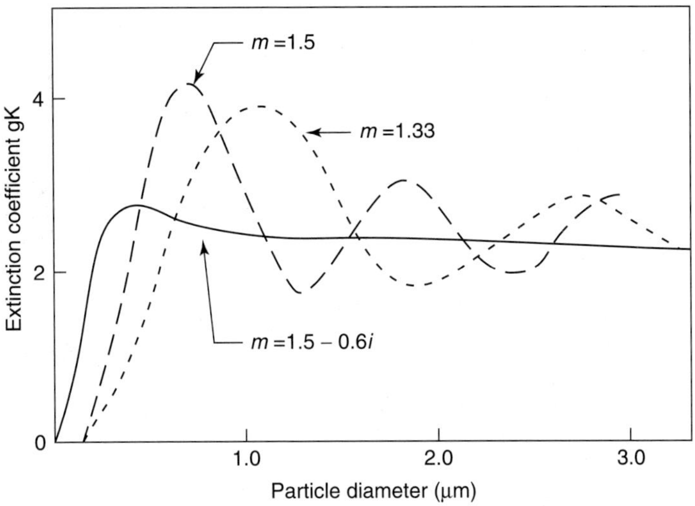

Figure 1.11 Extinction coefficient at various particle sizes with different refractive index.

1.3.3.4 *Acoustic scattering* The attenuation of sound is affected by a set of mechanical, transport and thermodynamic properties. The temperature-dependent properties are sound attenuation, speed of sound, coefficient of thermal expansion, thermal conductivity, and density and viscosity of the liquid. In addition, the shear rigidity of the particle is also required before a particle size can be measured by the instrument patented by Alba (US Patent 5,121,629 (1992)) and calculated from the theory of Allegra and Hawley (1972) (Figures 1.12 and 1.13 and Table 1.11).

1.3.3.5 *Photon correlation spectroscopy (PCS)* A coherent monochromatic laser light beam illuminates a representative sample of dispersed particles in a liquid. The scattered light intensity is measured by a photodetector placed at an angle (usually at 90°) to the light

Table 1.10 Salient points of particle optical counters.

Principle	Lambert–Beer Law and Mie theory
Size range	Scattered light 5 to $< 0.1\ \mu m$ at slow flow rates (0.1 CFM). In airborne systems, use an aerodynamic size ($\alpha = \pi x/\lambda$)
	Attenuated light not less than 3 μm due to variation in the extinction coefficient, K ($K = 1$–2 for $\alpha = 100$). Each particle is counted and its cross-section represents its size
Sample size	Dispersed powders μg. Single particles measured
Calibration	Certified reference material and also particle number calibration
Analysis time	Seconds/minutes
Training	Medium. Few (2) days
Strengths	Sub-micrometre measurement. Single particle measurement in liquids and gases
Limitations	Particles $< 5\ \mu m$ dependent on particle/fluid refractive index
	Low particle concentrations to minimise coincidence effects
ISO Standard	BS ISO 13323-3: 2000
Makers	Amherst Aerosizer

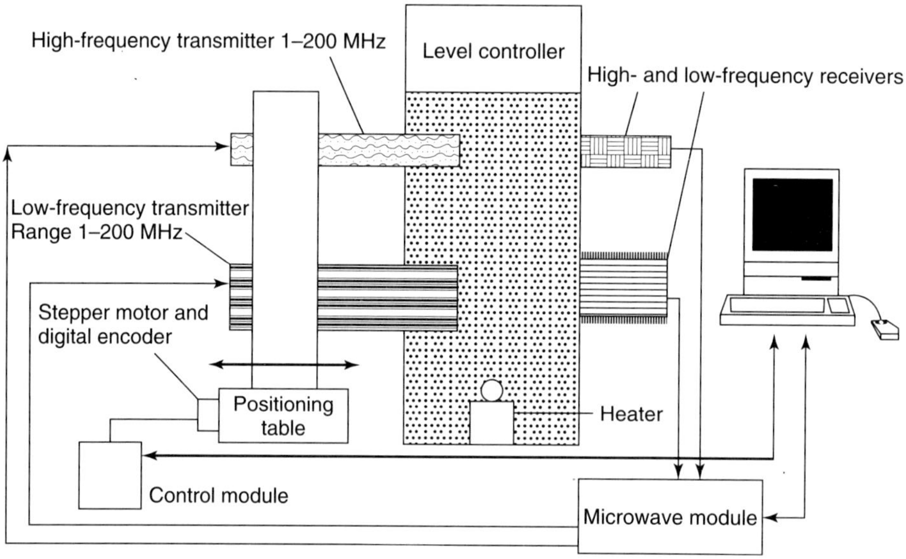

Figure 1.12 Schematic diagram of an acoustic spectrometer.

source (Figure 1.14). The time-dependent electrical signal is then time-delayed in a correlator to generate an autocorrelation function. Mathematical interpretation of the autocorrelation by the cumulants method of Koppel (1972) gives a polydispersity index. Photon correlation spectroscopy is also known as quasi-elastic light scattering (QELS) or dynamic light scattering (DLS) and is the measurement of the time-dependent fluctuations of the intensity

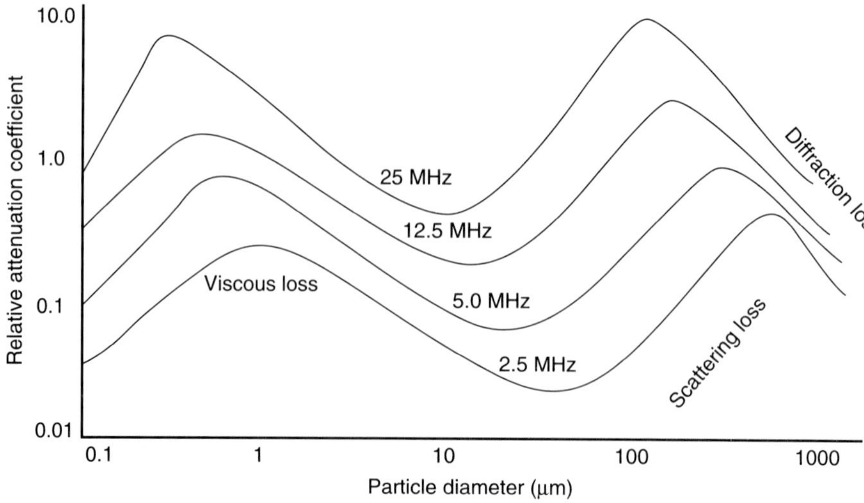

Figure 1.13 Relationship between acoustic attenuation, sound wave (MHz) and particle size for particle-size measurement.

Table 1.11 Salient points of acoustic scattering.

Principle	Acoustic size dependent on Allegra & Hawley theory Ultrasonic frequencies in the MHz range
Size range	0.01–1000 μm.
Sample size	Dispersed powders μg
Calibration	Fundamental. Certified reference materials also used
Analysis time	Seconds/minutes
Training	Medium. Few (2) days
Strengths	Measurement of suspensions up to 50% volume concentration
Limitations	Resolution of size distribution, cross-correlation
ISO Standard	BS ISO 13321-1: 1996
Makers	Malvern

of light scattered by particles or molecules in a dispersed system illuminated by a coherent (laser) light source. These time-dependent scattered light fluctuations arise from the Brownian motion, and possibly other forces, of the dispersed particles, colloids or macromolecules. The technique of PCS yields a diffusion coefficient which, together with the Stokes–Einstein equation, can evaluate a particle size termed an equivalent spherical diameter. Smaller particles diffuse faster than larger particles. From the scattered light-intensity fluctuations over time, an autocorrelation (intensity–intensity) function and photon–photon autocorrelation is computed. From a Fourier transformation of the decaying autocorrelation function, a frequency distribution can be obtained which, when analysed, gives a particle-size distribution. Since large particles scatter light at small angles and small particles at broader angles, PCS may have 'blind spots' at one angle to particles of a certain size. This may be due to the weakness of the scattered light from that particle at that angle. Information gathered at other

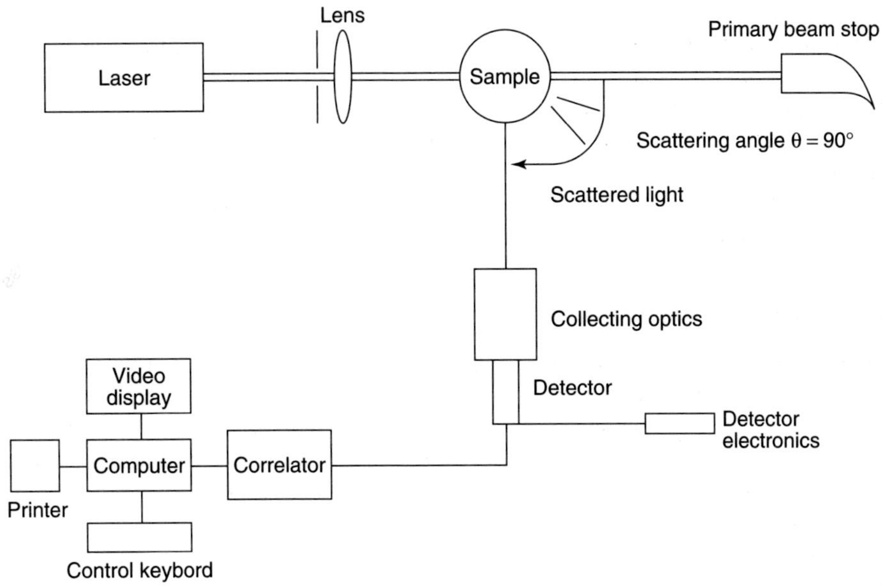

Figure 1.14 Schematic diagram of a photon correlation spectrometer.

Table 1.12 Salient points of photon correlation spectroscopy.

Principle	Brownian motion. Stokes–Einstein equation
Size range	1–1000 nm. Depends on laser/density
Sample size	Dispersed powders μg Sample concentration mg/ml
Calibration	Fundamental. Certified reference materials also used
Analysis time	Seconds/minutes
Training	Medium. Few (2) days
Strengths	Sub-micrometre measurement, polymer chains, microemulsions 10:1 size range
Limitations	Resolution of size distribution from cross-correlation computation. Shape affects the rotational and translation diffusion of particles
ISO Standard	BS ISO 13321-1: 1996
Makers	Brookhaven Instruments, Beckmann Coulter, Honeywell (formally Leeds and Northrup), Malvern Instruments, Otsuka Electronis, Particle Sizing Systems

angles will thus enhance the reliability of the particle size and size distribution computed – known as multiangle PCS (Table 1.12).

1.4 Surface area and porosity: General principles

An alternative to the characterisation of irregular particles by particle size and shape and one of the characteristics which is now becoming increasingly important, mainly due to the ease with which it can now be measured, is the measurement of the surface area of a powdered solid.

The ease of agglomeration and compaction, together with many other powder technological processes, can vary widely with materials which have chemically identical structures but which may possess a differing physical nature. Expressing the topographical nature of materials in terms of the energy, enthalpy, solubility or dissolution potential per unit surface area has thus enabled optimisation in the choice of preferred raw resources. Whenever a solid is in contact with a fluid, the interface between the two phases is dependent upon the surface area available to interact with the surrounding fluid, be it gaseous or liquid (ISO 1995b). Sorption is the preferred method for the measurement of the internal surface area of an assembly of particles, since the surface area calculated with scattered laser light instrumentation is an external geometric surface area computed by assuming a distribution of spherical particles.

1.4.1 *External and internal surface area*

Owing to the wide nature of powdered ingredients, there is no unique method for analysis or characterisation of the topography (surface) or the internal structure (pore/void distribution) in complex porous matrices. The internal surface areas of size-enlarged particles or powders (>1.0 m^2/g) are generally determined from the monolayer or multilayer adsorption region of a low-temperature nitrogen adsorption isotherm using the Brunauer, Emmett and Teller (BET) equation (Brunauer *et al.* 1935; BCR 1989). The multilayer and capillary condensation region of the low-temperature nitrogen isotherm can also be used to evaluate the size and shape of pores within particles or the voids between particles. The maximum relative pressure achievable with low-temperature adsorption equipment usually limits the magnitude of the size of the pore/void measured to a size not generally greater than 500 nm. Krypton adsorption is mainly used for low-surface-area powders ($<1.0\ m^2/g$).

Permeametry, achieved by the passage of gases slightly above atmospheric pressure through a packed bed of powder, evaluates the external area of particles. The external surface area of the packed bed of particles is measured in the square centimetre per gram range.

1.4.2 *Porosity*

The material matrix within an agglomerated or compacted mass of powder may have pores or voids which are generally in the size range 1.0 nm to 0.1 mm. These pore/void spaces can be sized by the physical adsorption of various vapours or gases at low (nitrogen, krypton and carbon dioxide) and ambient (water, butane or organic vapours) temperatures. Since the techniques chosen for the characterisation of porous materials are, to some extent, dependent upon the dimensions of the space to be measured and the mechanisms of pore space filling, it is expedient to classify pore space or porosity in terms of the magnitude of the space, with the caveat that the size limits are arbitrary and can be influenced by the interactions of the solid and fluid used in the measurement technique (Sing *et al.* 1985).

Pores exceeding a width of 50 nm are generally termed macropores, while pores with a width not exceeding 2 nm are micropores. Pores of intermediate sizes are termed mesopores. When the interparticulate sizes within agglomerates and compacts are greater than those readily measurable by low-temperature adsorption and application of the Kelvin equation (i.e. greater than 500 nm), the technique of mercury intrusion porosimetry can be used to evaluate intraparticle and interparticle space in the approximate size range 3.5 nm to 1.0 μm.

One of the difficulties in interpreting mercury porosimetry is the inability to distinguish between penetration into inter- or intra-particle space. However, determination of the pore space distribution by mercury penetration, because of the relative simplicity, versatility and speed of the technique, is currently used for many quality characterisations despite the hazardous nature of mercury.

1.4.3 *On-line and* in situ *particle characterisation*

Harold Heywood pronounced in 1967 (Heywood 1967) that, '*a size analysis procedure which can form part of a fully automated production unit will be needed to feed back information to the process*'. Thus, in the mid-1990s, a new era of rapid-response particle-process control sizing instruments began. On-line monitoring and process control do not, however, remove the problem of obtaining a representative powder sample but only change the essential sampling operation from a manual laboratory collection of particles to a possible automated particle collection. These on-line sample collection processes, followed by size or particle characterisation techniques, may have computer control, collect many more powder samples and possibly have a more rapid response time, but on-line particle characterisation still does not overcome the problem of collecting, removing and transporting the particulate sample to the particle-characterisation instrument.

Since time, effort and money can be wasted by analysis of a non-representative sample, the emphasis on size measurement/characterisation, with the advent of computerised information technology, has shifted from off-line batch to on-line and *in situ* analysis (Stanley-Wood

2000). Ultimately, the desired goal for many process and product engineers will be *in situ* particle characterisation and control. On-line measurements nowadays are largely devoted to modification and adaptation of off-line methods. Even if the on-line analysis has a quick response of only a few seconds or a fraction of a second, this will eventually be superseded by in-line and/or *in situ* particle-size and size-distribution measurements.

In-line and *in situ* methods of characterising particulate systems attract the attention of industry because:

(1) The physical characteristics of the particles in a chemical process influence both the dynamics of the process and the final product specification.
(2) In a dynamic continuous chemical process, there should be a continuous dynamic measurement which can control the particle characteristics of the powdered material in terms of particle size, particle-size distribution, shape, bulk density and other particle or powder characteristics.
(3) The characterisation of the mechanics and dynamics of the chemical process is also essential, initially to understand the process but eventually to control, improve and optimise the plant, in terms of safety and cost-effectiveness and ultimately to tailor-design the final product from a chemical process. In these scenarios, in-line and *in situ* characterisation will predominate.

Laser-based particle-characterisation instruments thus may be ideally positioned to achieve in-line or *in situ* particle-size analysis because of the manner in which particles are detected and the ease with which computational manipulation of the collected information, needed to record a desired particle size or process parameters, can be accomplished. With the advent of the production of purer and finer powders, the current methods, instrumentation and techniques used to size particles may soon necessitate the application of new and novel chemical, physico-chemical, electrical, optical and physical criteria to determine particle or droplet sizes. Whatever the choice of method used to size any irregularly shaped particle, it must be realised that particle characterisation is solely a means to correlate powder properties to the functionality of a manufacturing process, product usage or preparation.

1.5 Shape characterisation of irregular particles

Shape measurement, initially of academic interest only, began to find credibility when shape characteristics were incorporated into industrial and commercial specifications for powders and granulates. Nowadays, particle morphology is finding increasing importance in many aspects of powder flow, filtration, compaction and granulation.

Particle shape is a fundamental characteristic of a particle, in a manner similar to that of particle size. Various types of particle shape have been described in words and numerous general shapes subsequently defined (BSI 1958), but these verbal descriptions are hopelessly inadequate for incorporation into equations and calculations to ascertain the effect of particle shape on various particle systems, particle properties or the design of bulk powder-handling equipment.

Pilpel (1969) has shown that the shape of small particles obtained by comminution is dependent upon the method of size reduction. Micronized pigments are often less irregular and more spherical in shape than particles which have been repeatedly fractured and sheared

Table 1.13 Effect of micronisation on the shape of pigment particles.

Pigment	Measured surface area (m^2/g)	Calculated surface area (m^2/g)	Shape factor
Prussian blue	61.3	2.3	0.04
Prussian blue, micronized	58.9	12.7	0.23
Red iron oxide	7.6	3.4	0.44
Red iron oxide, micronized	7.3	4.9	0.68
Burnt amber	138	1.2	0.01
Burnt amber, micronized	94	5.0	0.05
Yellow iron oxide	17.6	1.4	0.08
Yellow iron oxide, micronized	17.7	2.5	0.14

by attrition. Table 1.13 lists the shape factors for a number of materials either micronized or fractured. This shape factor – the ratio of the calculated surface area obtained from electron microscopy to the BET measured surface area – becomes closer to unity when a particle has a more spherical form. The effect of size and shape on the flow and failure properties of penicillin powders, which had an angular prismoidal form, has been shown by Walton and Pilpel (1972) to affect the bulk powder characteristics of cohesion, tensile strength, internal friction and failure factors.

The historical method for shape analysis compares the irregularly shaped particle profiles with known geometric shapes such as a sphere, cube or rectangle. The shape of the irregular particles is then described in terms of length and width or simply as a ratio devised from image analysis. The development of techniques to characterise and measure the shape of irregularly shaped particles is shown in Figure 1.15 (Kaye 1981; Hawkins 1993).

1.5.1 *Dimensionless surface-volume conversion shape factors*

Herdan (1952) stated that any 'shape factor' should have a threefold function.

(1) It should be a factor of proportionality between the particle size determined by different methods of particle characterisation. Thus, the particle size determined by microscopical techniques could be compared with that determined by sedimentation.
(2) Shape factors were conversion factors for expressing the results of different particle measurement techniques as an 'equivalent sphere' parameter.
(3) Shape factors transformed the second and third power of a measured particle diameter into the parameters of particle surface or particle volume.

One of the inherent restrictions in the characterisation of particle size is that the measurement of a particle dimension is dependent upon the physical criterion used in the measurement technique. This illustrates the first of Herdan's functions. For example, an irregular particle characterised by sieve analysis, sedimentation or by an electrical sensing zone technique will not have identical diameters. The difference between the measured sizes can therefore be advantageous in shape recognition. The factor of proportionality between the two particle sizes determined by different analysis techniques can thus be regarded as a shape factor.

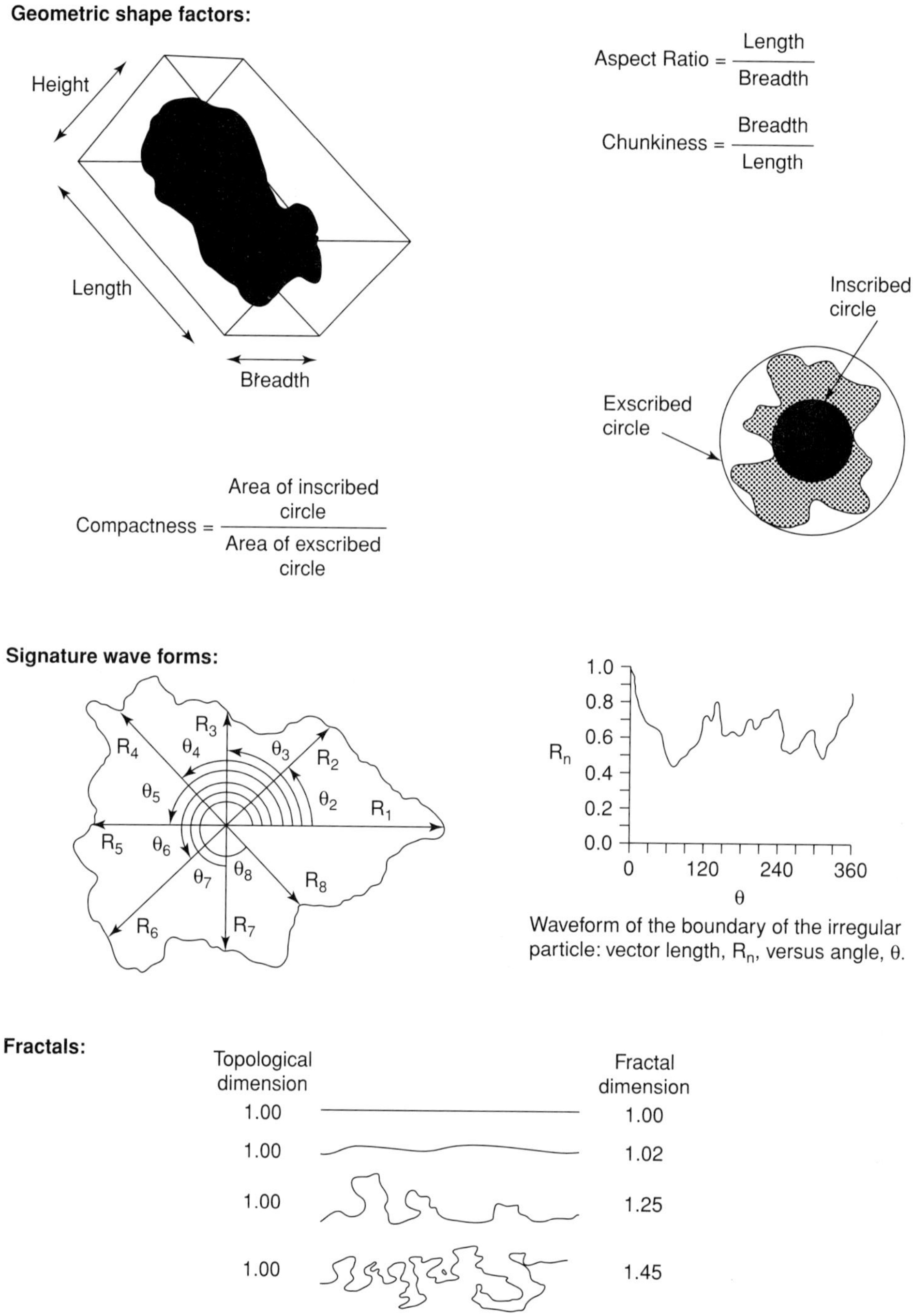

Figure 1.15 Techniques for the characterisation of shape for an irregular particle.

BS 3406 Parts 2, 3 and 4 (1963) stated that a 75 μm particle by sieve analysis may have a mean projected diameter of 105 μm or a mean Stokes diameter of 70.5 μm, while in BS 4359 Part 3 (1970), Appendix B, various calculated and experimental values of surface, volume and specific-surface coefficients (shape factors) are tabulated for various dimensional forms and irregular particle shapes. These shape factors being obtained from sieving, permeability, sedimentation, ESZ and light extinction methods of particle size and surface area measurement. In all cases, the numerical values for spherically defined forms are 3.14, 0.52 and 6.0 or, algabraically, α_s, α_v and α_{sv}, respectively. But when the same criteria are applied to non-spherical or irregular-shaped particles over a range of sizes, 0.5–2000 μm, these shape factor values are exceeded.

From Herdan's function 3 above, the following relationship for the volume of a particle holds:

Volume of particle, V = Volume coefficient (volume shape factor) multiplied by the cube of some characteristic dimension. (1.3)

This reasoning is also valid for the surface of a particle:

Surface of particle, S = Surface coefficient (surface shape factor) multiplied by the square of some characteristic particle dimension. (1.4)

Algebraically, these definitions can thus be written for irregular particles as:

$$V = \alpha_v x^3 \text{and } S = \alpha_S x^2, \tag{1.5}$$

where x is the size characteristic of the particle, and α a shape factor.

When the measured dimension is that of a sphere with the same surface area as an irregular particle, then by definition, the surface of that sphere is equivalent to πd_s^2 or πx_s^2. Likewise, from the volume viewpoint, the volume of a sphere, with a diameter, x_v, is, by mathematics, equivalent to $\pi x_v^3/6$. The surface and volume shape factors of the spherical particles α_s and α_v are thus numerically equal to 3.14 (π) and 0.52 ($\pi/6$) respectively, because of the mathematical relationship of surface and volume of spheres being πd^2 and $\pi d^3/6$.

The surface and volume of an irregularly shaped particle, however, can be expressed algebraically as $\alpha_{sa} x_a^2$ and $\alpha_{va} x_a^3$, respectively, where x_a is the diameter of the irregularly shaped particle measured by microscopy. These shape factors will not have the same numerical values as spheres; they will always be of a greater magnitude because the surface area and volume of any object are always minimal for a spherical object.

For many years, it was assumed that empirically (Stanley-Wood 1983), the ratio V/x^3, where V is the average particle volume and x the mean particle size, remained sensibly constant for various grades of irregular and non-uniformly shaped particles. Likewise, the ratio S/x^2, where S is the average particle surface and x the mean particle size, was deemed to remain sensibly constant for different sizes of the same material. It was thus possible, at times, to speak of an average particle shape within an assembly of particles, although variations in particle shape with size for many processed and powered materials can occur.

Measurement of the two-dimensional quantity of area can be obtained from the projected microscopic image of the particle while the three-dimensional quantity (volume) can be measured from either a sedimentation or ESZ technique. When the characteristic dimension of an irregularly shaped particle is the projected area diameter x_a, the surface of that irregularly shaped particle can thus be symbolised as $\alpha_{sa}x_a^2$, where the letter 's' in the shape factor α_{sa} denotes a surface shape factor, and 'a' is related to the projected area diameter, x_a. This can then be compounded by the symbolised particle-size measurement, x_a, obtained from microscopy, of the particle's projected area diameter squared. Although the same intellectual reasoning and logic can be applied to methods for particle volume measurement, these theoretical shape factors find little application in industry.

1.5.2 *Dimensionless geometric shape factors*

One of the first practical attempts to standardise particle shape measurement for individual particles was by the measurement of the length or diameter and the breadth or width of a particle and then, from the protocol outlined in BS 512 (1966), to designate different and arbitrary ratios of particle breadth to length, B/L, or particle thickness to breadth, T/B, to various three-dimensional particle shapes. Regular geometric shapes, such as discs, rods, blades and equidimensional particles, could then be assigned a numerical value corresponding to the appropriate shape of the body or particle rather than a literal description.

Heywood (1947) derived an expression for the shape of a particle in terms of the general geometrical form of an irregular particle from the ratios of length, L, breadth, B, and thickness, T (Figure 1.16). Heywood's elongation ratio has unfortunately also been defined, in different terminology, as an aspect ratio which has the ratio of length of an irregular particle divided by the width (W) or breadth of the particle profile at right angles to the length measurement ($L/B = L/W$).

To separate the dimensional proportions of the particle from the geometric shape of the particle, Heywood used the ratios of elongation, n (ratio L/B), and flatness, m (ratio B/T), together with a term defined as the equidimensional shape factor, α_o. The volume shape factor of an irregular particle, α_v, could therefore be represented in part by the α_o factor

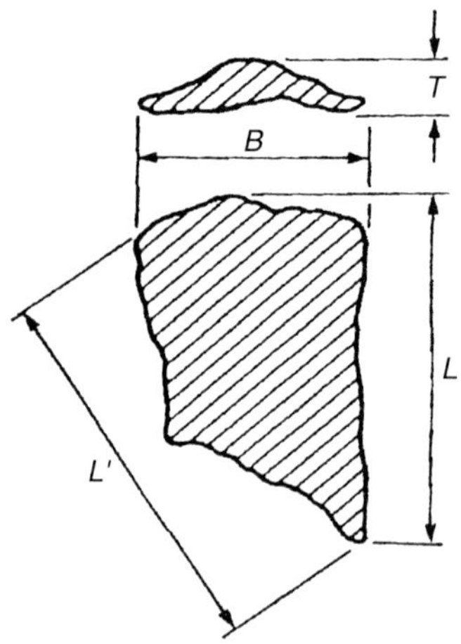

Figure 1.16 Heywood's elongation shape factor (n = ratio L/B) and Heywood's flatness shape factor (m = ratio B/T).

and also by the ratios of the elongation and flatness ratios, n and m, respectively, which characterised the influence of particle shape due to the dimensions of the particle to give:

$$\alpha_v = \alpha_o / m\sqrt{n}.$$

Values of L, B and T can be easily obtained by an automatic scanning microscope technique and software. This approach to shape recognition has been used by Mandelbrot (1957), Church (1968), and Dwyer *et al.* (1970) to obtain relationships for a collection of powders between measured particle and linear diameters and other particle parameters.

A disadvantage to Heywood's elongation ratio is that as an irregularly shaped particle becomes more elongated, and the breadth becomes smaller, the elongation ratio tends to an infinite value. Kaye and Clark (1992) have thus suggested that a more acceptable and possibly more practical ratio would be the chunkiness ratio, W/L or B/L, which is the reciprocal of Heywood's elongation ratio or the aspect ratio of the irregular particle. The chunkiness ratio thus gives a numerical range from zero to unity, instead of unity to infinity, as the breadth of the irregular particle decreases with length.

The comparison of an irregularly shaped particle with that of a more symmetrical body, such as a cube or sphere, which has a monodimensional value, also generates various shape factors. To these ends, measurement of the degree of asymmetry of an irregularly shaped particle to that of the symmetrical body was proposed by Wadell for petrological and geological powdered soil samples. Wadell (1932, 1933, 1935) defined the shape of an irregularly shaped particle as having either sphericity, ψ, or roundness, P.

Sphericity was defined as the ratio of the surface area of a sphere which has the same volume as an irregular particle to the actual surface area of the irregular particle. This can be expressed algebraically as:

$$\psi = \pi d_v^2 / \pi d_S^2 \tag{1.6}$$

but can also be expressed in a more practical manner as the working formula:

$$\psi = dc/Dc = \frac{\text{Diameter of a circle equal in area to the projected area of a particle in a stable plane}}{\text{Diameter of the smaller circle circumscribing the projection of the particle}}. \tag{1.7}$$

Wadell's roundness can be obtained by measurement of the radius of curvature (r) of each projection in the perimeter of an irregular particle. The sum of these radii can then be related to the radius of the maximum inscribed circle (R) by:

$$P = \sum (r/R)/N, \tag{1.8}$$

where N is the number of projections measured (Figure 1.17). As the corners of a particle are worn down, the radius of curvature, r, tends to R, and thus P tends to unity.

Wadell's roundness factor can be used to distinguish between the form of the particle and the roundness or surface perimeter, sometimes termed the roughness, of a particle. Thus, Wadell's roundness ratio is currently preferred over Wadell's sphericity because of the insensitivity of Wadell's sphericity ratio, which requires large changes in relative particle proportional dimensions of length and breadth to thickness, to produce significant changes in Wadell's sphericity, ψ (Table 1.14).

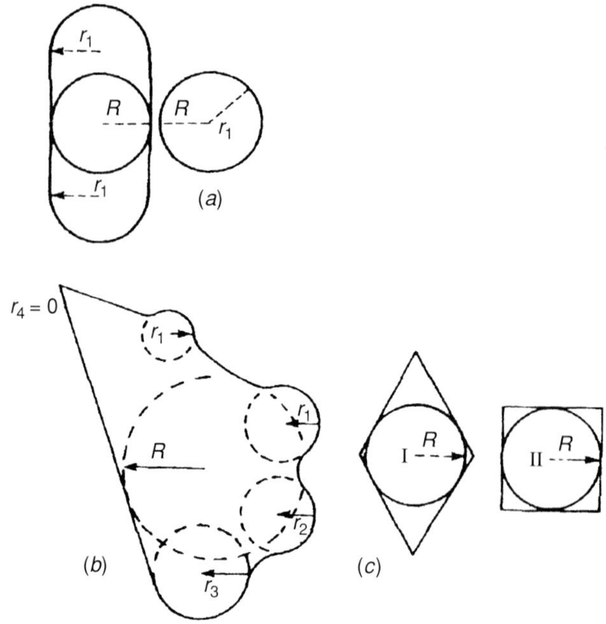

Figure 1.17 Wadell's shape factors for an irregular particle: (a) two particles with identical roundness ($P = 1$) but different forms; (b) Wadell's roundness for an irregular particle; (c) two particles with $P = 0$ but with different form.

Any shape factor which attempts to visualise and enumerate the inherent roughness of the surface of most real powdered materials is dependent on the measured fineness or surface area of particles, because the evaluation of the surface area or perimeter of an irregularly shaped particle changes with the measurement technique used to characterise an assembly of irregularly shaped particles. This can be illustrated by the sphericity value of a porous spherical particle which is relatively low in numerical value and may therefore be indistinguishable from the sphericity of a smooth but irregular and highly convoluted particle. The area and perimeter parameters from an irregular particle were used by Hausner (1966), who portrayed the complicated image of an irregularly shaped particle encompassed within a rectangle. The dimensional characteristics measured are then the side lengths of the rectangle, a and b, the projected area of the particle's image, A, and the circumference

Table 1.14 Shape coefficients for sections of a cube.

Relative dimensions of cube	α_o	α_s	ψ
Thickness = breadth	0.696	4.71	0.806
Thickness = breadth/2	0.348	3.14	0.761
Thickness = breadth/10	0.070	1.88	0.434
Thickness = breadth/100	0.007	1.60	0.110
Thickness = 0	0.0	1.57	0.0

of this area, C. Since the cross-sectional area and circumference of a spherical particle are $A = d^2\pi/4$ and $C = d\pi$, respectively, the relationship $C^2 = 4\pi A$ exists. Therefore, for a spherical particle, which gives the minimum surface area for a given volume, a Hausner shape factor can be defined as:

$$Z = C^2/12, \tag{1.9}$$

which has a value of unity for any spherical particle.

1.5.3 *Common industrial usage shape factors*

Prior to the practical recognition of the value of particle shape factors within industrial processing, such theoretical considerations, although of intellectual and academic interest, had little or no application in particle processing for many years. The practical application of shape analysis was addressed when Davies (1975) stated that since particle shape information was needed to describe the behaviour of particles in many industries in which particles were mixed, reacted, stored or transported, and although the main parameters still remained particle size, particle-size distribution, density, and surface interactions within complex particle processes, for the purposes of clarity, particle shape should be represented by:

(1) the general particle form, by an aspect ratio, a proportionality constant or a shape factor;
(2) the surface topography, by a roughness, angularity or roundness; or
(3) a combination of the above two categories.

These shape criteria compound the threefold functions of Herdan (see Section 1.5.1).

The methodology commonly used in shape characterisation nowadays is based upon either the computation of Fourier descriptive coefficients or dynamic shape factors together with fractal dimensional analysis. The particle shape is usually determined from sensing and analysing particle images. These images are then analysed in terms of the particle contour. Since automatic image analysers, with appropriate computer software, can now rapidly measure particle area, with and without porosity, perimeter, Feret diameter, longest chord length and number of particles (Figures 1.5 and 1.15), numerical values of the various shapes and forms of irregular particles can be readily obtained from ratios of these various image parameters. With the advent and development of automated microscope systems linked into and coupled with computers, the description of irregular particle shapes and characterisation of profiles by complex mathematical functions have now become available.

1.5.4 *Fourier transformation or signature waveforms of irregular particles*

There are three popular forms of the Fourier descriptors used in the analysis of an irregularly shaped particle profile (Hundal *et al.* 1997):

(1) Measurement and analysis of the radial dimensions, at various angles from the centre of gravity to the perimeter. This method, although used successfully for convex particles, is not applicable for particles which have contours which fold back on themselves, i.e. concave particles.

(2) Boundary co-ordinates in the complex plane, also known as the Granuland method.
(3) Angular bends as a function of arc length, also known as the Zahn–Roskies Fourier. This has been used in conjunction with computer neural networks to classify shapes by pattern recognition, a classification which cannot be achieved by the use of ordinary statistical methods.

Schwartz and Shane (1969) were the first to describe a two-dimensional profile of an irregularly shaped particle by a geometric radial signature waveform which could then undergo Fourier analysis (Flook 1982). The signature waveform was generated by locating, at a central position within the particle profile, a vector which touched the outline profile of the particle. The magnitude of the vector, at various angles when moved around the irregular profile, could either produce a graph of magnitude against angle or be treated as a continuous periodic harmonic waveform (Figure 1.15). The complex particle signature waveform can then be broken down by Fourier analysis (Beddow *et al.* 1976; Flook 1982) into a list which commonly used the first five harmonics, in a Fourier series of the simple harmonic wave, to uniquely characterise the shape of the irregular particles. A list of the Fourier harmonic series within the complex signature waveform gave a dimensionless shape description of the particle. The central location of the vector can be, as in the work of Schwartz and Shane, the centre of the smallest circle inscribing the particle, or can be at the centre of gravity of the particle (Erlich & Weinberg 1970; Meloy 1977). The higher harmonics (>5) were believed to give an indication of the texture of the irregular particle.

Kaye (1978) reviewed the work on Fourier shape analysis and found that the geometric signature method was, in reality, only of use for relatively rounded materials, such as geological fluvial sediments and spray-condensed metals or spray-dried products. Kaye, following the work of others (Austin *et al.* 1990), used the signature waveform methodology in combination with slip-chord analysis, to separate the edges of an irregular particle image and therefore introduced a new concept to the armoury of shape analysis, termed the angular facet signature waveform, which is now used in the shape characterisation of abrasives and comminuted materials (Kaye 1993).

Although computing power has increased, the amount of computing power required to perform these analyses, coupled with the complicity of the analysis, has confined this approach to research and thus has found only limited acceptability in industry.

1.5.5 *Fractal geometry of rugged irregular particles*

Mandelbrot (1977) applied fractal geometry to describe the ability of a convoluted line to fill two-dimensional space. A straight line, on a flat piece of paper, without any 'ups and downs' and without 'wiggles' has a topological dimension of unity. As the 'straight' line becomes more irregular in shape, wiggles up and down on the flat piece of paper and has an 'irregular wavelike' form, this line will occupy more two-dimensional space than a straight line. The fractal dimension thus becomes greater than one because of this irregular waveform or ruggedness of the line (Figure 1.15). The extended irregularities of the boundary of an irregular particle can then be likened to a convoluted line. Therefore, fractal geometry and the fractal dimension of the boundary of irregularly shaped profiles can be applied, as an alternative mathematical method, to obtain an index of ruggedness or texture of irregularly shaped particles. The length of a convoluted curve can be estimated from the product of the number of known fixed-size dimensional steps (λ) required to traverse a convoluted curve.

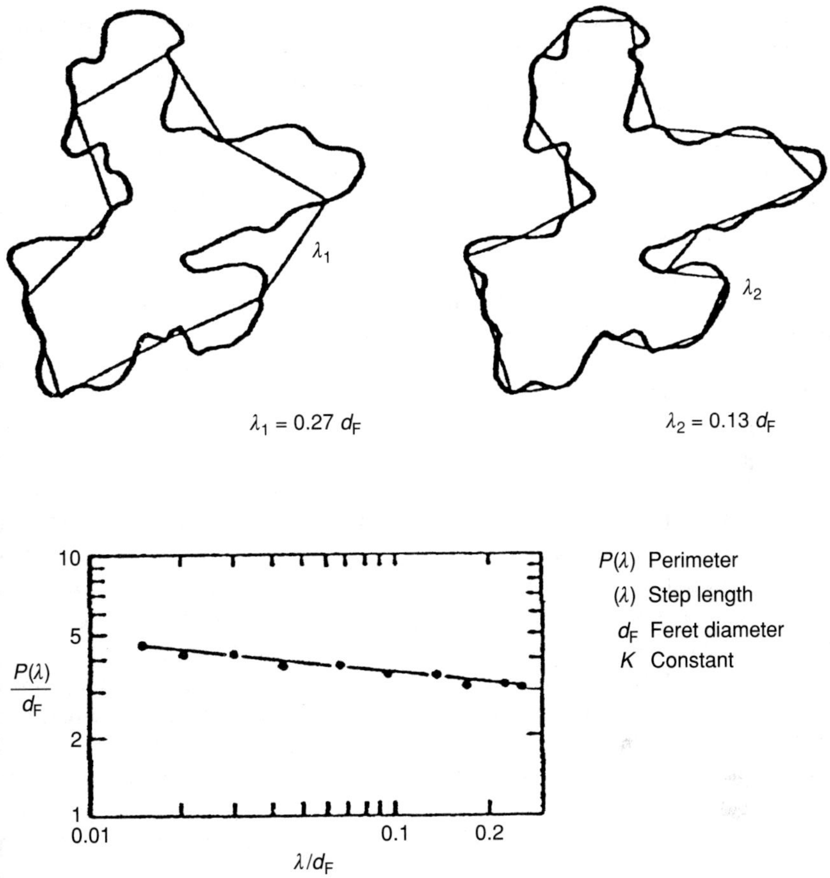

$$P(\lambda) = K\lambda^{(1-D)}$$

$$\ln(P) = (1 - D)\ln\lambda + \ln K$$

Figure 1.18 Fractal geometry of a rugged irregular particle. The two fractal step sizes, λ_1 and λ_2, are expressed in terms of the Feret diameter, d_F, of the irregular particle: 0.27 d_F and 0.13 d_F, respectively. The log graph shows the normalised values of the perimeter, $P(\lambda)/d_F$ vs. the normalised step length, λ/d_F to give a slope of $(1 - D)$, which is the fractal dimension of the rugged irregular particle.

As the size of the fixed step dimension decreases, the closer the zig-zag stepped curve will follow the convoluted curve being measured (P) until, eventually, a constant value of the length of the convoluted curve is reached (Figure 1.18).

Mandelbrot has shown that for a fractal curve, a plot of the logarithm of the estimated length of a convoluted curve, P, against the logarithm of the step size, λ, used gave a linear relationship which extrapolated to an infinite length. The slope of the log-log plot, however, gives a quantity (1-D) which is defined as the fractal dimension of the rugged boundary curve of an irregular particle. The Mandelbrot fractional dimension is a fractal which is solely a number between 1 and 2, and which can be used to describe the structure or shape of an irregular profile. Flook (1978) applied fractal dimensions to characterise the texture and structure of geometrically constructed model particles and the profiles of carbon black aggregates. He found, for carbon black aggregates, that the plot of logarithmic perimeter length against logarithmic step size showed a departure from linearity at large and small step sizes. At small step sizes, the fractal dimension – from the slope of the log-log plot – was dominated by the structure of the sub-unit, while at large step sizes the fractal dimension was due to the coarse structure or shape of the agglomerate. Flook concluded that it was now a practical proposition to use fractal dimensional measurements to characterise the textured and structured profiles of a wide variety of materials.

1.5.6 *On-line shape characterisation and measurement of particle shape*

In-line and on-line techniques for the characterisation of industrial powder processes and product engineering continue to attract the attention and demands of industry because particles passing through any physio-chemical process impact upon the performance of the process and the specification of the product. For most operations, the two most important parameters in process monitoring and control are particle size and morphology. With recent advances in sensor technology, it is now possible to implement both size measurement and shape morphology characterisation simultaneously either using two separate units, one for size and the other for shape (Bonifazi *et al.* 1995; Scott *et al.* 1998), or from a single unit (Castellini *et al.* 1993; Galai, Migdal Haemek, Israel; Lasentec, Redmond, WA). With laser-light illumination, both a particle-size distribution and a morphological particle shape characteristic can be obtained from a flowing stream of particles, using either a time-of-flight (TOF) or a time-of-transition (TOT) technique.

Although light-scattering techniques are not commonly used for particle-shape classification, there is within the scattering signature encoded information from which it should be possible to infer the shape of particles. By changing the geometric shape of the photodetectors, from a series of semi-circular concentric detectors to an array of wedge-shaped detectors in a segmented ring, the shape of particles may be measured (Heffels *et al.* 1994). To record the particle shape information from scattered light, the classical laser diffraction detection system of semi-ring-shaped detectors, which usually collects the radial spatial frequencies or intensities of the particle field, has to be changed to include wedge-shaped photodetectors. This detection system then collects encoded particle-shaped information in the form of the azimuthal spatial-intensity frequencies. The influence of shape on size measurement by light blockage and the characterisation of plate-like particles has been appraised for rapid-response instrumentation and in modified on-line light techniques (Chin *et al.* 1988).

1.5.7 *Particle image contour analysis*

Morphology is best investigated by automated image analysis because the increase in reproducibility of a set of shape factors, obtained from automated data collection, is greater than the variation seen when recorded data are obtained from human image analysis. Non-automated systems are prone to operator fatigue and subjective visual recognition.

It has been recognised that three steps are necessary to describe the morphology of particles:

(1) There must be means to visualise the particle image, since shape is fundamental to an image. An image, with the introduction of digital cameras, can now contain lines of picture elements (pixels) where each pixel can correspond to an initial analogue image.
(2) Image treatment. Since only a relatively small number of particle images are to be characterised, it is essential that a representative sample be viewed. The particles should be separate and well dispersed to yield as clear a resolution of the particle as possible.
(3) Shape characterisation. Since irregularly shaped particles tend to take up the most stable position on a flat surface, the morphology of imaged particles is now termed the bidimensional shape and based on the particle silhouette or contour line.

Pons *et al.* (1999) have defined two classes of shape descriptors, the macroshape descriptor in which measurements of the silhouette surface, S, the maximum Feret (F_{max}) and minimum Feret (F_{min}) diameter are measured together with the geodesic length (L_g) and Euclidean thickness (E) (Figure 1.19). Table 1.15 shows the main macroscopical shape descriptors from these measurements where the equivalent circular diameter (D_{eq}) is equal to $2(S/\pi)^{1/2}$.

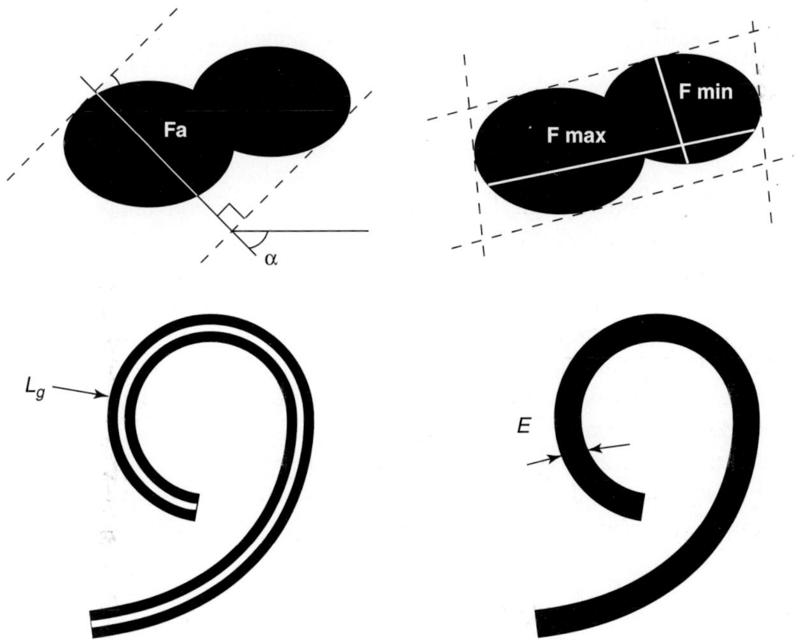

Figure 1.19 Dimensions of convex and elongated concave particles to calculate shape descriptors (after Pons *et al.*). F_{min}, minimum Feret diameter; F_{max}, maximum Feret diameter; F_a, Feret diameter, at angle α; L_g, longest geodesic length; E, Euclidian thickness.

Table 1.15 Main macroscopical shape descriptors.

Name	Formula	Comments
Form factor	$P^2/4\pi S$ or $4\pi S/P^2$	Compares the surface of the object (S) to the surface of the disc of the same perimeter (P)
Roundness	$4S/\pi F_{\max}^2 = D_{eq}/F_{\max}$	Compares the surface of the object (S) to the surface of the disc of the same diameter ($F_{\max}$)
Aspect ratio	$F_{\max}/F_{\min}$	
Elongation	L_G/E	Aspect ratio substitute for a very elongated particle (fibre)
Curl	$F_{\max}/L_G$	For very elongated objects
Convexity	P/P_C	P_C is the perimeter of the convex bounding polygon
Solidity	S/S_C	S_C is the surface of the convex bounding polygon
Compactness	$\sqrt{\text{Roundness}}$	
Extent	$S/(F_{\max} F_{\min})$	

The other class of shape descriptor is the mesoshape descriptor, which can give a more detailed appreciation of shape because comparison is made with a reference shape. These mesoshape descriptors have found applicability in the investigation of abrasion of particles or the erosion of, and change in, the silhouette of a mathematically defined polygon (Pons *et al.* 1999).

With the advent of dynamic image data collection, the particle image contour can be represented as a fractal dimensional shape factor or an aspect ratio (or its reciprocal chunkiness), or transformed by Fourier analysis. It is now currently believed and accepted that a combination of both size and shape measurements will lead the way forward to the characterisation of mixed particles species, complex designed particles and varied sized, shaped, mixed and complex multi-functional agglomerated systems (Davies 1996).

1.6 Particle hardness

Particle hardness is a powder characteristic which always has to be considered when powdered solids are transported by mechanical conveying, flowing from hoppers and silos, or when subjected to any in-plant movement.

If particles are harder than the plant equipment, erosion and degradation of plant surfaces will ensue, whereas if particles or agglomerates have a weak nature or constitution and undergo any movement, either within the powder (shearing) or by any means of transportation, the initial particle or granule size may be reduced. Particle breakdown tends to be a disadvantage in most processes and may ultimately lead to a hazardous environment or even cause a dust explosion.

The measurement of hardness of small particles can be approached from either a physical or scientific viewpoint, which is generally complicated and time-demanding, or alternatively particle hardness can be assessed from an industrial, quick and easy, but reliable, measurement protocol.

1.6.1 *Particle and material hardness by indentation: General principles*

Static indentation hardness testers are mainly used for measurement and assessment of the yield of metals used in the construction of plant. The form of the test piece is generally a flat plate or slab, but at times, indentation tests can be used on large particles.

Table 1.16 National and international standard test methods of the assessment of hardness.

BS 1881: Part 4: 1970	Brazilian test. Methods of testing concrete for strength
BS 1881: Part 4: 1983	Brazilian test. Methods of testing concrete for strength
BS 1881: Part 5: 1970	Methods of testing hardened concrete for qualities other than strength
BS 1881: Parts 114, 115, 116, 117, 118, 119, 120: 1983	Methods of testing concrete for strength
BS EN ISO 6508-1: 1999	Metallic materials – Rockwell hardness test – Part 1: Test method (Scales A B C D E F G H K N T)[a]
BS EN ISO 6508-2: 1999	Metallic materials – Rockwell hardness test – Part 2: Verification and calibration of testing machines (Scales A B C D E F G H K N T)[a]
BS EN ISO 6508-3: 1999	Metallic materials – Rockwell hardness test – Part 3: Calibration of reference blocks (Scales A B C D E F G H K N T)[a]
EN ISO 6507-1: 1997	Metallic materials – Vickers hardness test – Part 1: Test method
EN ISO 6507-2: 1997	Metallic materials – Vickers hardness test – Part 2: Verification and calibration of testing machines
EN ISO 6507-3: 1997	Metallic materials – Vickers hardness test – Part 3: Calibration of reference blocks
ISO 6506: 1981	Metallic materials – Hardness test – Brinell test
ISO 3876: 1983	Hard metal – Vickers hardness test
ISO 6507-1: 1997	Metallic materials – Vickers hardness test. Part 1. Test method
ISO 3738-1: 1982	Hard metal – Rockwell hardness test
ISO 6508: 1997	Metallic materials – Rockwell hardness test. Part 1. Test method
ISO 4545: 1993	Metallic materials – Hardness test – Knoop hardness testing
ISO 2039-1: 1997	Determination of plastics. Part 1. Ball indentation

[a] Scales A, C & D for a diamond cone indenter; Scales B, F & G for either a 1.5875 or 3.175 mm ball; Scales N & T on a superficial test for convex cylindrical surfaces.

Macro-indentation hardness tests are achieved from knowledge of the applied load (kilogram) to a hardened metal probe, which penetrates and indents into the substrate of the material. Subsequent measurement of the ensuing indent diagonals (millimetre) evaluates the degree of hardness of the material. These tests have been used for many years and have been standardised by the authorities which publish British Standards (BS), European Standards (EN) and International Standards (ISO), or the combination of British, European and International test protocols (Table 1.16).

Micro-indentation is, however, a relatively new protocol arising from miniaturisation of the loaded indenter, which can operate over a range of lower loads or forces (0.1–5.0 N). Miniaturisation of the hardness probe or indenter enables particulate materials to be tested. Assessment of the hardness or yield of particles in sizes above 100 μm can thus occur because the indenter has a size of a few micrometres. The micro-indentation diagonals are generally measured from a magnified image of the indent. The relationship between the logarithmic value of the applied load (P) and the logarithmic average length of the indent diagonal (d) lies between two asymptotes of slope 2, indicating that hardness is a function of indent diagonal squared (Wyatt & Dew-Hughes 1974).

1.6.2 *Macro- and micro-indentation hardness numbers*

The Vickers static indentation test uses a diamond pyramid indenter, which has an included angle of 136°. The diamond pyramid number (DPN) or Vickers hardness number (H_v) is calculated from the applied load, P (kg), divided by the actual area of the indent impression (mm^2). The value is independent of the depth of penetration because all indentations are geometrically similar. The load is automatically applied for 10 s and the diagonal (d) of the indent measured from the following algebraic relationship:

$$H_v = \text{DPN} = P/\text{pyramidal area of indentation.} \tag{1.10}$$

Since the angle of indenter is 136° between opposite faces and 146° between opposite edges, the pyramidal area of Vickers indent is greater than the observed flat plane projected area of the indentation seen by image magnification, by the ratio 1:0.9272.

Thus:

$$\begin{aligned} H_v = \text{DPN} &= 0.9272P/\text{projected area of indentation} \\ &= 1.8544P/d^2 \text{ (kg mm}^{-2}\text{)}, \end{aligned} \tag{1.11}$$

which becomes:

$$H_v = 1.89032 \times 10^6 F/d^2, \tag{1.12}$$

where F is the applied force in Newton's (F = Pg, where P is applied load and g is the gravitational constant) and d the diagonal (mm) (Figure 1.20).

Since hardness can be expressed in terms of the area of indentation, a valid logarithmic index may also be obtained with a ball indent (Knoop and Shore ISO 2039-1: 1997), which has a similar protocol and method of test as seen with a Vickers pyramidal indent (ISO 6507-1: 1997). Brinell (ISO 156: 1982) Vickers (ISO 716: 1986) and Rockwell are valid macro-indentation tests and currently used extensively in the metallurgical industries (Table 1.16).

The micro-indentation hardness of particles can be evaluated in a similar fashion to that of macro-indentation, the diagonal of the indent being measured in micrometres, with the aid of an optical microscope and the Vickers number calculated from Mott (1956):

Precise microhardness measurements by determining the depth of the indent via computer-controlled monitoring of the position of the indenter to the surface are now achievable. Scanning indentation mechanical microprobes (SIMM) can also test the hardness of complex coatings and inhomogeneous structures (National Physical Laboratory, Teddington, UK).

1.6.3 *Particle fracture: Griffith crack fracture and stress-intensity factors*

Dependent upon the degree of force applied to a solid surface, via an indenter, radial cracks may originate from the sides or corners of the indent impression in addition to the impression created by the pyramidal indenter. These indentation cracks can be used, in addition to the indent size, to calculate hardness numbers and thus characterise the mechanical properties of discrete or compacted particles.

In terms of stress intensity factors, Marshall and Lawn (1979) stated that in fracture mechanics or the study of the phenomena of instability within a particulate system undergoing cracking, two mechanisms may be in operation, dependent upon the degree of plasticity and elasticity within the system.

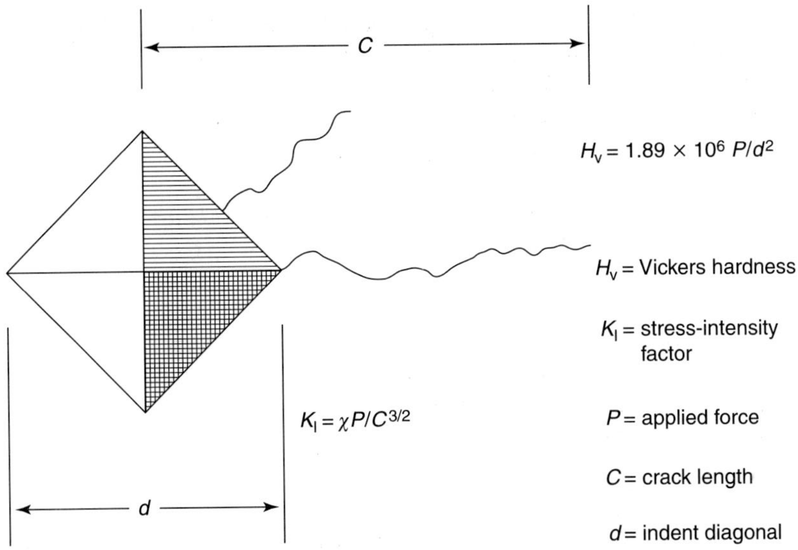

Combination of stress-intensity factors

$$K_{Ic} = \phi\, \sigma_a C^{1/2} + \chi P/C^{3/2}$$

(a) Stress-intensity factor due to crack arrest

$$K_{Ia} = \phi\, \sigma_a C^{1/2}$$

ϕ = crack geometry constant [approx. 1.0]
σ_a = applied tensile stress
C = crack length

(b) Stress intensity from crack initiation due to indentation

$$K_I = \chi\ P/C^{3/2}$$

χ = dimensionless indenter parameter. A function of elastic modulus E and hardness H
P = applied force on indenter

Figure 1.20 Hardness and stress intensity factors from indentation and crack length.

The classical work of Griffith (1970) suggested that the size of a flaw, crack or void within a system remained stationary up to a critical load. Upon increasing the load, the particulate system could then become unstable and fracture into radial cracks. This material characteristic – commonly termed toughness – can be expressed mathematically in terms of a stress intensity factor, K_I.

$$K_I = \chi P/C^{3/2}, \tag{1.13}$$

where P is the applied load or force on the indenter, C is the crack length from an indent side, and χ is a dimensionless indenter parameter which determines the intensity of the indentation stress field. For an ideal elastic-plastic material, χ is equivalent to $(E/H)^{1/2}$

$(\cot\theta)^{2/3}$ where θ is the indenter half angle, which, for a Vickers diamond indenter, has a value of 72°, and E and H are the elastic modulus and hardness of the material, respectively.

The application of an applied tensile stress (σ_a), obtained from the indent area and applied force, creates a significant degree of precursive extension of the particle material which precedes the formation of fracture or beginning of the crack phase, in all indentation experiments.

This contribution to particle fracture can be mathematically expressed as an additional stress intensity factor, K_{Ia}, which is due to crack arrest, where

$$K_{Ia} = \phi\sigma_a C^{1/2}, \tag{1.14}$$

where ϕ is a crack geometry constant, which is of the order of unity and the other symbols as above. It should be noted that K_{Ia}, the stress-intensity arrest factor, is obtained from the crack length, and not from the indent diagonal.

The resistance to particle fracture can thus be summed as:

$$K_{Ic} = \phi\sigma_a C^{1/2} + \chi P/C^{3/2}, \tag{1.15}$$

where K_{Ic} is known as the critical stress-intensity factor which confusingly is also known as particle toughness. The indentation crack lengths are measured from one side of the indent impression to the crack end, which is conventionally termed the crack half length.

While commercial equipment is available for the measurement of Vickers, Rockwell, Brinell, and Shore Numbers to characterise the nature of large crystalline particles containing location slip planes within the crystal habit and for granulates with internal porosity, these indentation techniques do not lend themselves easily to the determination of the 'hardness or toughness' of particulate products.

1.6.4 *Mohs' scale of hardness*

In the late eighteenth century and early nineteenth century, attempts were made to categorise and to place in some kind of order the many metals, ores and natural minerals available to metallurgists and the scientific and engineering arena. To these ends, Mohs (1820, 1825a) began to assemble known non-metallic and metallic materials into various classes, orders, genera and species to, in his words '*. . . enable students to discriminate minerals on principles similar to those of botany and zoology*'.

The physical properties investigated, described and classified were:

(1) Odour
(2) Taste
  (a) Astringent
  (b) Sweetish
  (c) Saline
  (d) Alkaline
  (e) Cooling
  (f) Bitter
(3) Electricity
(4) Magnetism
(5) Specific gravity
(6) Hardness

(7) State of aggregation
  (a) Brittle: Those particles which lose their coherence with the aid of a knife
  (b) Malleable: From a malleable mineral, slices of the material may be detached, as from metallic lead
  (c) Sectile: Intermediate stage between malleable and brittle; particles lose their connection but do not allow separation by slicing
  (d) Ductile: Can be wrought into sheets or wire. [On the] application of force, particles change but do not lose their connection
  (e) Flexible: Do not resume their former situation but are neither ductile nor malleable
  (f) Elastic: Resume their former situation.

Of these seven physical properties of minerals, hardness is the major physical property which has stood the test of time, mainly due to a scale devised by Mohs.

The mechanism of particle hardness, or degree of abrasion, can be visualised by the 'scratchiness' of one material upon another. Mohs stated that, '*Although the degrees of hardness among minerals is easily discovered by the simple experiment of scratch, the establishment of an accurate scale is attended with great difficulty*'. Mohs (1825b) selected ten minerals ranging from talc, with a Mohs hardness of unity, to diamond, with a scratch hardness of 10, to generate a scale. In the Mohs scale, the material with a hardness factor of 1 (talc) could not scratch a material with a hardness factor of 2 (gypsum) ending with diamond (Mohs factor 10). When the Mohs scale was originally proposed, diamond could not be scratched by any other material (Table 1.17 and Figure 1.21).

Table 1.17 Mohs hardness scale.

Mohs' hardness scale	Material	Comments and explanation
1	Talc	Very soft; can be powdered between fingers
2	Gypsum (hexahedral rock salt)	Moderately soft; can scratch lead
3	Calcite (rhombohedral lime)	Can scratch a fingernail
4	Fluorite	Can scratch copper
5	Apatite	Can scratch a knife blade with difficulty
6	Orthoclase feldspar (or prismatic feldspar)	Can scratch a knife blade
7	Quartz (rhombohederal)	
8	Topaz (prismatic)	All particles harder than Mohs Number 6 will scratch glass
9	Corundum (silicon carbide)	
10	Diamond (octahedral)	

There are, however, a number of caveats to the Mohs scale, which were highlighted in Mohs' original table translated from German by William Haidinger (Mohs 1825a, b).

(1) 'The interval[s] between members of the scale are not of the same magnitude. Between 5–6 is greater than it should be'.
(2) '[It is] very difficult to ascertain the perfect quality of the interval between the different degrees of hardness'.
(3) 'The force applied in this experiment [should] be always the least possible'.

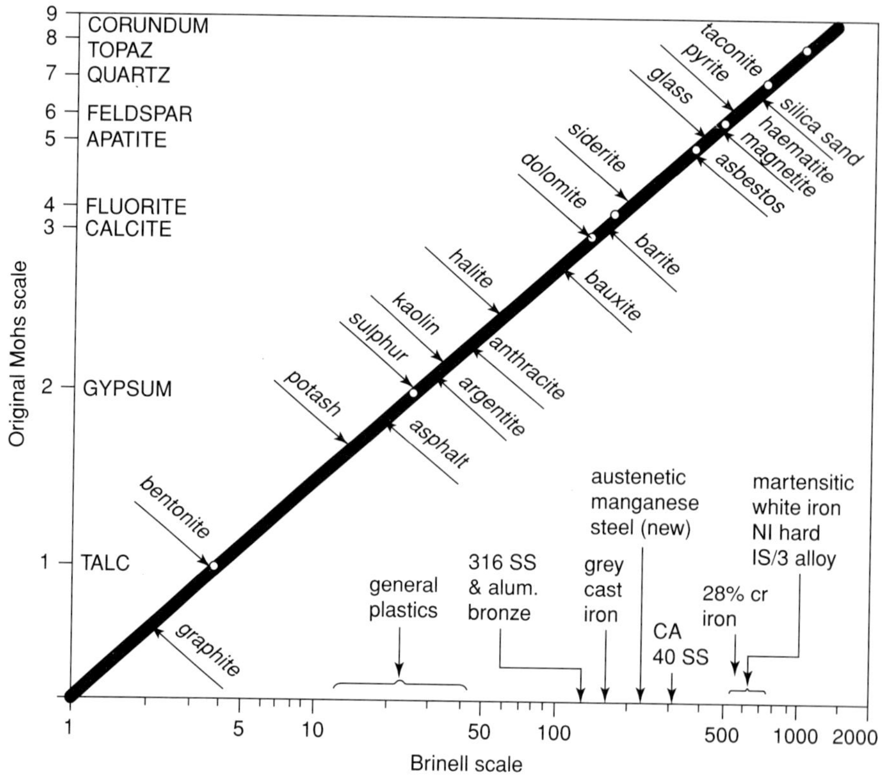

Figure 1.21 Relationship between Brinell Hardness and Mohs scale of hardness.

1.6.5 *Abrasion, attrition and fragmentation*

Wear on particles can be regarded as abrasion, as opposed to erosion by particles on plant. Abrasion is associated with the blunting of the corners and the 'smoothing off' of particle edges, together with the removal of protuberances from the surface of particles to give a more rounded profile. The term 'attrition' tends to be used when a particle divides into two or more smaller particles. If the particles divide into sizes that are only slightly smaller than the initial size, this can be regarded as fragmentation, the body of the particulate material being shattered.

The physical quality of particles, pellets and other agglomerated material may be assessed by tumbling the powdered material in a rotating drum. The broken down material of a known size can then be quantified (ISO 3271: 1975 *Tumble Drum* and ISO 3271: 1995 *Iron ores – Determination of tumble strength*). Alternately, the compression of a bed of powder – a compression test – as outlined in ISO document 283 (1974) from ISO Technical Committee, Subject Committee 3, may be used.

1.6.5.1 *Abrasive wear* Whenever particles are moving in plant, either by gravity, mechanical conveying or being mixed by paddles or spirals, abrasive wear can occur to both plant surfaces (erosion) and also abrasion to the outer profile of particles. A grindability factor for ores and coal has been described in ISO 5074: 1994. *Hardcoal. Determination of Hardgrove grindability index*, while the protocol for the determination of the abrasiveness of coal is described in ISO 12900: 1997. *Coal. Determination of abrasiveness.* In addition, particles and agglomerates may also be subjected to attrition and fragmentation if the product is friable, friability being the tendency of particles to break or fragment during handling or under pressure.

The initial stages of abrasive wear are believed to begin by plastic deformation of the plant surface resulting in groove formation and thus the 'roughing up' of surfaces. Prolonged abrasion results in the separation or generation of particles from the containing walls and possible degradation of particle sizes and shapes. Prolonged periods of abrasive wear can lead to a decrease in the elastic modulus and possible onset of fatigue failure within plant.

The way in which particles flow over plant surfaces, the pattern of flow in mechanical equipment as well as particle–particle interactions, other than Jenike type shear, is, however, poorly understood. A number of variables, however, can be listed to predict abrasive wear (Johanson & Royal 1982).

1.6.5.2 *Particle size* The volume of wear increases as particle diameter increases up to a critical size. With all particle sizes the attack angle of particles, either with the plant wall or any internal structures to aid production, agglomeration, size enlargement, size reduction and mixing, is also of importance.

1.6.5.3 *Particle-size distribution* The closer particles pack together the lower the voidage within the powder mass becomes and therefore the greater the wear. This may be due to the increase in the proximity of particles and probably the greater number of particle–particle contacts. Increase in packing tends to increase when powder masses have a wide distribution of sizes, with the smaller particles fitting between and into the voids generated by the larger particles.

1.6.5.4 *Particle shape* Increase in the angularity of particles increases wear because of localised stresses between particle and wall surfaces. If the particles suffer abrasion and become 'rounded', they might roll over each other and/or the walls, thus decreasing wear.

1.6.5.5 *Particle hardness* The abrasive particles must be harder than the wall surface for wall wear to occur. The transition of plastic deformation to tearing of surfaces to ultimate fatigue is a function of material hardness.

1.6.5.6 *Moisture and temperature* Increase in moisture tends to increase the cohesiveness of the powdered material. This may have two effects:

(1) Wear rate may be greater if the rolling of particles is decreased.
(2) However, if the degree of moisture is increased appreciably, this can act as a lubricant and therefore decrease the rate of abrasion.

Elevated temperatures can affect the hardness of both particles and wear surfaces (Corder & Thorpe 1987).

1.6.5.7 *Attrition* Attrition is a common problem in bulk powder handing and the process industries. Like abrasion, little is known about the mechanisms by which attrition occurs. Some success has been achieved, however, in assessing the relative attritability of materials by using devices which mimic the industrial process to be monitored (Beinrose & Bridgwater 1987; British Materials Handling Board 1987):

(1) small-scale grinding (Bond 1961)
(2) fluidised beds (Forsythe & Hertwig 1949)
(3) shear attrition cells (Paramanathan & Bridgwater 1983)
(4) impact fracture analysis (Ghadiri & Yuregir 1987; Yuregir *et al.* 1986).

Three methods which have found application in the assessment of industrial attrition (Knight & Bridgwater 1985) are:

(1) an annular attrition shear cell
(2) single particle compression
(3) a test where powder is blown inside a spiral air classifier

In the first and third methods, attrition was measured by determining the mass of attritable particles passing through a 500 μm sieve. The compression test strengths were measured by compressing individual particles between two anvils. Single-particle fracture mainly

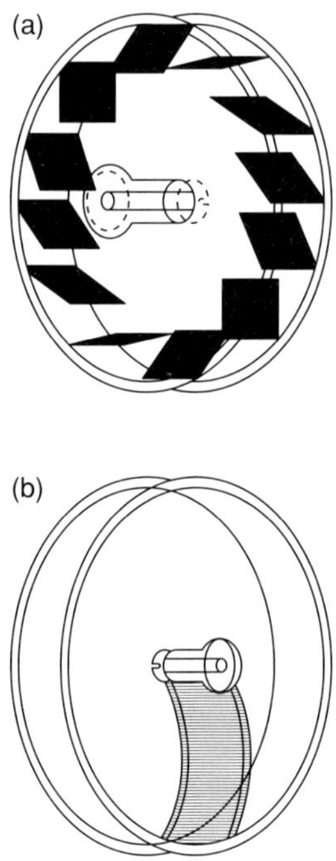

Figure 1.22 Schematic of tumbling drums used in the pharmaceutical industry: (a) abrasion drum; (b) friability drum.

occurred diametrically across the particle placed between the two plates. The diametric compaction strength, σ_f, using the theory of Frocht (1948), can be expressed as:

$$\sigma_f = 2F/Dt\pi, \tag{1.16}$$

where F is the applied load (Newtons), D the particle diameter (mm) and t the thickness (mm) of the particle. This test was originally developed for an indirect tensile strength test for drilled cores of concrete and is also known as the 'Brazilian test' (BSI 1983) (Table 1.16).

The single compression test does not, however, provide a reliable test for attrition, but the annular attrition cell and spiral classifier showed good agreement on an empirical correlation basis between laboratory test and industrial processes. These two techniques may thus be used to compare different methods in assessing powder attrition.

1.6.5.8 *Friability* This term tends to be commonly used in the pharmaceutical industry, and although it can be applied to particles and agglomerates, it was initially used to assess the breakdown or abrasion of tableted surfaces. Figures 1.22a and 1.22b show the equipment used to create the circular motion to tumble material. The amount of breakup within the apparatus, after any desired period of time, can then be determined by the mass of the broken down material which then gives the degree of friability of different products. This can be sieved and sized in a similar manner used in the ISO Tumble Drum test ISO 3271: 1975 and 1995.

References

Ah Chin, A.D., Bulter, P.B. & Luerkens, D.W. (1988) *Powder Technol.*, **54**, 99–105.
Allegra, J.R. & Hawley, S.A. (1972) *J. Acoust. Soc. Am.*, **51**, 1545–64.
Allen, T. & Nelson, R. (1996) *Powder Technol.*, **88**, 11.
Austin, L., Yekeler, M., Dumm, T. & Hogg, R. (1990) *Part. Part. Syst Charact.*, **7**(4), 169–254.
BCR (1980) *BCR Information EUR 6825 EN Certification report on reference materials of defined particle size: Quartz BCR No. 66, 67, 68, 69, 70.* Commission of the European Communities, Office for Official Publications of the European Communities, Luxembourg.
BCR (1989) *The certification of nitrogen bet specific surface areas of six materials in the range 0.1 to 8* m^2g^{-1}. Community Bureau of Reference (BCR), Commission of the European Communities, Brussels (via Joint Research Centre, Institute for Reference Materials and Measurements (IRMM), Geel, Belgium).
Beddow, J.K., Vetter, A.F. & Sissan, K. (1976) *Powder Metall. Int.*, **8**(2), 69.
Beinrose, C.R. & Bridgwater, J. (1987) *Powder Technol.*, **49**, 97–126.
Bond, F.C. (1961) *Br. Chem. Eng.*, **6**, 378–85.
Bonifazi, G., Francini, F., Guamierri, V., Longobardi, G., Massacci, P. and Recinella, M. (1995) *Part. Part. Syst. Charact.*, **12**, 318–323.
British Materials Handling Board (1987) *Particle Attrition State of the Art Review*. Trans Tech Publications Inc., Germany.
British Standards (1993) *Glossary of terms relating to particle technology. BS 2955 1993.*
Brunauer, S., Emmett, P.H. & Teller, E.J. (1935) *J. Am. Chem. Soc.*, **60**, 309.
BSI (1958) *BS 2955.* BSI Standards, London.
BSI (1983) *British Standard, 1881 Part 4.* BSI Group, London www.bsi-global.com
Castellini, C., Francini, F., Longobardi, G. & Pampaloni, E. (1993) *Part. Part. Syst. Charact.*, **10**, 7–10.
Church, T. (1968) *Powder Technol.*, **2**(1), 27–31.
Corder, G.D. & Thorpe, R.B. (1987) Tribolology. In: *Particulate Technology* (eds B.J. Briscoe & M.J. Adams). Adam Higer, Bristol, UK.
DallaValle, M.J. (1943) *Micromeritics. The Technology of Fine Particles*. Pitman, London.
Davies, R. (1975) *Powder Technol.*, **12**, 111–124.
Davies, R. (1996) *Powder Technol.*, **88**, 191–196.

Donald, M.B. (1989) *Elizabethan Copper, The History of the Company of Mines Royal 1568–1605*. Reprinted by Michael Moon, Whitehaven, UK.
Dwyer, J.L., Manalon, D.A. & Merbon, R.R.A. (1970) *Proceedings of Second International Conference on Particle Size Analysis*. Society for Analytical Chemistry, London.
Erlich, B. & Weinberg, B. (1970) *J. Sediment. Petrol.*, **40**, 205.
Flook, A.G. (1978) *Powder Technol.*, **21**, 295–298.
Flook, A. (1982) In: *Proceedings of the 4th Particle Size Analysis Conference, Loughborough University of Technology. September 1981* (eds N. Stanley-Wood & T. Allen). Wiley Heyden, London.
Forsythe, W.L. & Hertwig, W.R. (1949) *Ind. Eng. Chem.*, 41, 1200–6.
Frocht, M.M. (1948) *Photoelasticity*, Vol. 2. Wiley, New York. Also *Proceedings of the Symposium on Photo-elasticity*, Chicago 1961. Pergamon, Oxford, 1963.
Ghadiri, M. & Yuregir, K.R. (1987) Tribolology. In: *Particulate Technology* (eds B. J. Briscoe & M. J. Adams). Adam Higer, Bristol, UK.
Griffith, A.A. (1970) *Phil. Trans. Roy. Soc.* **A221**, 163.
Hausner, H.H. (1966) *Proceedings of the Conference on Particle Size Analysis*. Society for Analytical Chemistry, London.
Hawkins, A. (1993) The shape of powder – particle outlines. In: *Series on Materials Science and Technology* (ed. G.W. Greenwood). Research Studies Press/Wiley, Taunton, UK/New York.
Heffels, C.M., Heitzmann, D., Dan Hirleman, E. & Scarlett, B. (1994) *Part. Part. Syst. Charact.*, **11**, 194.
Herdan, G. (1952) *Small Particle Statistics*. Butterworth, Sevenoaks, UK.
Heywood, H. (1937) *Chem. Ind.*, **15**, 149–54.
Heywood, H. (1947) *Symposium on Particle Size Analysis*. Suppl. 25, 14. Institution of Chemical Engineers, London.
Heywood, H. (1954) *J. Imp. College, Chem. Eng. Soc.*, **8**, 25–33.
Heywood, H. (1966) *Proceedings of the 1st Particle Size Analysis Conf. September 1966*, pp. 355–59.
Heywood, H. (1967) *Proceedings of the 1st Particle Size Conference, September 1966.*
Hundal, H.S., Rohani, S., Wood, H.C. & Pons, M.N. (1997) *Powder Technol.*, **91**, 217–27.
ISO (1990a) *ISO 3310-1:1990 Test Sieves – Technical requirements and testing – Part 1: Test sieves of metal wire cloth.*
ISO (1990b) *ISO 3310-3:1990 Test Sieves – Technical requirements and testing – Part 3: Test sieves of electro-formed sheets.*
ISO (1995a) *ISO 9276-1:1995 Representation of results of particle size analysis. Part 1: Graphic representation.* International Organisation for Standardisation, Geneva.
ISO (1995b) *BSI:1995 BS 4359 Part 1: 1996 ISO 9277:1995 Determination of the specific surface area of powders. Part 1. BET method of gas adsorption for solids (including porous materials).* BSI Standards, London and Commission of the European Communities, Office for Official Publications of the European Communities, Luxembourg.
ISO (1998) *ISO 9276-2:1998 Representation of results of particle size analysis. Part 2: Calculation of Average Particle Sizes/Diameter and Moments from Particle Size Distributions.* International Organisation for Standardisation, Geneva.
ISO (1999a) *BS ISO 9044:1999. Industrial woven wire cloth – Technical requirements and testing.*
ISO (1999b) *ISO 3310-2:1999 Test Sieves – Technical requirements and testing – Part 2: Test sieves of perforated metal plate.*
ISO (2001) *BS ISO 13317-1:2001. Determination of particle size by gravitational liquid sedimentation methods – Part 1: General principles and guidelines.* BSI Standards, London.
Johanson, J.R. & Royal, T.A. (1982) *Bulk Solids Handling*, **2**, 517–23.
Kamack H.J. (1951) *J. Anal. Chem.*, **23**, 844.
Kaye, B. (1981) *Direct Characterisation of Fine Particles*. Wiley, New York.
Kaye, B. (1993) *Part. Part. Syst. Charact.*, **10**(3), 99–110.
Kaye, B. & Clark, G. (1992) *Part. Part. Syst. Charact.*, **9**(1), 1–82.
Kaye, B.H. (1978) *Powder Technol.*, **21**, 1.
Knight, P.C. & Bridgwater, J. (1985) *Powder Technol.*, **44**, 99–102.
Koppel D. (1972) *J. Chem. Phys.*, **57**, 4814–20.
Mandelbrot, B. (1957) *Science, NY*, **156**, 636.
Mandelbrot, B.P. (1977) *Fractals, Forms, Chance and Dimension*. WH Freeman, San Francisco.
Marshall, D.B. & Lawn, B.R. (1979) *J. Mater. Sci.*, **14**, 2001–12.
Meloy, T.P. (1977) *Powder Technol.*, **17**, 27.
Mohs, F. (1820) *The Characters of the Classes, Orders, Genera and Species or the Characteristic of Natural History System of Mineralogy*. Printed for W & C Tait, Edinburgh. British Library, London.
Mohs, F. (1825a) *Treatise on Mineralogy or the Natural History of the Mineral Kingdom by Fredrick Mohs (Professor in Mining, Academy of Freiburg).* Translated from German with considerable additions by William

Haidinger, FRSE. Vols 1, 11 and 111. Printed for Archibald Constable & Co., Edinburgh and Hurst, Robinson & Co, London. British Library, London.
Mohs, F. (1825b) *Treatise on Mineralogy or the Natural History of the Mineral Kingdom by Fredrick Mohs (Professor in Mining, Academy of Freiburg)*. Translated from German with considerable additions by William Haidinger, FRSE. Vol. 111. Printed for Archibald Constable & Co., Edinburgh and Hurst, Robinson & Co, London. British Library, London.
Mott, B.W. (1956) *Micro-Indentation Hardness Testing*. Butterworth, London.
Oliver, R. & Ford, L.J. (1987) *Proceedings of the Institute of Briquetting and Agglomeration (IBA)*, p. 20. IBA, Erie, PA.
Paramanathan, B.K. & Bridgwater, J. (1983) *Chem. Eng. Sci.*, **38**, 207–24.
Pilpel, N. (1969) *Paint Mfn.*, July 23.
Pons, M.N., Vivier, H., Belaroui, K., Bernard-Michel, B., Cordier, F., Oulhana, D. & Dodds, J.A. (1999) *Powder Technol.*, 103, 44–57.
Groves, M.J. & Wyatt-Sargent, J.L. (eds) (1970) *Proceedings of Particle Size Analysis 1970*. Society for Analytical Chemistry, London.
Rumpf, H. (1990) *Particle Technology*. Chapman & Hall, London.
Schwartz, H.P. & Shane, K.C. (1969) *Sedimentology*, **13**, 213–31.
Scott, D.M., Boxman, A. & Davies, R. (1998) CD-ROM, pp. 165–189. Institution of Chemical Engineers, Rugby, UK.
Scott, D.M., Boxman, A. & Jochen, C. (eds) (1998) *Part. Part. Syst. Charact.*, **15**, 47–50.
Sing, K.S.W., Everett, D.H., Haul, R.A.W., Moscou, L., Pierotti, R.A., Rouquerol, J. & Siemieniewska, T. (1985) Reporting physisorption data for gas/solid systems with special reference to the determination of surface area and porosity. *Pure Appl. Chem.*, **57**(4), 603–19.
Singer, C., Holmyard, E.J., Hall, A.R. & Williams, T.T. (1954–1956) *History of Technology*, Vols 1–5. Clarendon Press, Oxford.
Stanley-Wood, N., Sarrafi, M., Mavere, Z. & Schaefer, M. (1993) *Adv. Powder Technol.*, **4**(1), 33–40.
Stanley-Wood, N.G. (1983) *Enlargement and Compaction of Particulate Solids*. Butterworths, London.
Stanley-Wood, N.G. (2000) Particle size analysis: Introduction. In: *Encyclopaedia of Analytical Chemistry; Applications. Theory and Instrumentation* (ed. R. A. Myers), Vol. 6, pp. 5301–37. Wiley, Chichester, UK.
Trachtler, A. & Mesch, F. (1998) *Part. Part. Syst. Charact.*, **15**, 30–35.
Wadell, H. (1932) *J. Geol.*, **40**, 443.
Wadell, H. (1933) *J. Geol.*, **41**, 310.
Wadell, H. (1935) *J. Geol.*, **43**, 250.
Waldie, B. (1980) *Size Control in Particle Growth Processes*. Industrial Research Fellowship Report. Institution of Chemical Engineers, Rugby, UK.
Waldie, B. (1989) *Proc. 5th Int. Symp. Agglomeration, Brighton*. Institution of Chemical Engineers, Rugby, UK.
Walton, C.A. & Pilpel, N. (1972) British Pharmaceutical Conference, Keele, paper 3. *J. Pharm. Pharmac.*, **24**, 10–16.
Whitby, K.T. (1959) *ASTM Special Technical Publication No. 234*.
Wyatt, O.H. & Dew-Hughes, D. (1974) *Metals, Ceramics and Polymers*. Cambridge University Press, Cambridge.
Yuregir, K.R., Ghadiri, M. & Clift, R. (1986) *Powder Technol.*, **49**, 53–57.

Dedicated to Styran Honeydew – **Silver** – 12.10.1989 to 30.8.2004

2 Bulk property characterisation

DON McGLINCHEY

2.1 How do powders behave?

How do powders behave? Are they like liquids or solids or something different altogether? Observing and describing how collections and assemblies of large numbers of individual particles behave and respond to various stimuli is not easy. When the number of particles becomes large, it becomes all but impossible to determine the behaviour of the bulk material from knowledge of the properties of the individual particles making up the collective. We have to take a different tack and approach the bulk material as a substance in its own right. We then test and model the bulk material, deriving properties of the bulk such as bulk density and bulk strength, which are quite different from the properties of the constituent particles. Some powders show elastic properties under certain loads, that is they will 'spring' back to their original shape if the load is removed. Increasing the load to some threshold or yield value will cause the bulk material to deform plastically, in which case the deformation will be irreversible. This is typical elasto-plastic behaviour. Some materials show an almost non-detectable movement before plastic deformation: rigid-plastic behaviour. This behaviour is similar to solid materials, however a major difference is that for a granular material, significant deformation can occur under hydrostatic loads. Under certain conditions, a granular material will fail or shear along definite shear planes rather like blocks of solids sliding over each other. With enough energy input, some powders and granular materials will flow and exhibit characteristics of a fluid such as viscosity but only for as long as the energy keeps being added. Over the next few pages, I hope to show that these behaviours can be described and that powders and granular materials can be characterised by various tests and model parameters.

2.2 Introduction to characterisation

2.2.1 *Bulk density*

Bulk density is probably one of the most common and widely used of the bulk characteristics. It is used in hopper design to determine wall loading; it is used to size volumetric feeders, such as screws and rotary valves; and it is used to estimate 'flowability' and used in many other ways. It is rather unfortunate, then, that such a useful characteristic is not a constant for a given material. The bulk density of a material is simply the mass of material divided by the volume that it occupies. The density of the particles themselves can be taken as constant; however, the complication comes because the amount of 'space' between the particles depends on how the material has been handled before the measurement is taken. The volume that a unit mass of product can occupy can change by 50% between the material being in a compressed state to being in a very loose state. Cement, for example, has a compacted bulk density of 1400 kg/m^3 and an aerated bulk density of

1000 kg/m^3. It is obviously important that the correct bulk density value is selected for any calculation.

The full expression for bulk density is

$$\rho_b = \frac{\text{Mass}_{\text{solids}} + \text{Mass}_{\text{spaces}}}{\text{Volume}_{\text{solids}} + \text{Volume}_{\text{spaces}}}. \tag{2.1}$$

For dry bulk solids, the void spaces would usually contain air or some other gas, the density of which can be taken as negligible compared with the density of the solid particles, so we can approximate

$$\rho_b = \frac{\text{Mass}_{\text{solids}}}{\text{Volume}_{\text{total}}}. \tag{2.2}$$

We can relate this to another common characteristic, voidage or void fraction, ε, which is the percentage of the total volume not occupied by particles:

$$\varepsilon = \frac{\text{Volume}_{\text{spaces}}}{\text{Volume}_{\text{total}}}. \tag{2.3}$$

Again, assuming air or gas in the void spaces and taking particle density as ρ_p, we can write:

$$\rho_b = \rho_p (1 - \varepsilon). \tag{2.4}$$

The range of values voidage can take can be illustrated by considering a static heap of mono-sized spheres. If the spheres are in a regular hexagonal packing, the classic 'cannon ball' stack, the voidage would be 26%, whereas if they were in regular cubic packing, the voidage would increase to 48%. However, this does not represent the loosest packing even for large smooth identical spheres. The cannon ball stack gives each ball six contact points, but simple static mechanics requires only two contact points below the centre of gravity of the ball for equilibrium. Therefore, it is possible to have a stable structure with far fewer contact points and a resulting increase in voidage (see Duran 2000). If the particles are irregular in shape, have a size distribution and in some way cohere to one another, the packing arrangement can be very loose, and so the voidage can be very large.

Measurement of bulk density is in theory quite simple, requires only a knowledge of material mass and volume and is generally based on one of two techniques. The first is to weigh out a quantity of material using a simple balance and put this in to a calibrated cylinder in much the same way as you would a liquid. If the particulate material is poured into the cylinder, the volume taken up would be of the material in a loose or poured state; the associated bulk density would be commonly described as 'poured bulk density'. If this same cylinder is then tapped or dropped from a small height on to the bench several times, the volume would likely decrease and the new value of bulk density would be the 'tapped bulk density'. Similar techniques can be used to determine aerated bulk density from a fluidising column or compacted bulk density from a material placed under load.

The second technique is to fix the volume of the bulk material by filling a cup-like vessel to overflowing and then level with a straight edge. This is then weighed on a balance and the bulk density calculated. This circumvents some of the problems of trying to estimate the actual level of powder in a cylinder with a surface which is typically anything but flat and see through a glass that has become coated in powder (Table 2.1).

Table 2.1 Typical bulk density values for a few materials.

Material	Tapped bulk density (kg/m^3)	Poured bulk density (kg/m^3)	Particle density (kg/m^3)
Iron powder	3410	3360	7200
Aluminium powder	1220	1095	2650
Cement	1400	1100	2700
Nylon pellets	680	680	1140

One possible complication with bulk density measurements is the effect of the porosity of the particles themselves, and to avoid ambiguity, it is worthwhile stating whether the bulk density value is inclusive or exclusive of closed pores.

2.2.2 *Flowability – bulk density*

Bulk density measurements have been used to give some qualitative prediction of the 'flowability' or 'handleability' of a bulk solid, that is some estimate of the likely ease or difficulty in dealing with these materials. One such predictor is the often-quoted Hausner ratio (H), which is defined as:

$$H = \frac{\rho_b \text{ tapped}}{\rho_b \text{ aerated}}. \tag{2.5}$$

Another close relative is Carr's Compressibility Index:

$$\text{CI } (\%) = \frac{\rho_b \text{ tapped} - \rho_b \text{ aerated}}{\rho_b \text{ aerated}} \times 100, \tag{2.6}$$

which ranks materials as:

(1) 5–15%: free–flowing to excellent flow – granules
(2) 12–16%: free–flowing to good flow – powders
(3) 18–21%: fair to passable powdered granule flow
(4) 23–28%: easy fluidisable powders – poor flow
(5) 28–35%: cohesive powders – poor flow
(6) 33–38%: cohesive powders – very poor flow
(7) >40%: cohesive powders – very very poor flow

These relatively quick and easy measurements can be effective in giving some indication as to how powders will likely behave in solids handling plant but are by no means comprehensive, and a little caution should be exercised if relying only on this information.

2.2.3 *Angle of repose*

The angle of repose is another parameter that is used to determine the flowability of a bulk solid. The angle of repose is defined as the angle of the free surface of a heap of particulate material to the horizontal plane. We are faced with the same problem as we were with bulk density: the angle of repose is not a constant for a given material and is dependent on the method of heap formation. There are again two measurements which are commonly

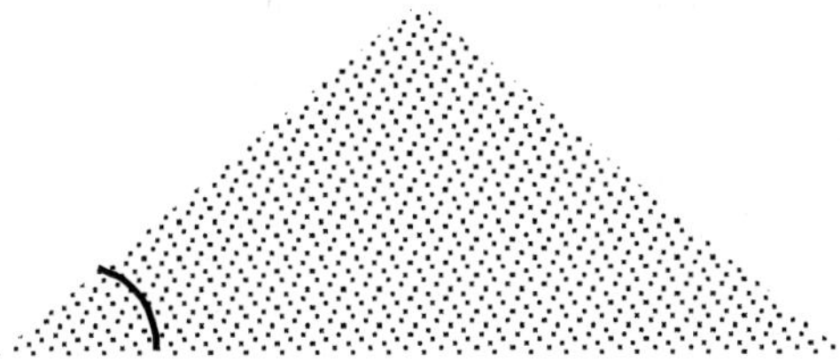

Figure 2.1 Poured angle of repose.

quoted: the poured angle of repose and the drained angle of repose. The poured angle of repose is the angle measured from a heap formed by pouring material on to a flat surface to the horizontal (Figure 2.1). The drained angle of repose is the angle measured on the internal conical face of a material which has been formed when drained from an orifice of a flat bottomed container to the horizontal (Figure 2.2). A third angle of repose which you may come across is the dynamic angle of repose, which is the angle to the horizontal of the free surface formed in a relatively slowly rotating drum (Figure 2.3).

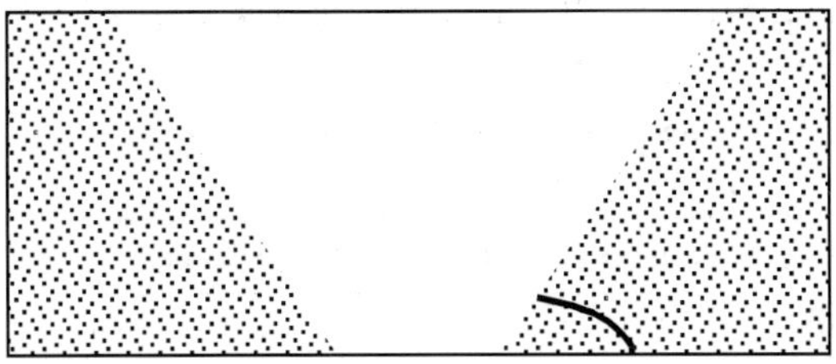

Figure 2.2 Drained angle of repose.

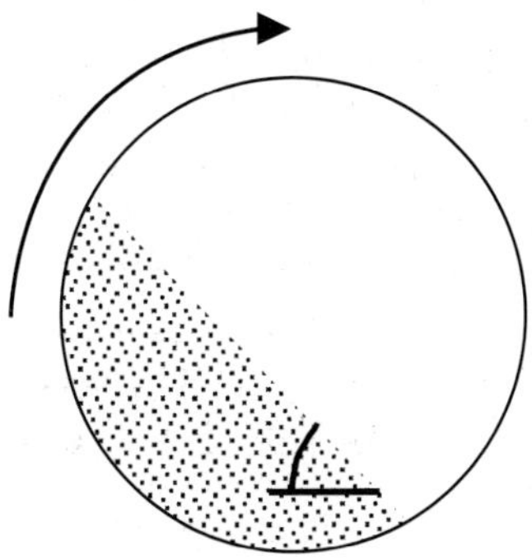

Figure 2.3 Dynamic angle of repose.

There are several things we should be aware of when using angles of repose. The first is that the angle formed will depend on the details of the formation process. For example, the fall height for the poured angle or the orifice size for the drained angle will influence the

Table 2.2 Angles of repose of three materials (based on Duran 2000).

Material	Shape	Poured angle (°)	Drained angle (°)	Dynamic angle (°)
Tapioca	Spherical	30	37.5	32
Sand	Angular	37	39	36.5
Coal	Angular	37.5	41	34

angle. Therefore, the angle measured is not independent of the measuring apparatus. The second is that a different angle will be found for the same material tested using the three different techniques (Table 2.2). The measurements can only be reliably made when using powders that are free flowing to slightly cohesive and are fairly homogenous. Materials that are a mixture of components or have a wide size distribution will give angles that are difficult to determine and have a low repeatability. There are also some uncertainties based on the fundamental physics of the problem, relating to stress history and avalanche behaviour, discussed in Duran (2000). Despite these difficulties, the angle of repose in whatever form can be a useful tool to rank materials. As a rough guide, the relationship between the angle of repose and flowability often follows the structure in Table 2.3.

Table 2.3 Angle of repose as a 'rough guide' to flowability.

Angle of repose (°)	Flowability
25–30	Very free flowing
30–38	Free flowing
38–45	Fair flowing
45–55	Cohesive
>55	Very cohesive

This classification allows us to make some judgement on the likely flow behaviour of a material but has very limited use for equipment selection and design, and in particular, it is a mistake to use the angle of repose in an estimate of the wall angle required for the converging section of a hopper. However, the angle of repose can be used in some cases to estimate the surcharge in a storage vessel or is useful in estimating the ground area requirements when forming a stockpile.

An interesting development of the rotating drum device is the Aero Flow device, developed by Brian Kaye, where the avalanche behaviour is studied using the covering and uncovering of elements of an array of photodiodes as the drum revolves. The signals from these detectors are processed to create a 'fractal fingerprint', which is typical of the avalanching behaviour. The time between successive avalanche events is used to characterise the behaviour with poorly flowing materials demonstrating more scatter in their avalanche times than freer flowing powders. The tests are relatively quick, and the Aero Flow can be a useful tool in comparative measurements and quality control in some circumstances. More detail can be found in Kaye (1997).

2.3 Introduction to theory and models

The stress–strain behaviour of particulate materials has been investigated mainly by two branches of engineering, and hence there are two main sources of information: one from geotechnical engineers involved in the study of soil mechanics and one from process engineers in the study of material handling problems. There are obvious similarities and the application of soil mechanics techniques to the problems of materials handling has a long history going back over a hundred years to Janssen (1895) and his work on bin design. There are, however, significant differences: the geotechnical engineer generally works at pressures and strains that reflect their particular interest, i.e. high pressures of the order of megapascals with relatively small strains, whereas the process engineer is interested in the behaviour of materials at much lower stresses of kilopascals and with much larger plastic strains. The whole inspiration is also different; the geotechnical engineer generally wishes to find the point at which a material will fail and start to deform and then design to a point that ensures that failure does not occur, whereas a process engineer wishes to find the point at which a material fails and starts to flow and then will design to ensure material failure and continued flow.

The behaviour of particulate material in response to applied stresses or strains imposed during testing has been modelled in many ways. Several of the more popular are discussed in the following sections. Several important topics such as triaxial stresses and strains, and the concept of critical state are described along with the models proper.

2.3.1 *The Ideal Coulomb material*

In this simplest of powder models, we treat the granular material not as a collection of individual particles but as a continuous substance that behaves much like the familiar solid. Nedderman (1992). If we take a sample of powder and apply some force to it, as shown in Figure 2.4, the material may respond in many different ways. If the material acts as an

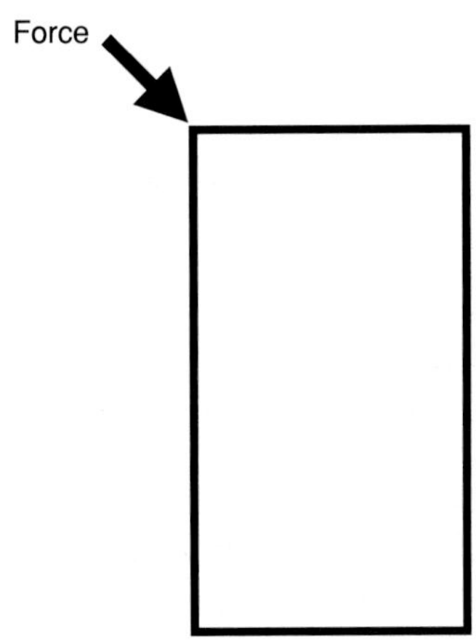

Figure 2.4 Force applied to a powder sample.

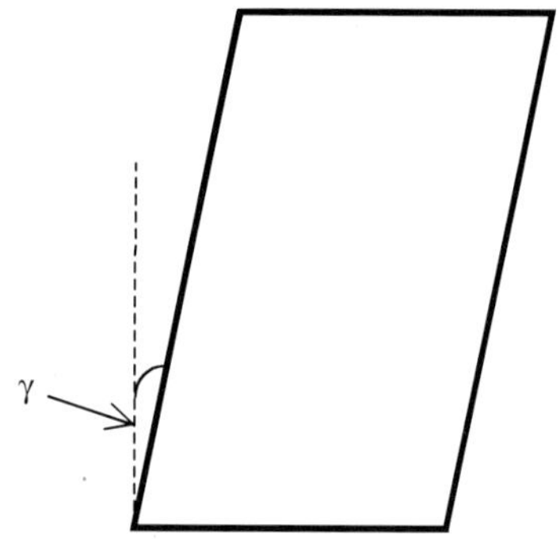

Figure 2.5 Deformation of sample.

elastic solid, the applied force will result in a deformation of the block by some measurable strain, γ, as shown in Figure 2.5.

If, however, our block is a sample of a bulk solid, this elastic (recoverable) deformation is very small before a critical strain is reached when the block divides into two, and these two smaller blocks slide over each other (see Figure 2.6). For very many bulk solids, it transpires that the shear stress, τ, on the plane that divides the two blocks is independent of both the extent and the rate of the relative movement. This behaviour can be described as plastic.

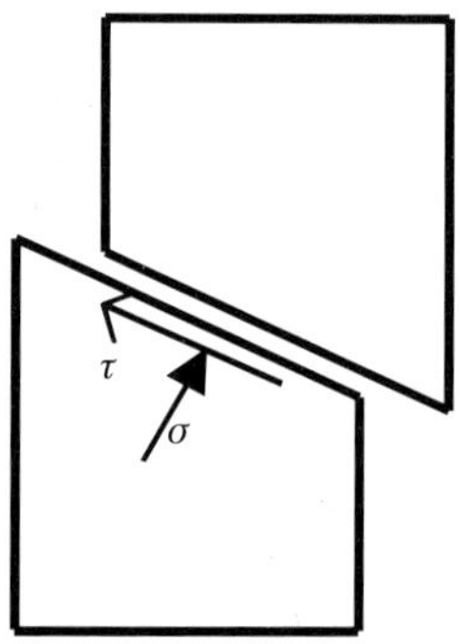

Figure 2.6 Sample failure.

The small elastic deformation can be important in practice, and the region of slip can be anywhere between ten to a hundred particles wide, but as a simplifying assumption we take the initial elastic deformation to be negligible and assume that the width of the failure zone is also negligibly small. We now have a situation where we have two rigid blocks sliding over one another on a narrow plastic, failure or yield plane, and this gives the basis for a simple rigid-plastic failure model.

In this simple model, the shear stress, τ, is dependent on the normal stress, σ, acting on the plane. There is a linear relationship between these two stresses such that

$$\tau = \mu\sigma + c, \tag{2.7}$$

where μ is the coefficient of internal friction, and c is the cohesion of the material. This is a form of the Coulomb yield criterion, and materials that obey this relationship are known as Ideal Coulomb materials. Most coarse-grained materials have a very small value of cohesion and are generally referred to as 'cohesionless' materials.

The values of internal friction and cohesion for a material are determined experimentally where the friction is determined by analysis of the results and commonly quoted as an angle, ϕ, such that:

$$\tan\phi = \mu, \tag{2.8}$$

and the Coulomb yield criterion is often written as:

$$\tau = \sigma \tan\phi + c. \tag{2.9}$$

Angles of internal friction typically range from about 20° for rounded objects to about 50° for angular products (Table 2.4), and cohesion varies between extremely small values for coarse granular products to values of about 50 kPa for a stiff clay.

Table 2.4 Angle of internal friction of various materials.

Material	Angle of internal friction (°)
Wheat	15–30
Wheat flour	28–42
Granulated sugar	30–40
Quartz sand	33–40
Alumina	27–44
Fly ash	25–35

The cohesion of a material may be affected by many factors such as surface condition, moisture content and electrostatic charging. For example, interfacial forces and capillary pressure at freely movable liquid surfaces:

(1) liquid bridges
(2) capillary forces at the surface of aggregates filled with liquid

Attraction forces between solid particles:

(1) Molecular forces
    (a) Van der Waals forces
    (b) Chemical binding forces (valence forces)
(2) Electrostatic forces
(3) Magnetic forces

Interlocking bonds:

Rumpf (1962, 1975) discussed the strength of agglomerates, and the following figure has been produced to illustrate the relative magnitude of the forces which may cause particles to cohere (see Figure 2.7).

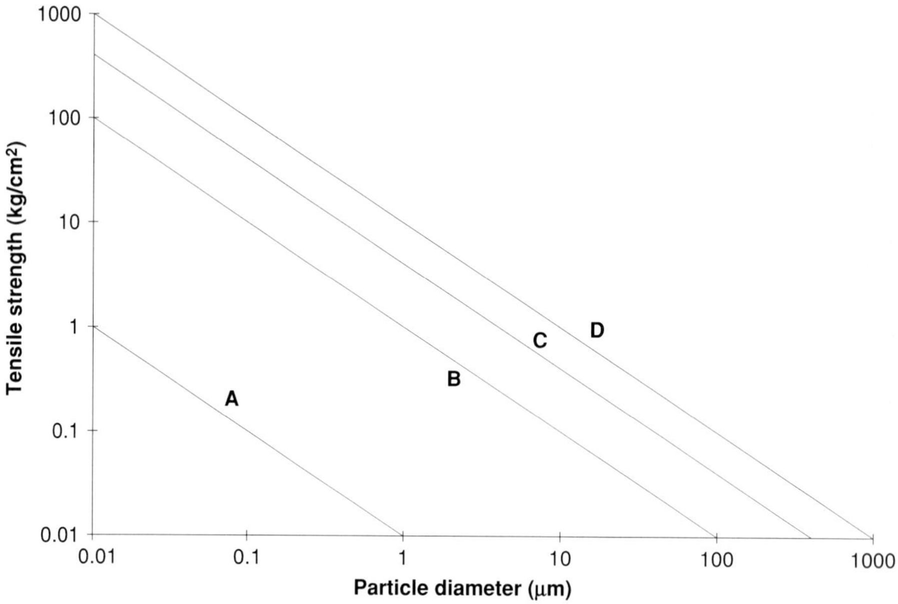

Figure 2.7 Tensile strength vs. particle diameter.

Relative magnitude of bonding forces:

(A) van der Waals forces calculated at a distance of 3.3 nm
(B) van der Waals forces in the presence of adsorbed liquid layers with zero inter-particle distance
(C) pendular bridges between particles (water)
(D) capillary bonding (water)

The influence of particle size on tensile strength is obvious, and in general finer powders will exhibit larger values of cohesion than coarser powders.

In contrast to the mechanisms involved with fine powders, fibres as well as flat and bulky materials can interlock or weave and fold about each other. Such interlocking occurs only infrequently in assemblies with disperse particle characteristics because the shape of their elements does not meet the above requirements. The strength of the resulting bulk material depends on the type of interaction and the material characteristics. It could be argued that this is not really 'cohesion', but it can result in similar (if not worse) flow problems in the processing plant.

The Coulomb yield criterion allows us to relate normal and shear loads to the frictional and cohesion properties that characterise particular materials. This relationship forms the basis of a very useful mechanical design, for example in the design of storage silos or screw conveyors.

2.3.2 *Moisture content*

This may be an appropriate point to say a little about one of the most common sources of cohesion: moisture. I live close to a beach, and my little girl loves to build sandcastles. I would like to share our first-hand experience on the effect of moisture content. At the top of the beach, the sand is dry. Sandcastles cannot be built successfully here. The sand on its own has no cohesion, and sandcastles are no more than piles of sand with shallow sloping walls: low angles of repose. Moving down the beach a little towards the sea, the sand is damp; the liquid bridge bonds give some cohesion, but the sea water has started to dry up, forming salt crystal bonds between some sand grains. This makes the sand difficult to dig, and when the salt bonds are broken by force, the resulting sandcastles are poor. The sand has an inhomogeneous distribution of liquid bonds, and fracture planes are easily established in the drier areas. Moving closer still to the sea, the sand is damp, and the tide has not been out long enough for significant evaporation of the sea water. The moisture in the important (for making sandcastles) first few spades full is reasonably consistent. The sand exhibits cohesion and, for a time, can even sustain vertical free surfaces. Here, the best sandcastles can be made, with seashell turrets and feather flags, if desired. If you venture too close to the sea, the sand becomes wet, the moisture content is too high for strong bridges, and areas will be saturated. The sand loses its strength, and castles slump.

There are other things to note on a beach. Observe what happens when you walk on wet sand: areas around your foot may appear drier (caused by stress dilatancy). Little pools may form where your heels have been (caused by plastic deformation). Look at how the dry sand is picked up by the breeze, blown around and deposited.

The implication of our trip to the seaside is clear for bulk solids handling: a free-flowing powder will become cohesive and may present problems if moisture is present. The difficulty in handling is usually a function of moisture content with flowability getting worse with increasing moisture until a peak is reached; then, flowability improves as moisture increases until the material reaches the state of a slurry. Allowing damp material to dry out can make matters much worse if solid bridges are formed. This is in fact a standard technique for making large stable agglomerates.

The quantity of moisture in a bulk sold is usually expressed as a percentage by mass relative to the dry mass of the bulk solid. One detail to note is the difference between free and inherent moisture. Moisture which is only on the surface of the particles is generally referred to as free moisture and that which is 'trapped' in the particle as inherent moisture. The simplest method of determining moisture content is to allow a sample of material to air-dry naturally and measure the resulting loss in mass. One common method of obtaining the total moisture content, free plus inherent, is to heat a sample in an oven maintained at $105 \pm 2°C$ until equilibrium conditions are reached. There are more sophisticated techniques such as those based on the chemical reaction between calcium carbide and moisture, which form the basis of quick on-site tests, or the use of microwave radiation for on-line measurement. Whichever technique is used, care must be taken in the selection of a test sample, which can be tricky as the moisture content may not be homogeneous throughout the bulk material.

2.3.3 *Wall friction*

The friction between a bulk material and a solid surface such as a hopper wall or a chute is described by much the same Coulomb relationship we have seen before, where the stress acting normal to the powder and wall surface is related to the shear stress to cause sliding at the powder wall interface, and is given by:

$$\tau = \mu_W \sigma \tag{2.10}$$

or

$$\tau = \sigma \tan \phi_W. \tag{2.11}$$

Knowing this relationship, we can say something about the stress conditions required to ensure that a powder will slide against a wall. This is a vital part of silo design and is used in many other occasions for design and troubleshooting problems of hang-up in the process plant.

2.3.4 *Material flowability factor*

A comparison of the relationship between consolidation stress and the stress required to fail the unconfined material or unconfined yield stress is used as a measure of flowability. If we take another look at Figure 2.4 and assume that the stress required to compress this material to the shape shown can be taken as the major principle consolidation stress, σ_1, then the stress that results in a failure of the unconfined sample represents the unconfined yield stress, σ_c. The same results can be obtained indirectly from shear cell data. Powders and granular material show a range of behaviours that define the dependence of σ_c on σ_1. If we were to perform a series of tests at various consolidation pressures and measure the unconfined yield strength, also denoted as compressive strength, we might obtain results as shown in Figure 2.8, which would represent an idealised flow function.

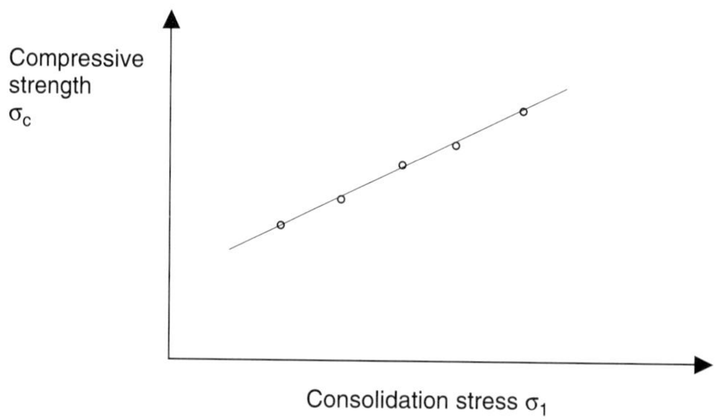

Figure 2.8 Compressive strength vs. consolidation stress.

The flow function will vary from material to material, as illustrated in Figure 2.9. Here, we see three products: A, B and C. The flow function of product A lies above that of B and C, and so product A has the greatest strength and is the least free-flowing. The flow function of product C lies almost along the horizontal axis; this material has no cohesive strength and is completely free-flowing. Product B is therefore somewhere between C and A in terms of flow behaviour.

Jenike's (1961, 1964) protocol characterises flowability as the ratio of the major principal stress, σ_1, at steady state flow to the unconfined yield strength, σ_c:

$$ff_c = \frac{\sigma_1}{\sigma_c}. \tag{2.12}$$

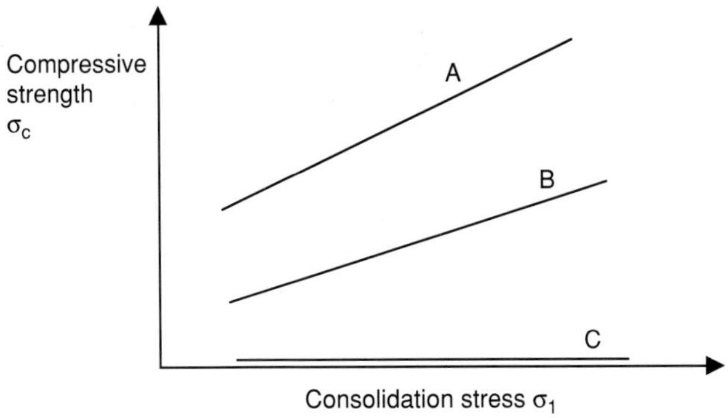

Figure 2.9 Flow functions of materials: A, B, C.

Table 2.5 Flowability from Jenike tests.

		ff_c	<	1	Hardened
1	<	ff_c	<	2	Very cohesive
2	<	ff_c	<	4	Cohesive
4	<	ff_c	<	10	Easy flowing
10	<	ff_c			Free flowing

Jenike's classification was extended by Tomas and Schubert (1979) to give the values shown in Table 2.5.

Unfortunately, in practice, the relationship between σ_1 and σ_c of many materials is non-linear and may span a range of ff_c values. It is important, therefore, to be sure of the stress conditions under which any quoted values of ff_c were obtained and that these match the stress condition of interest.

2.3.5 *Time consolidation*

When considering the behaviour of a powder, an important factor that must not be overlooked is the time the material is likely to be at rest and under load, for example in a storage silo for a few hours or perhaps a few weeks. In general, there is a tendency for the strength of a powder to increase over time if allowed to consolidate. There may be several reasons for this. One is that there is a de-aeration and rearrangement on the particle level; here, the material settles into a more compact and stronger configuration. There may be some problem due to an increase in moisture content of the bulk solid, for example from moist warm air cooling overnight, resulting in condensation. Or there may be some chemical reaction taking place, forming solid bridges between individual particles, which can result in very strong materials for which it is very difficult to initiate flow. These effects can be investigated by many of the testers, which are described later.

2.3.6 *The Cam-clay model*

A granular material may be described as frictional blocks of material sliding over one another as in the Coulomb model. A different approach to describing the deformation is to treat the material as a continuum and apply the theories of elasticity and plasticity. Models of this type were developed by soil mechanists Schofield and Wroth (1968) to describe the behaviour of soils and were written in terms of triaxial stresses and strains.

2.3.6.1 *Triaxial stresses and strains* The stresses and strains used in soil mechanics and in which terms the soil mechanics models are generally expressed are equations of triaxial stresses and strains. It is possible to divide a Cartesian compression which has only principal components into one spherical and up to three deviatoric tensors

$$\begin{bmatrix} \sigma_1 & & \\ & \sigma_2 & \\ & & \sigma_3 \end{bmatrix} = p\begin{bmatrix} 1 & & \\ & 1 & \\ & & 1 \end{bmatrix} + q_1\begin{bmatrix} 0 & & \\ & 1 & \\ & & -1 \end{bmatrix} + q_2\begin{bmatrix} -1 & & \\ & 0 & \\ & & 1 \end{bmatrix} + q_3\begin{bmatrix} 1 & & \\ & -1 & \\ & & 0 \end{bmatrix} \tag{2.13}$$

$$p = \frac{1}{3}(\sigma_1 + \sigma_2 + \sigma_3) \tag{2.14}$$

$$q_1 = \frac{1}{3}(\sigma_2 - \sigma_3) \tag{2.15}$$

$$q_2 = \frac{1}{3}(\sigma_3 - \sigma_1) \tag{2.16}$$

$$q_3 = \frac{1}{3}(\sigma_1 - \sigma_2), \tag{2.17}$$

where σ_1, σ_2 and σ_3 are the principal stresses, p is the spherical stress, and q is the deviatoric stress. A useful illustration of this approach is to examine the Cam-clay model. Many models have in common the concepts of Cam-clay, such as critical state and a plastic yield surface, and were developed from a study of work on the material. Schofield and Wroth (1968) derived the Cam-clay model from work equations applied to a triaxial test sample suffering very slow drained increases in stress δq on q and $\delta p'$ on p', where q is the deviatoric stress and p' the average effective stress.

The concept of effective stress originates from the field of soil mechanics, where a soil is usually tested in a fully saturated condition, i.e. all the void space between the particles that go to make up the soil structure or skeleton are filled with water. The soil is then treated as a two-phase continuum with the total stress made up of the stress contribution on the skeleton of solid soil particles and the pressure of the interstitial fluid, termed the pore water pressure.

The total stress component, σ, normal to any plane in a soil media is divided into two parts; the pore pressure, u, and the effective stress component, σ', which must be considered to be effectively carried by the structure of soil particles. A basic supposition is that the mechanical behaviour of the effective granular structure depends on all the components of the effective stress and is quite independent of u. Of course, in a dry granular bulk solid, the pore pressure is zero. In a partly saturated soil or granular material, the situation is more complex, in that the system has three phases: the solid particles which make up the soil skeleton, and the pore fluid which now has both water and gas (usually air) phases.

The complete description of the Cam-clay model is lengthy and beyond the scope of this chapter, but one of the most important results of this model is the concept of critical state deformation. There are several important observable behaviours that are described by

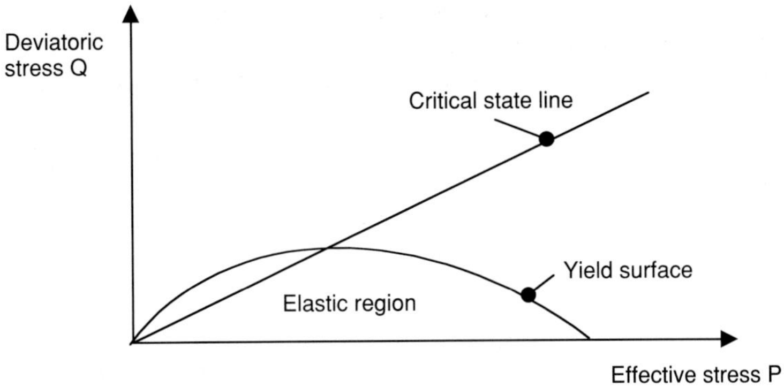

Figure 2.10 Simplified picture of the Cam-clay model in stress space.

the Cam-clay model and illustrated in Figure 2.10. There is a limit region in stress space where strain is recoverable; that is, the material exhibits elasticity. There is a predictable limit to the elastic region bounded by the yield surface beyond which the material deforms plastically, and strain is not recovered; that is, the material does not 'bounce back' to its original shape. The yield surface 'moves' during slow deformation in such a way as to approach the critical state line. There is an effective stress which will cause the material to yield and deform even in the absence of a deviatoric stress. A granular material will deform under a hydrostatic pressure. This is unlike a true solid such as a block of steel, which will not deform in the absence of a deviatoric stress.

2.3.6.2 *The concept of critical state* The critical state concept, as described by Schofield and Wroth (1968), states that granular materials, if continuously distorted until they flow as a frictional fluid, will come to a well-defined critical state, determined by two equations

$$q = Mp' \tag{2.18}$$

$$\Gamma = v + \lambda \ln p'. \tag{2.19}$$

The constants M, Γ, and λ represent basic granular material properties, q is the deviatoric stress, p is the spherical stress, and v is the specific volume. Equation 2.18 determines the magnitude of the deviatoric stress needed to keep the soil flowing continuously as the product of a frictional constant, M, with the effective pressure. Equation 2.19 states that the specific volume occupied by a unit volume of flowing particles will decrease as the logarithm of the effective pressure increases.

If a granular material is in a state looser than critical, then during deformation, the effective granular structure will give way (contract); if the material is in a state denser than critical then during deformation, the effective granular structure will expand (dilate).

2.3.7 *Stress dilation*

This property of a granular material to dilate (or contract) when stressed was noted in an article by Reynolds (1885), who observed that 'a strongly compacted granular material placed in a flexible envelope invariably sees its volume increase when the envelope is deformed. If the envelope is unstretchable but deformable, no deformation is possible until the applied force breaks the envelope or fractures the granular material'. There are several key points in this short statement which have important consequences for those involved with handling bulk solids. The first is that the material must be 'strongly compacted' (denser than critical state), so that we do not expect to observe dilatancy of a loosely packed bulk solid. In fact, we might well expect to find that this loose (looser than critical state) material contracts under stress. There are a couple of ways to try to visualise what is happening to result in the observed behaviour; perhaps one of the simplest is to follow the explanation given by Duran (2000). Skipping the algebra, we represent the granular media as four disks arranged as shown in Figure 2.11. The only restriction imposed by the model is that the four disks remain in contact at all times during the deformation. The lines connecting the centres of the four disks form a parallelogram that changes shape when the particles move relative to one another. If we consider the enclosed area, denoted A, as the disks 1 and 3 move closer and disks 2 and 4 are forced apart, Duran shows that this area increases to a maximum (dilation) then reduces again (contraction). A real bulk solid of non-spherical particle shape with some size distribution is not as easy to visualise or analyse, but it takes no great leap of imagination to see that Reynolds' dilatancy behaviour is dependent in some way on the packing and so bulk density of the material being stressed.

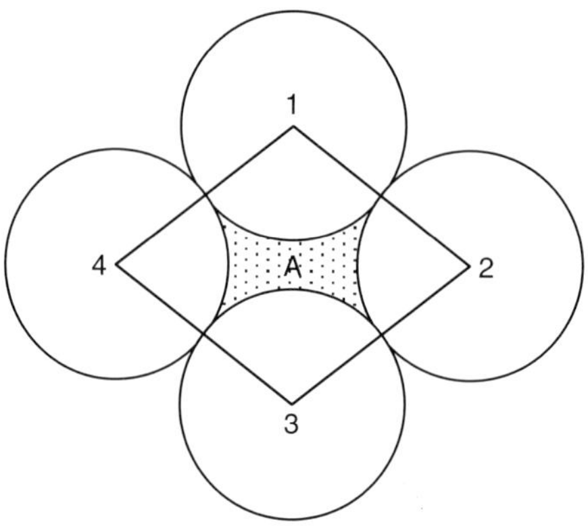

Figure 2.11 Simple picture of dilantancy.

So, what are the implications for bulk solids handling? Go back and look at the second part of Reynolds' statement; 'If the envelope is unstretchable . . . '. How many solids handling storage vessels or processing plant form a stretchable envelope? Screw conveyors that have become blocked or even some pneumatic conveying line blockages where material has become compacted can be extremely difficult to clear. Beware also of consolidating material in storage, for example by operating vibrating discharge aids when the hopper outlet is closed.

The application of a model such as the Cam-clay model, which was developed to model the behaviour of a clay type material, to a dry granular bulk solid may be questioned. Crouch *et al.* (1994) reported that elastoplasticity-based models have been successful because of their flexibility and their simplified treatment of the complex microscopic slip and rolling mechanisms that control the deformation. While it was recognised that sands and clays have quite dissimilar microstructures, their macroscopic behaviours (ignoring viscous effects) show many common features such as:

(1) extensive plastic straining prior to failure yet no clearly defined yield point;
(2) significant hardening associated with compaction and mild softening associated with dilation, both under shear;
(3) the strong influence of the mean effective stress on the shear strength.

The Cam-clay type models have considerable success in describing the behaviour of soils, but these models are not routinely applied to the field of industrial bulk solids handling. One of the reasons for this may be the difficulty in obtaining data in effective–deviatoric stress space necessary to determine parameters such as M or λ. The standard test apparatus, the triaxial cell, used in soil mechanics labs is not suitable for application to dry bulk solids in suitable pressure ranges. However, modified triaxial cells have been used and are described later.

Several key points can be gleaned from a review of scientific literature. It can be seen from the currency of elasto-plastic models that they provide a useful method of describing the behaviour of particulate materials. The number of variations and modified versions of these models reflects the difficulty in realising a model that quantitatively describes material behaviour, and so in the context of this book, the terminology of these models, i.e. elastic space, yield surface, mean and deviatoric stress, etc., shall be used to describe the general behaviour of a bulk material in a qualitative manner.

The elasto-plastic models require macro-properties of the material, and so the behaviour of different materials can be predicted, provided that the correct parameters are ascertained. These models are continuum models and cannot be related to properties of individual particles.

2.3.8 *Rheology*

Another way to describe or characterise the bulk behaviour of a bulk solid is to describe it not as a solid at all but as a fluid. This is justified in situations where the granular material is in a state of continuous deformation or flow and may be useful as a design strategy in dynamic environments. Typical applications of this approach are fluidised beds or high-energy powder mixers.

We are again attempting to model the behaviour of a granular material by a relationship between an applied stress and the resulting deformation. If we take an element of powder (behaving like a fluid) and subject it to a shear stress, it will deform by an amount dependent on the nature of the material (Figure 2.12).

Provided that the deflection increment, Δx, is small we can write:

$$\text{The stress } \tau = \frac{F}{A}. \tag{2.20}$$

$$\text{The strain } \gamma = \frac{\Delta x}{\Delta y}. \tag{2.21}$$

$$\text{And the strain rate } \dot{\gamma} = \frac{\delta \gamma}{\delta t}. \tag{2.22}$$

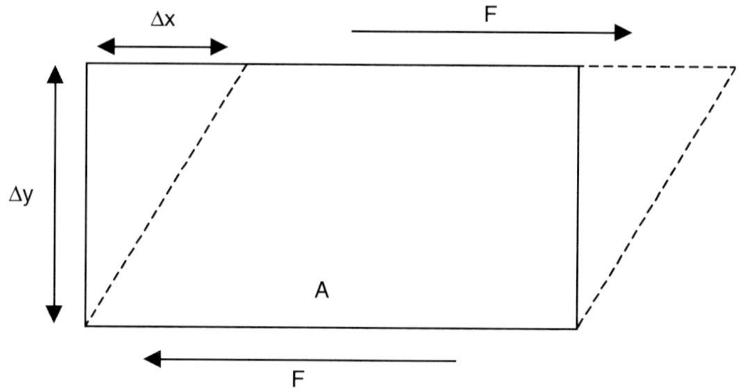

Figure 2.12 Deformation of a 'fluid' element.

The total rheological description of a material depends on the relationship between these quantities.

Under steady-state conditions, the applied shear stress will be related to the strain rate through a property of the material termed viscosity.

$$\tau = \eta\dot{\gamma}. \tag{2.23}$$

If the viscosity is a constant value, then the powder will be behaving as a Newtonian fluid, and the slope of the shear stress/strain rate line is linear and is the Newtonian viscosity, μ, which is constant regardless of strain rate. All gases and many liquids exhibit this behaviour.

There are three other general material responses. Shear thinning fluids or pseudoplastics exhibit a viscosity which decreases with increasing shear strain rate. That is, the faster you stir, the easier it gets. Many paints, which are basically particles of pigment in a polymer solution, behave in this way. This is generally a desirable quality for a paint. When you are brushing the paint on to a surface, it experiences shear, and its viscosity decreases, allowing the paint to flow and cover the surface, but when the brush stroke is finished, the viscosity increases again and prevents it from flowing under the influence of gravity. Shear thickening or dilatant fluids exhibit a viscosity which increases with increasing strain rate. This behaviour is shown by some particle suspensions such as sand in water. Here, the particles can rearrange themselves to yield a suspension which shows an increased resistance to flow. Bingham plastics are materials where a finite stress or yield stress is required before any flow takes place. Examples of Bingham plastic materials are toothpaste, blood, coal slurries and fluidised beds of glass beads.

All four stress/strain rate behaviours are illustrated in Figure 2.13. To add to the complicated stress behaviour, there are also time-dependent effects; a thixotropic material exhibits a viscosity that decreases with time at a constant strain rate. A material with a viscosity which increases with time at a constant strain rate is known as rheopectic.

2.3.9 *Internal structure of a granular material*

As we have seen, the behaviour of a particulate material may be described by the use of models of the type discussed in the previous sections, i.e. continuum models e.g. Cam-clay.

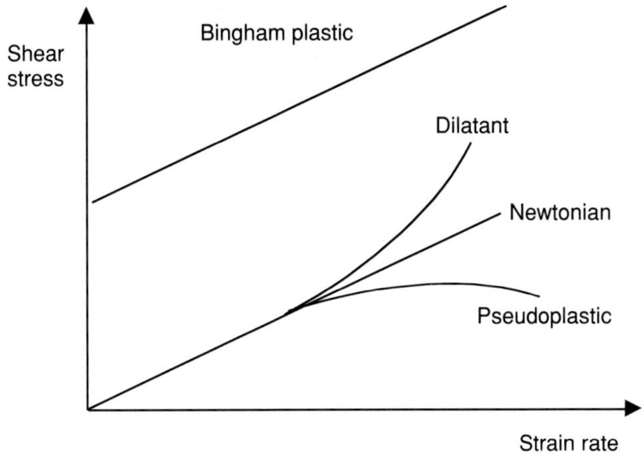

Figure 2.13 Stress–strain rate behaviours of classes of materials.

The material's behaviour can also be pictured at the particulate level in terms of the internal structure. This provides an alternative qualitative description of some aspects of the material behaviour disclosed in this chapter.

2.3.9.1 *The concept of 'stressed columns'* Troadec *et al.* (1991) stated that the stress–strain behaviour of granular material was extremely complex and was dependent on the structure of the medium, the nature of the particle contacts between particles and their geometry, as well as the direction and magnitude of applied pressure and any boundary conditions. The relationship between the structure of a packed bed and its mechanical properties was studied by Troadec *et al.* (1991) using a two-dimensional packing of cylinders made from Plexiglas and rubber. These experiments revealed that deformations were governed by a characteristic length much larger than that of the cylinders and was of the order of the length of the bonds in the network of strongest stresses, giving an axial loading the appearance of 'stressed columns' of particles. The macroscopic stress–strain behaviour of the packed bed was different from the microscopic behaviour applicable to two single cylinders in contact. The network of strongest stresses, i.e. the geometry of the stressed columns and therefore the mechanical behaviour of the packed bed, depended on the geometrical structure of the packing. This behaviour was also affected by disorder due to the non-uniformity in contacts between cylinders. This concept of stress networks can be reasonably assumed to hold for three-dimensional systems.

Applying a similar concept, Dunstan *et al.* (1988) proposed a minimal form of local collapse as a mechanism of slow steady-state deformation, complementing the familiar mechanism of potentially expansive sliding and rolling. Dunstan *et al.* suggested that granular material may be modelled as a changing collection of spacially dispersed crooked columns which are supported by distributed lateral forces arising from the minor principal stress and which transmit all the major principal stress through the material. During the very brief existence of each of these transitory columns in an infinitesimal strain increment, the column stores elastic energy, just as a solid column would, and on collapse the stored energy is released and dissipated.

This transient column concept was taken up by Arthur *et al.* (1985, 1991), who studied experimentally, in the directional shear cell, the effects of rotation of principle stress directions on the initiation of flow in particulate materials. Arthur *et al.* provided a simple picture of columns of particles roughly aligned with the major principle stress direction carrying the load imposed by the major principal stress through the mass of granular material in a very non-homogeneous manner. These columns are transient; an increment of strain will cause several to become unstable and the associated column loads to be shared among the remaining columns and new ones. This is a material in which underlying mechanisms of deformation are associated with local instability and the elastic loading of transient columns rather than a homogeneous frictional interparticle sliding, which is the traditional interpretation.

2.3.10 *Summary of powder mechanics models*

As we have seen, there are at least three basic ways to describe and characterise the mechanical behaviour of powders and granular materials:

(1) To treat the powders as a solid, which deforms by sliding over itself in shear planes. We can characterise a material by its relationship between normal stress and shear stress via a friction coefficient. This approach is also used for the movement of powder against a solid surface such as a hopper wall, but when a powder shears internally, there is a second factor which can be measured and is characteristic of a material under given conditions and that is the cohesion of a material. The concept of cohesion is not normally applied when considering two truly solid blocks sliding against each other but plays an important role in describing powder behaviour. The usual way to present this model is through the use of Mohr's circle analysis, as demonstrated in standard texts.

(2) To treat the powder as an elasto-plastic solid. In this treatment, there is an elastic region in stress space bounded by the yield line which can be described by theories such as the Cam-clay model, which also includes the very important concept of critical state and describes the critical state line in stress–bulk density space. The full description using this model requires details of tri-axial stress and volumetric strains, which can prove a difficulty for dry bulk solids.

(3) To treat the powder not as a solid but as a fluid and use the approach of relating the stress state not to strain but to strain rate by some notional viscosity.

The value of any model may be judged from two criteria:

(1) Descriptive: The ability of the model to describe the current system and how the material will respond to changes in the system.

(2) Predictive: The ability of the model to predict how the material will behave in a different system or how a different material will behave in the same system.

All three approaches have shown themselves to be successful in describing and predicting bulk solids behaviour under representative conditions. How, then, do we practically go about obtaining the parameters for these models and so characterise our powder or granular material?

2.4 Shear testers

There is an array of powder testers commercially available or residing in research establishments around the world. The difficulty is deciding which of these devices is the one (or two) required to solve a particular problem or perform a useful function. Several of the more common powder testers are briefly described with an indication of their key features starting with probably the most well known, the Jenike Shear Tester.

2.4.1 *The Jenike cell*

The Jenike cell was developed by Andrew W. Jenike as a tool in a system for hopper design (Jenike, 1961, 1964). The Jenike cell is widely used in industry and is well established as a tool for safe and reliable silo design. In fact, the Jenike testing technique forms the basis for the Standard Shear Testing Technique, as published by the European Federation of Chemical Engineers (1989). In its simplest form, the cell consists of a circular fixed base, a shearing ring which rests on the top of the base, and a cover for applying the load (see Figures 2.14 and 2.15).

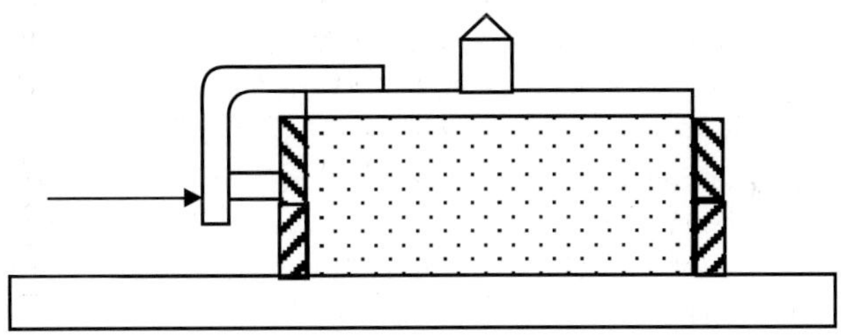

Figure 2.14 Simple form of Jenike type shear cell.

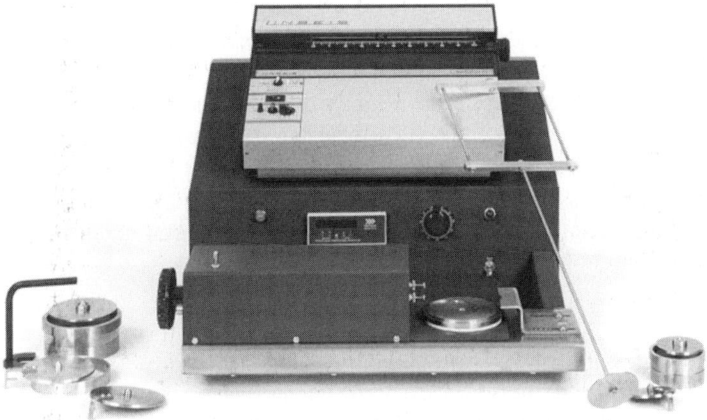

Figure 2.15 Jenike Cell. Courtesy of Jenike and Johanson Inc.

Table 2.6 Jenike cell dimensions.

	Standard Jenike	Small Jenike	Standard metric
Diameter (mm)	95.25	63.5	95
Base height (mm)	12.7	9.525	13
Ring height (mm)	15.875	11.113	16
Material	Stainless steel or aluminium	Aluminium	Stainless steel or aluminium

The dimensions of the Jenike cell are given in Table 2.6. The actual dimensions are not believed to be critical, so long as the proportions are maintained. Verlinden (2000).

The full test procedure can be followed from the SSTT or from the manual supplied with a new Jenike cell. There are a few important stages to these tests: preshear, consolidation and shear. During the first phase, the sample is loaded under a normal stress, σ_P, and subjected to shear stress until a steady state is reached. One of three behaviours will be noted: the material may have been under-consolidated, over-consolidated or critically consolidated. As the total travel of the relative displacement of the shear cell is of the order of 6 mm, a steady state must be reached in the first 4 or 5 mm. The material must be in a critically consolidated state at the outset; this is achieved in a trial and error process by a sequence of loading and twisting a ‘pre-consolidation’ cover. This stage of the test is necessary to attain steady-state flow across the entire cross-section and in the same direction as the intended test shear. The importance of this on powder response to shear is discussed by Schwedes (1999). When consolidation is complete, the shear load is removed, the normal stress lowered to σ_S, and the shear force reapplied until a failure plane has developed. As the sample would be over-consolidated for the new stress regime, failure is indicated on the record of shear stress passing through a maximum value.

This value gives the ‘end point’ on the failure curve (the curve of normal stress vs. shear stress). A series of tests are performed for material prepared to the same consolidation load but then sheared under a range of loads lower than the consolidation load. A number of quantities which are expressions of different aspects of the resistance of the powder to flow may be determined from this failure curve; for example, unconfined yield strength, angle of internal friction and apparent cohesion. Since the maximum shear displacement for the standard Jenike cell is about 6 mm, the largest particle size that can be practically tested is approximately 1 mm. The test cell in a slightly different form is used to determine wall-friction properties. The test procedure for silo design purposes is fairly lengthy and ideally requires a trained and competent operator to carry out the tests. There are simplified tests suggested for use in ranking powders’ flowability and for quality-control purposes. Time consolidation effects can be investigated by leaving the material under load for a desired time before performing the shear test.

2.4.2 *The triaxial tester*

The triaxial tester is used extensively in soil mechanics, and the test is based on cylindrical samples that are usually twice as long as their diameter. A sample is placed on a pedestal within an enclosing cell so that it can be subjected independently to an all-round water pressure and to a vertical force acting through a piston. The sample is squashed either by cell pressure or by a motor which drives it upwards against a fixed beam. As it is being tested, the pressure of the water in the cell can be measured either by a gauge or by a transducer: it is usually kept constant. Drainage through the top cap can be prevented (an undrained test) or, if allowed (a drained test), measured by a volume-change transducer. The force caused by the compression of the sample can be measured by a load cell in the piston. The sample is enclosed in an impermeable rubber membrane, which is held against the top cap and the pedestal by O-rings. It is commonly expected that a continued deformation will eventually lead to the so-called critical state, where no further volume changes occur. However, in triaxial compression, this critical state is not usually obtained within the range of homogeneous deformations.

Ooi and Rotter (1991) conducted a series of tests on dry wheat using a triaxial testing apparatus that was modified to use air as the cell fluid. Instead of using water in an enclosing cell to confine the sample, the confining pressure was obtained by reducing the pressure inside the sample. Shinohara *et al.* (2000) used a triaxial device to investigate the effects of particle shape on internal friction values, and Tripodi *et al.* (1994) used a triaxial test on cohesive powders for a Cam-clay type model. However, these testers have not had a significant impact on industrial practice so far.

There is another form of triaxial tester known as the 'True Tri-axial Tester', or the cubical triaxial tester, which has six plane walls that bound the sample (Figure 2.16). These walls are arranged in such a way that independent movement in the *x*-, *y*-, and *z*-directions is possible. This design has the potential to completely determine the stress and strain on the sample, which would be a great tool for fundamental study; however, few have actually been constructed, and there is little published data (Schwedes, 1999).

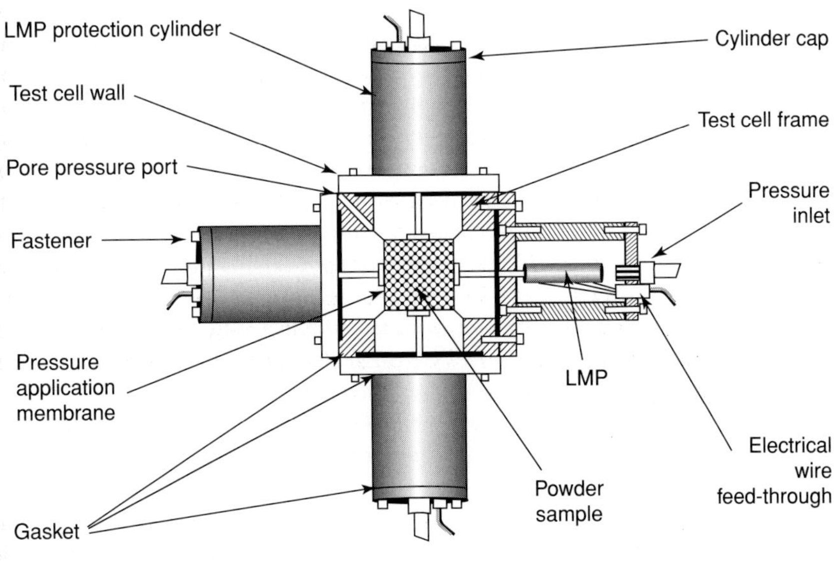

Figure 2.16 Penn State Cubical Triaxial Tester. Courtesy of Prof. V. Puri, Penn State University.

2.4.3 *The ring shear cell*

The annular, or ring shear, cell overcomes the problem of the limited travel of the Jenike cell. The most widely used ring shear cell was developed by Schulze and is shown schematically in Figure 2.17.

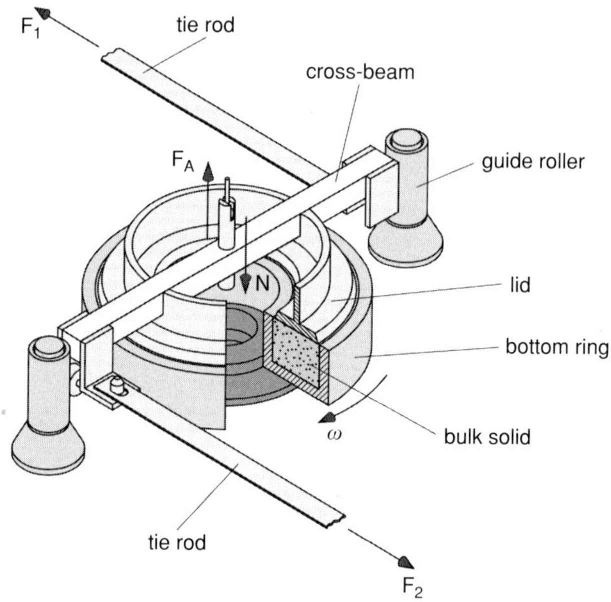

Figure 2.17 Schulze ring shear tester. Courtesy of Dr Dietmar Schulze.

The material under test is contained in an annular trough connected to a central drive shaft. A shoe with recesses in the underside is allowed to rest on the material. This shoe is freely pivoted concentric to the annular trough and is restrained by two tie rods connected to a cross-beam which present and indicate the shear force. The material is trapped in the recesses, and hence shear takes place between the lower face of the shoe and the material in the trough. The shoe is counterbalanced or has weight added to provide the consolidating loads and shear loads. Consolidation is carried out by shearing the material under load until a steady state is reached. The shear force is removed by reversing the cell, the weight of the shoe is adjusted to the desired shearing load, and the cell is moved forward again until shear takes place, indicated by a peak in the shear stress. With the cell driving forward, the shoe weight is adjusted back to the consolidation load and shearing continued until the steady force is recorded as before. The cycle is then repeated for a series of shear loads. The results obtained from this device are similar to those of the Jenike cell, are processed in the same way and have been shown to give similar values. There are automated versions of these machines commercially available which, combined with the test procedure, give less time-consuming and less operator-dependent results. Again, like the Jenike tester, there are simplified tests that can be used for quicker quality-control tests; a modification enables the determination of wall friction (see Figure 2.18) and can be used to perform time consolidation trials.

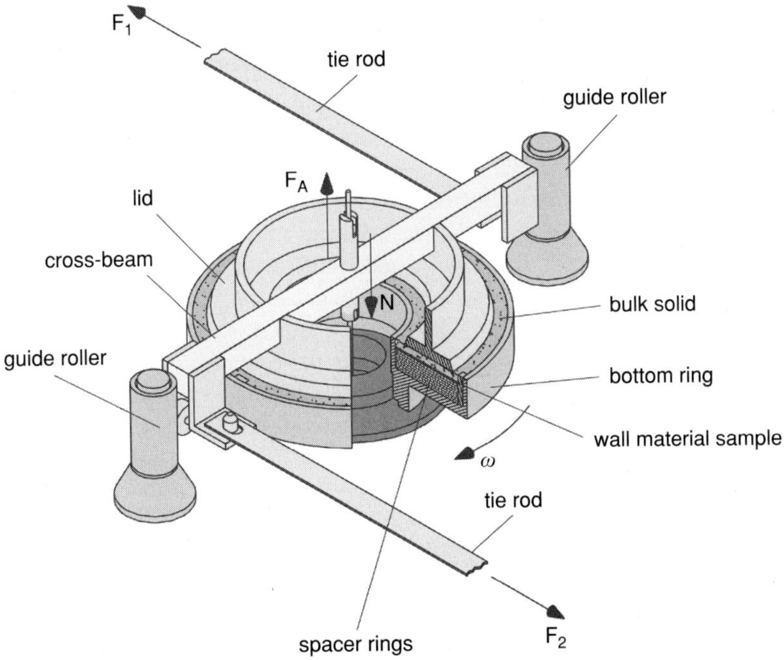

Figure 2.18 Schulze ring shear tester for wall friction tests. Courtesy of Dr Dietmar Schulze.

2.4.4 *Rotational split-level shear tester*

The rotational split-level shear tester, or torsional shear tester as it is also known, is a rotational shear tester with a circular cross-section. The best-known version of this device is that developed by Peschl. The design is conceptually very simple, as illustrated in Figure 2.18. The operation principle is similar to that described for the ring shear cell (see Figure 2.19).

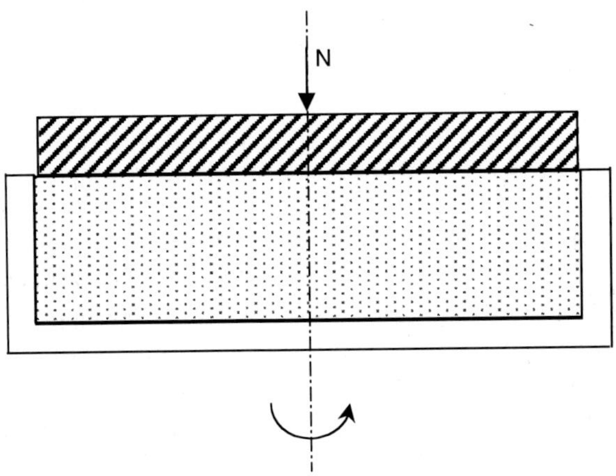

Figure 2.19 The torsional shear tester.

In this design, there is no shear in the centre, and we might expect the dependency of shear stress on radius to have an influence on performance. Several comparative studies have been undertaken between Jenike, Ring Shear Cell (Schulze) and the torsional cell, which show rather ambiguous results. Some results have shown good agreement between the test methods, whereas others have shown significant differences, and in the trials that showed a discrepancy, the torsional shear tester results would underestimate the strength of the material with respect to the results from other methods. These results would then suggest that a material had a better flowability than results from a Jenike test. The use of the torsional shear cell for silo design must therefore be undertaken with caution. Whether there was a discrepancy or not may have been due to the influence of the bulk solid under test, but the details of how this would happen have not been explained. The torsional shear tester does have several strong points: its simplicity and ability to be automated are appealing features for quality-control tests. If automated, the results will be largely operator-independent, and the device can be operated by any competent person. The small volume of material required for a test and the ability to apply low normal loads may make this an attractive option for the pharmaceutical industry, for example, where these shear cells have an established market.

2.4.5 *Biaxial shear testers*

In a biaxial shear tester, the sample is constrained in the lateral x- and y-direction by four plates, which have a degree of independent travel. The plates in the z-direction are fixed and restrict any deformation in this direction. The use of rubber membranes and grease to avoid friction between the material and the walls means that the applied stresses on the boundaries of the sample are principal stresses. The cross-section of the sample can therefore take different rectangular shapes, and these shapes can be reached through a variety of stress–strain paths. Testers of this type are used at the Technical University of Braunschweig, and a different design, the ‘Flexible Wall Biaxial Tester’, is used at the Telemark Institute of Technology of Porgrunn and the Technical University of Delft. Such testers are capable of straining a sample and measuring the stress response or alternatively stressing the sample and measuring the strain response to the applied loading. Various consolidation and shearing combinations can be performed, and studies of anisotropy, relaxation, creep and time-dependent effects can be undertaken. Models of stress–strain relationships, constitutive models, can have their parameters calibrated. All this makes these devices extremely useful for research into some of the more fundamental aspects of powder behaviour. The tests, however, are very time-consuming, and if a complete flow function is required for silo design, a standard Jenike test or ring shear test would be more efficient particularly if time-consolidation effects have to be measured. The relative slowness also makes the application to quality control unattractive.

A device of this type was used by Friese and Schwedes (1995), in an investigation of the behaviour of cohesive powder, and by Dunstan *et al.* (1994), in an investigation of powder strength after changes in principal stress direction. Arthur *et al.* (1991) and Dunstan *et al.* (1994) used a development of the biaxial tester, the directional shear cell, which allows both normal and shear stresses to be applied to the sample faces. The biaxial tester and the directional shear cell are mechanically complex, and it would be difficult to adapt them for industrial use.

2.4.6 *Uniaxial testers*

The usual approach here is to form a compact of material in a cylindrical mould by applying a consolidation load, $\sigma_{1,c}$; remove the load and retaining walls of the mould then apply an increasing load in one direction, normally from the top, until the sample fails at a stress equivalent to the unconfined yield strength, σ_c, of the material in that condition. This presupposes that a stable compact can be formed, which restricts these tests to cohesive materials. Depending on the application, this may not be seen as a major drawback of these testers, since most of the industrial flow problems occur with cohesive powders. However, studies with biaxial testers have shown that the relationship between σ_c and $\sigma_{1,c}$ cannot be regarded as equivalent to the Jenike flow function. In fact, the curve $\sigma_c = f^n\ (\sigma_{1,c})$ lies below the flow function and so overestimates flowability.

There are several problems with trying to use uniaxial testers for design information:

(1) The inevitable wall friction in the mould means that the consolidation deviates from the ideal case.
(2) The consolidation does not bring about a steady-state condition, as occurs in the Jenike and Ring shear cells.
(3) Removal of the retaining walls may be a problem if the material adheres to the cell surfaces.
(4) As a sample must be consolidated with sufficient load to form a stable compact, no tests are possible at low stresses.

The main use for these testers is likely to be for quality control, and there are many advantages in this choice. Their simplicity of design and operation lends to 'in-house' versions being used in a range of industrial situations as well as in many test laboratories. Tests can be performed quickly, and the procedure can be mostly automated, so reducing operator dependency (see Wiche *et al.* 2003).

2.4.7 *The Johanson hang-up indiciser*

The Johanson hang-up indiciser is one instrument from a set of three semi-automatic testers developed by Jerry Johanson for system (hopper) design (Johanson 1991). The principle of the Johanson hang-up indiciser (see Figures 2.20 and 2.21), is that a cylindrical sample of material is compressed by a two-piece piston until a pressure is reached that is a function of the material's bulk density, which is monitored during compression, and a value for 'bin diameter', which is input by an operator. When this consolidation pressure has been reached, both sections of the top piston and the base piston are removed. The inner top piston is then lowered until the sample fails. The instrument's internal software then calculates a hopper outlet size based on the measurement of compression force and force at failure, and reports back either an Arching Index or a Ratholing Index, depending on the test selected.

The theoretical basis for the determination of an outlet size from the information measured is rather contentious, and the instrument has faced some serious criticism from the academic community, namely that: the states of stress during compaction and failure are not homogeneous and undetermined; the results are dependent on wall friction and the cell geometry; and the results are not in agreement with other testers. Comparative tests with

Figure 2.20 Johanson hang-up indiciser.

the hang-up indiciser and the Jenike cell have been published by Bell *et al.* (1994) and by Marjanovic *et al.* (1998), which show that the unconfined yield strength values found when using the hang-up indiciser are likely to be lower than those found using the Jenike Tester. However, there is more to silo design than finding the unconfined yield strength. The tests performed by the indiciser and the values these instruments generate are based on the wealth of experience gained by Jerry Johanson since his early days working with Andrew Jenike on silo design. This is both the strength and the weakness of the Indiciser. The test is rapid and automated, and can be performed with little operator training and so is useful as a ranking tool.

2.4.8 *Rheological testers*

A simple apparatus for studying the rheological properties of a fluid is shown schematically in Figure 2.22. The sample is contained between two concentric cylinders. The outer cylinder is stationary, and the inner cylinder is rotated. This creates a velocity profile in the fluid of zero velocity at the outer cylinder to a velocity matching that of the inner cylinder due to

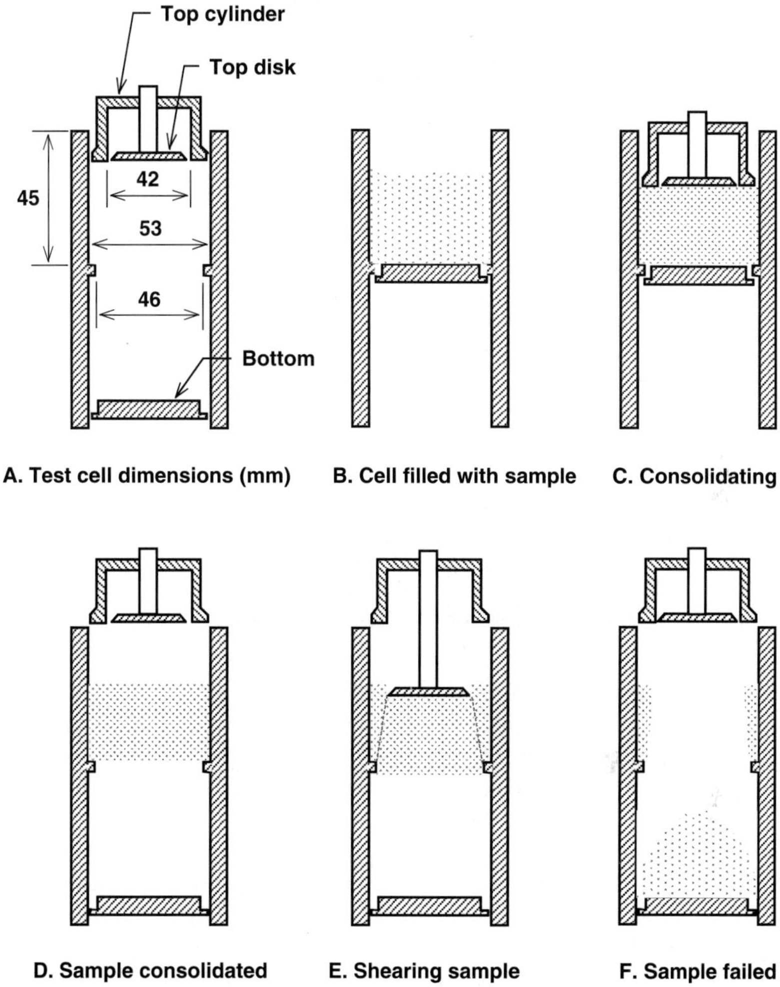

Figure 2.21 Johanson hang-up indiciser test sequence.

the condition of no slip at the surfaces. By measuring the force by means of a transducer on the drive shaft, it is possible to develop the relationship between stress and strain rates for a particular material. Variations and adaptations of this basic design have been used with powders, as have adapted ring shear testers (Barois-Cazenve *et al.* 1999; Klausner *et al.* 2000; see Figure 2.22).

2.4.9 *Other testers*

There are many more testers available than those already discussed which measure one or several characteristics of a bulk solid. Some of these are proprietary, such as the Johanson indiciser system or the Hosokawa tester; others which give, for example, some measure of wall friction or angle of repose can be obtained from various suppliers, or versions can

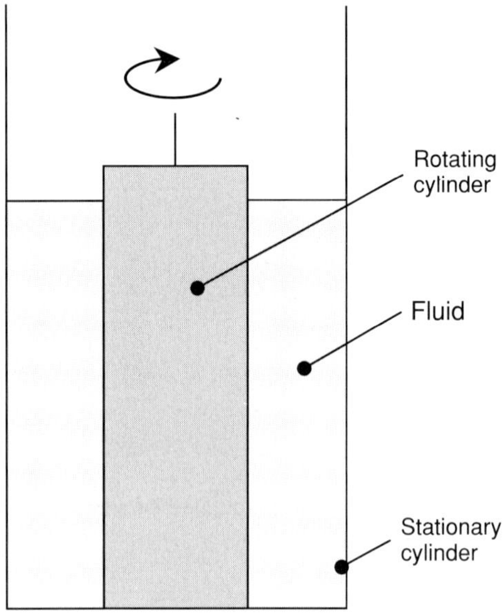

Figure 2.22 Schematic rheological property tester.

be manufactured in-house. There are testers which attempt to directly measure cohesion or tensile strength and testers to measure compressibility or voidage or specific surface. In addition to the properties we have reviewed, bulk density, powder mechanics and rheology, there are devices or methods that can give some quantification of what we might call process characteristics.

This would include things like dustiness, segregation potential, bulk thermal or electrical properties, reflectivity and perhaps we can include moisture content measurement in this category. There are simply too many to discuss individually. There are also some relative newcomers to the powder-testing market which have generated some interest and may prove to be very useful engineering tools, such as the powder flow analyser device from Stable Micro Systems. These testers rotate specially designed paddles or blades in a sample of material and record the torque or resultant force. The shape of the blades and the direction of travel allow the material to be densified or aerated, and the blade can rotate at various speeds. These tests are very rapid, and these instruments show considerable promise for use in quality control (see Figures 2.23 and 2.24).

2.4.10 *Testers summary*

There are many different situations where powders and granular materials need to be (or at least should be) characterised in terms of their bulk properties, but these would all fall into one of four categories:

(1) comparison with other powders
(2) study of fundamental behaviour

Figure 2.23 Powder analyser test cell. Courtesy of Stable Microsystems.

(3) design of handling and storage facilities
(4) simulation of specific problems

If you are clear about what you need to measure and why, then the choice of the correct test apparatus and methodology will be that much easier. Table 2.7 gives a quick overview of the testers discussed and carries with it all the usual limitations of such general approaches. If you are unsure about any particular application, the tests you require are too specialised, or the equipment too expensive, please get expert help. Powder behaviour is complex and at times counter-intuitive; performing the correct test under the correct conditions could prove crucial to the successful operation of process plant. Table 2.7 has four criteria: testers that are perhaps best for research establishments concerned with investigating fundamental powder mechanics; testers that are intended for the design of silos; those that are useful for making comparative measurements between powders; and those instruments or instrument systems that provide information on several aspects of powder behaviour, termed here multi-facet testers.

In using Table 2.7, find a suitable tester for your proposed purpose, read the comments in the body of the text, and then find out some more from the references or seek expert advice before spending good money. Table 2.7 should be seen as a guide only; for example, there has been fundamental research undertaken using uniaxial testers, or if you really wanted to, you could use the triaxial tester to determine parameters for silo design.

Figure 2.24 Powder flow analyser. Courtesy of Stable Microsystems.

Table 2.7 Tester characteristics summary.

Tester	Fundamental study	Silo design	Quality control	Multi-facet
Jenike	Yes	Yes	Yes	No
Ring shear cell	Yes	Yes	Yes	No
Torsional shear cell	Yes	Yes	Yes	No
Triaxial shear cell	Yes	No	No	No
Uniaxial compression tester	No	No	Yes	No
Johanson indicizer system	No	Yes	Yes	Yes
Powder flow analyser	No	No	Yes	Yes
Hosokawa powder characteristics tester	No	No	Yes	Yes

2.5 Segregation

Segregation occurs on many occasions with many materials in solid-handling plant and equipment, has a large influence on the ability to obtain a representative sample from bulk and can result in erroneous results from characterisation tests (Harnby *et al.* 1992; de Silva 1997; Bates 1998). It is such a common occurrence because opportunities arise for a material to segregate every time the bulk is set in motion. The effects of segregation can lead to changes in the material's appearance, texture, flow behaviour and so on, and may result in production difficulties, safety hazards, plant shutdowns, etc. The problems arising from segregation have important consequences for a broad spectrum of industries, such as chemical, pharmaceutical, food, steel manufacture and the utilities. Any assessment of segregation must be based on an understanding of the fundamental segregation processes, a correct sampling regime and a meaningful analysis of results.

Segregation or a separation of one fraction of a material from others in the bulk is due to some difference in particle properties and a force to 'drive' the fractions apart. The most common difference which results in segregation is a difference in particle size. However, a material may segregate based on a difference of density, shape, surface texture, elasticity, electrostatic charge and other physical characteristics. Each of these physical attributes plays varying roles in the segregating processes.

2.5.1 *Segregation mechanisms*

Sifting or percolating occurs when smaller particles in the bulk material are able to move through a lattice of larger particles. If the particle bed is static, the difference in particle size required for the sifting process to occur purely driven by gravity would generally be greater than around 10. If, however, the whole bulk is in motion, the opportunity for smaller particles to move into the voids between their larger neighbours is greatly increased, and even a small size difference can result in significant segregation (see Figures 2.25 and 2.26).

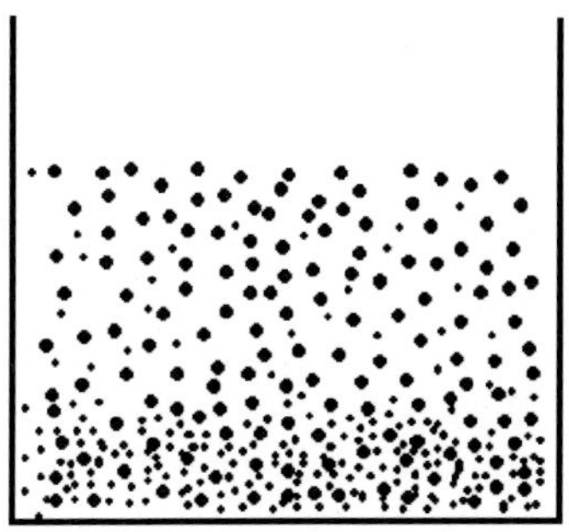

Figure 2.25 Percolation.

Figure 2.26 Sifting.

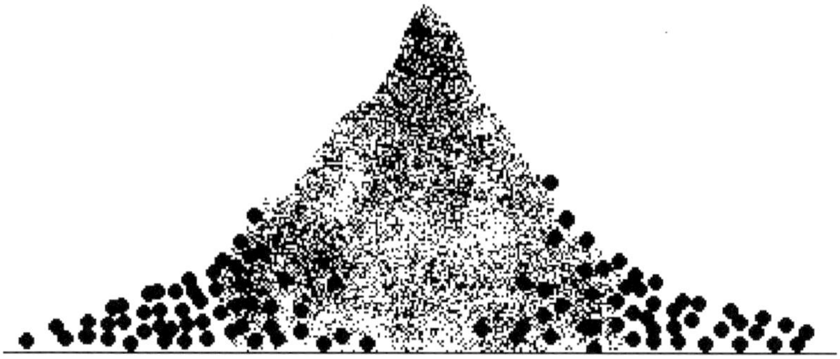

Figure 2.27 Angle of repose differences.

The angle of repose of a material is itself dependent on a variety of the constituent particles' physical properties. In general, coarse, rounded granules will have a lower angle of repose than a finer, more cohesive material. If both these components are present (even as a single material with a significant size distribution), then some operations, such as pouring the material into a heap, will result in the material segregating (see Figure 2.27).

Floating/sinking can occur in a material with a significant fines content which becomes highly dilated. This high dilation or fluidisation of the fines results in these fines behaving like a 'liquid' through which coarser or more dense material can sink. Elutration is somewhat akin to sifting, except that the fines fraction is driven upward through the lattice of larger particles by a driving force which is capable of overcoming gravitational effects. This force is generally provided by an up current of air from, for example, a dryer. Transverse gas flows can cause material to separate based on differing aerodynamic properties of particles within the bulk. Generally, the finer particles will be blown 'off course' to a greater degree than particles of a larger size. Up-thrusting occurs when a bed of material is subject to a series of 'shock' forces, e.g. vibration. The jostling of particles allows smaller particles to move beneath a larger particle and, once there, to consolidate with alike neighbours. The large particle therefore cannot move back down to its original position. As this process is repeated, larger particles move upward within the bed.

Lateral segregation is usually based on a particle size where fines sift through a continually changing repose slope. Fines generally concentrate in the central (axial) region of the rotating bed (see Figures 2.28 and 2.29).

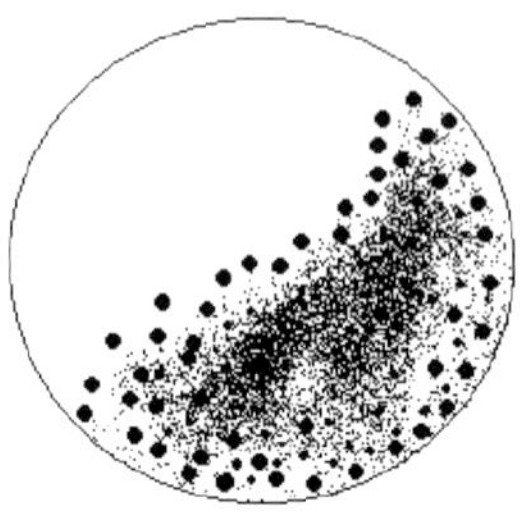

Figure 2.28 Lateral segregation.

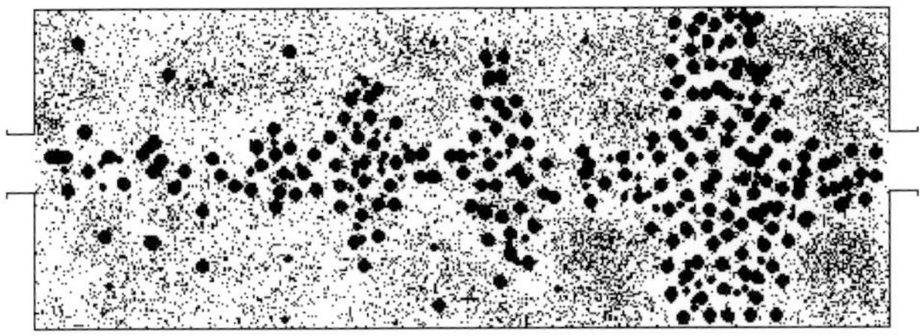

Figure 2.29 Transverse segregation.

Transverse segregation along the length of the rotating bed is again usually based on particle size. One possible explanation for this is that fines can pass through a matrix of larger particles but this leads to concentrations of fines in 'bands' through which further fines cannot pass.

Rolling, bouncing and capture are involved in dynamic situations where typically large numbers of particles are in motion. An example of material falling to form a heap serves well to illustrate these processes.

Segregation based on rolling relies on the fact that larger, more rounded particles will roll further down a heap of material than will smaller or more angular particles. This is mostly a function of a particle's size, shape and density.

Bouncing: particles of different size, shape or elastic properties will 'bounce' varying distances when impacted against an object or each other.

Capture: here, smaller particles can be more easily 'trapped' in the surface voids of a powder bed than larger particles.

Trajectory: segregation occurs because particles have different initial velocities or their velocities are a function of their physical properties. If material is discharging from a pneumatic conveying system, the size fractions may exit the pipeline at different velocities and will be slowed according to their inertia and aerodynamic properties.

2.5.2 *Taking samples*

Storage in hoppers and silos can produce considerable segregation dependent on the filling technique and the discharge pattern of the silo. A few general comments may be helpful. If possible, sample from the entire cross-section of the outlet and over the period of discharge. This can be achieved by using a correctly designed sample divider or splitter.

The discharge pattern of a material from a hopper can play a large part in either introducing or mitigating segregation problems. The correct choice will be a matter of individual circumstance, but it is generally thought that a mass flow discharge pattern is preferred to funnel flow. The sequence of filling and emptying the silo must also be subject to some thought, as for example the level of fill can influence the discharge pattern. Watch out for the effects of discharge aids, e.g. air currents or vibration.

For mixers, there are two situations to consider: (1) taking a sample from within the mixer, which can cause many problems with processing difficulties, physical access and a material distribution which changes with time; (2) taking samples as a mixer discharges, which can give a better indication of the material going into the downstream process as the method of emptying the device is important because an adequately mixed product may suffer segregation upon discharge. In that case, the same recommendations for sampling from a discharging silo would apply.

Material on a belt conveyor will be prone to segregation across its section. Initial loading can result in segregation similar to that of a poured heap; subsequent travel with successive small disturbances as the belt passes over the idlers can result in fines sifting to the lower layers and onto the belt. These same disturbances can also have the same effect as vibration in generating an upthrust of larger particles through a bed of fines. Screw conveyors, scrapers and en masse conveyors generally work at low speeds and, if operating correctly, inline segregation is not generally as problematic. With pneumatic conveyors, material will segregate in a pipeline due to differences in aerodynamic properties. In the dilute phase where air velocities are typically in the range 20–30 m/s, the differences in acceleration and velocity lead to size fractions travelling along the conveying line at different rates. This in-line segregation is generally only of practical importance at the beginning and end of the conveying cycle. However, the velocity difference may be important at the discharge from the line. Conveying in the dense phase with a low velocity would substantially reduce this effect, but radial segregation can still be present.

All these mechanisms must be considered when designing a sampling regime and in the characterisation test procedure to minimise segregation. Once the gross sample has been obtained from the bulk, care is still needed when dividing down to a test sample and when filling any of the testers' cells. A large scatter in test results may indicate a segregating material.

2.5.3 *Characterising segregation*

Most of the meaningful attempts to describe segregation numerically have a statistical foundation where the segregation is treated as some form of variance from the population mean. A critical aspect of any assessment is the sampling regime. Particulate solids are notoriously difficult to sample in a way which avoids some form of bias, and improper sampling will render the analysis results meaningless.

The sample collection and analysis techniques traditionally used in, say, particle size analysis may not be valid to describe segregation. The purpose of any investigation must be clearly defined. Does it relate to the end use of the product or the effect on the process? It is important to establish the scale of scrutiny at which any examination is made. This is related to the idea of a 'critical volume' in which the components are present in the correct proportions but are not necessarily well mixed. Sample size must be quoted in all reporting of results. The number of samples required depends on the application and the specifics of the plant. This becomes a trade-off between accurate description and economic viability. The consequences of being wrong are an important consideration.

The results of the analysis are generally subject to statistical treatment. This is open to the challenge that segregation processes are not random, and therefore statistics cannot be applied to their results. Although this argument has some basis, the usefulness and general acceptance of some form of statistical analysis outweighs any hesitation about their use. There are many techniques which will give an 'index' to quantify the level of segregation. An example of the use of an analysis of variance (ANOVA) technique is now given based on Rollins *et al.* (1995).

To apply this technique, samples must be taken in a particular manner. The entire volume of material under investigation must be notionally divided into a number of regions, and a number of samples taken from within each region. If a binary mixture is assumed and the samples measured to find the relative proportion of component, then the variance found within each region can be compared with the variance of all the samples. This gives an F-value which is related to the magnitude of segregation, where:

$$F = \frac{\dfrac{\sum_{j=1}^{N_R} (\hat{P}_j - \hat{P})^2}{N_R - 1}}{\dfrac{\sum_{j=1}^{N_R} \sum_{i=1}^{N} (Y_{i,j} - \hat{P}_j)^2}{N_R \cdot N - N_R}}, \tag{2.24}$$

where the numerator is the between-regions sample variance; the denominator is the within-region sample variance; N is the number of samples taken from each region; N_R is the number of regions; $\hat{P}$ is the average measured composition for all regions; $\hat{P}_j$ is the average measured composition for the jth region; and $Y_{i,j}$ is the ith measured composition from the jth region.

References

Arthur, J.R.F., Dunstan, T., Dalili, A. & Wong, R.K.S. (1991) The initiation of flow in granular materials. *Powder Technol.*, **65**, 89–101.

Arthur, J.R.F., Koenders, M.A. & Wong, R.K.S. (1985) Anisotropy in particle contacts associated with shearing in granular media. *Acta Mech.*, **64**, 19–29.

Barois-Cazenve, A., Marchal, P., Falk, V. & Choplin, L. (1999) Experimental study of powder rheological behaviour. *Powder Technol*, **103**, 58–64.

Bates, L. (1998) *User Guide to Segregation*. British Materials Handling Board, Marlow, UK.

Bell, T.A., Ennis, B.J., Grygo R.J. *et al.* (1994) Practical evaluation of the Johanson hang-up indicizer. *Int. J. Bulk Solids Handling*, **14**, 117–25.

Crouch, R.S., Wolf, J.P. & Dafalias, Y.F. (1994) Unified critical-state bounding-surface plasticity model for soil. *ASCE J. Eng. Mech.*, **120**, 2251–71.

Dunstan, T., Arthur, J.R.F., Dalili, A., Ogunbekun, O.O. & Wong, R.K.S. (1988) Limiting mechanisms of slow dilatant plastic shear deformation of granular media. *Nature*, **336**, 52–54.

Dunstan, T., Jamebozorgi, M. & Akbarian-Miandouab, S. (1994). Powder strength after changes in principal stress direction. *Powder Technol.*, **81**, 31–40.

Duran, J. (2000) *Sands, Powders and Grains – An Introduction to the Physics of Granular Materials*. Springer, New York.

Friese, H. & Schwedes, J. (1995) *Conference Proceedings Partec 95, Nuremberg*.

Harnby, N., Edwards, M.F. & Neinow, A.W. (1992) *Mixing in the Process Industries*, 2nd edn. Butterworth-Heinemann, London.

Janssen, H.A. (1895) Versuche uber getreidedruck in silozellen (on the measurement of pressures in grain silos). *Zeitschrift Des Vereines Deutscher Ingenieure*, 1045–9.

Jenike, A.W. (1964) *Storage and Flow of Solids Bulletin 123*. The University of Utah, Engineering Experimental Station.

Jenike A.W. (1961) *Gravity Flow of Bulk Solids Bulletin 108*. The University of Utah, Engineering Experimental Station.

Johanson, J.R. (1991) *Bulk Solids Flow Indices – A Simplified Evaluation System*. JR Johanson, San Luis Obispo, CA.

Kaye, B.H. (1997) Characterizing the flowability of a powder using the concepts of fractal geometry and chaos theory. *Part. Part. Syst. Charact.*, **14**, 53–56.

Klausner, J.F., Chen, D. & Mei, R. (2000) Experimental investigation of cohesive powder rheology. *Powder Technol.*, **112**, 94–101.

Marjanovic, P., Geldart, D. & Orband, J.L.R. (1998) Techniques for assessing powder flowability – a comparison. In: *Proceedings of the World Congress on Particle Technology 3, Brighton, UK.*
Nedderman, R.M. (1992) *Statics and Kinematics of Granular Materials.* Cambridge University Press, Cambridge.
Ooi, Y.E. & Rotter, J.M. (1991) Mechanical behaviour of wheat under storage stress histories. *IMechE,* **C418**, 181–86.
Reynolds, O. (1885) *Phil. Mag. Ser.* 5, **50**.
Rollins, D.K., Faust, D.L. & Jabas, D.L. (1995) A superior approach to indices in determining mixture segregation. *Powder Technol.*, **84**, 277–82.
Rumpf, H. (1975) *Particle Technology.* Chapman & Hall, London.
Rumpf, H. (1962) The strength of granules and agglomerates. In: *Agglomeration* (ed. W.A. Knepper). Wiley, New York.
Schofield, A.N. & Wroth, P. (1968) *Critical State Soil Mechanics.* McGraw-Hill, London.
Schwedes, J. (1999) Testers for measuring flow properties of particular solids. In: *Proceedings of the International Symposium on Reliable Flow of Particulate Solids III, Prosgrunn, Norway.*
Shinohara, K., Mikihiro, O. & Golman, B. (2000) Effect of particle shape on angle of internal friction by triaxial compression test. *Powder Technol.*, **107**, 131–36.
de Silva, S.R. (1997) *Mixing and Segregation in Industrial Processes – A Review.* IFPRI Report SAR, pp. 12–20.
Standard Shear Testing Technique for Particulate Solids using the Jenike Cell (1989) Institution of Chemical Engineers, Rugby, UK.
Tomas, J. & Schubert, H. (1979) *Conference Proceedings of Partec 97, Nuremberg.*
Tripodi, M.A., Puri, V.M., Manbeck, H.B. & Messing, G.L. (1994). Triaxial testing of dry cohesive powder and its application to a modified cam-clay constitutive model. *Powder Technol.*, **80**, 35–43.
Troadec, J.P., Bideau, D. & Dodds, J.A. (1991) Compression of two dimensional packings of cylinders made of rubber and plexiglas. *Powder Technol.*, **65**, 147–51.
Verlinden, A. (2000) *Experimental Assessment of Shear Testers for Measuring Flow Properties of Bulk Solids.* Ph.D. dissertation, University of Bradford, UK.
Wiche, S.J., Roberts, A.W. & McBride, B. (2003) A flowability tester for bulk solids flow measurement. In: *4th International Conference for Conveying and Handling of Particulate Solids, Budapest.*

3 Characterisation for hopper and stockpile design

ALAN W. ROBERTS

3.1 Introduction

Throughout the world, bulk materials-handling operations perform a key function in a great number and variety of industries. While the nature of the handling tasks and scale of operation vary from one industry to another and, on the international scene, from one country to another according to the industrial and economic base, the relative costs of storing, handling and transporting bulk materials are, in the majority of cases, very significant. It is important, therefore, that handling and processing systems be designed and operated with a view to achieving maximum efficiency and reliability.

Over the past four decades, much progress has been made in the theory and practice of bulk solids handling and powder processing. The procedures for the design of handling plant, such as storage bins, gravity reclaim stockpiles, feeders and chutes are well established and follow four basic steps:

(1) determination of the strength and flow properties of the bulk solids for the worst likely flow conditions expected to occur in practice
(2) determination of the bin, stockpile, feeder or chute geometry to give the desired capacity, to provide a flow pattern with acceptable characteristics and to ensure that discharge is reliable and predictable
(3) estimation of the loadings on the bin and hopper walls and on the feeders and chutes under operating conditions
(4) design and detailing of the handling plant including the structure and equipment.

The general theory, characterisation procedures and design methodologies pertaining to gravity flow of bulk solids are well documented, a selection of references being included at the end of this chapter. Of particular note is the foundation work presented in the three University of Utah Bulletins, namely Bulletin 108 of Jenike (1961), Bulletin 116 of Johanson and Jenike (1962), and Bulletin 123, Jenike (1964). The basic principles of this subject area, as presented in this chapter, are based largely on the work of Arnold *et al.* (1982) and Roberts (1991, 1998).

3.2 Modes of flow in bins

3.2.1 *Mass-flow and funnel-flow*

Following the definition by Jenike (1961, 1964), the two principal modes of flow are mass-flow and funnel-flow. These are illustrated in Figure 3.1. In mass-flow, the bulk solid is in motion at every point within the bin whenever material is drawn from the outlet. There is flow of bulk solid along the walls of the cylinder (the upper parallel section of the bin) and the

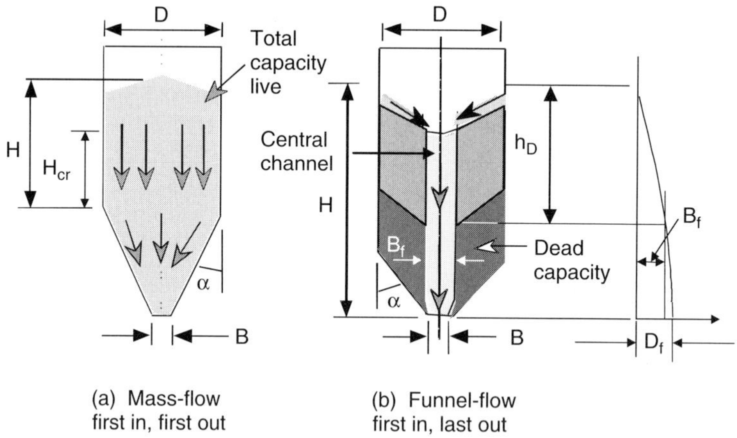

Figure 3.1 Principal modes of flow.

hopper (the lower tapered section of the bin). Mass-flow guarantees complete discharge of the bin contents at predictable flow rates. It is a 'first in, first out' flow pattern; when properly designed, a mass-flow bin can remix the bulk solid during discharge should the solid become segregated upon filling of the bin. Mass-flow requires steep, smooth hopper surfaces and no abrupt transitions or in-flowing valleys. The hopper slope is defined by the hopper half-angle, α. For flow to occur, the opening dimension B should be at least large enough to avoid blockages due to the formation of stable, cohesive arches, as well as 'mechanical' arches caused by interlocking of larger particles over the opening. Normally, the opening dimension is sized to allow the required discharge rate to be achieved.

Funnel-flow occurs when the slope or half angle, α, of the hopper walls is too large and the surface roughness of the hopper walls is too high to achieve mass-flow. It is a 'first in, last out' flow pattern with the bulk solid sloughing off the top surface and falling through the vertical flow channel that forms above the opening. The flow channel, which is governed by the outlet dimension B, opens out to a slightly larger dimension just above the outlet. The walls of the flow channel are generally steeply sloped in a slightly diverging manner but in some cases may even be vertical. The dimension B_f defines the cross-sectional diameter or width of the flow channel at a particular height above the outlet.

Flow will continue until the level of the bulk solid in the bin drops an amount H_D equal to the drawdown. At this level, the bulk strength of the contained material is sufficient to sustain a stable pipe or 'rathole' of dimension $D_f = B_f$ as illustrated in Figure 3.1(b). The dimension D_f is referred to as the critical rathole dimension for which the rathole is just on the verge of collapsing. D_f depends on the unconfined yield strength of the bulk solids which, in turn, depends on the degree of consolidation of the stored bulk solid. The variation of D_f with height of bulk solid is illustrated in Figure 3.1(b). Once the level defined by H_D is reached, there is no further flow, and the material below this level represents 'dead' storage. This is a major disadvantage of funnel-flow. For complete discharge, the bin opening needs to be at least equal to the critical rathole dimension determined at the bottom of the bin corresponding to the bulk strength at this level. However, for many cohesive bulk solids and for the normal consolidation heads occurring in practice, ratholes measuring several metres are often determined. This makes funnel-flow impracticable.

Funnel-flow is generally erratic and gives rise to segregation problems. It has the advantage of providing wear protection of the bin walls, since the material flows against stationary material. However, since it is a 'first in, last out' flow pattern, it is unsatisfactory for bulk solids that degrade with time. It is also unsatisfactory for fine bulk solids of low permeability. Such materials may aerate during discharge through the flow channel, and this can give rise to flooding problems or uncontrolled discharge.

3.2.2 *Types of mass-flow bins*

Mass-flow bins are classified according to the hopper shape and associated flow pattern. The two main hopper types are conical hoppers, which operate with axi-symmetric flow as in Figure 3.2(a), and wedge-shaped hoppers, in which plane-flow occurs as in Figure 3.2(b). In plane-flow bins, the hopper half-angle, α, will usually be, on average, approximately 8–10° larger than the corresponding value for axi-symmetric bins with conical hoppers. Also, the opening width, B, is approximately half the diameter of the corresponding conical hopper. Therefore, plane-flow bins offer a larger storage capacity for the same head room than axi-symmetric bins, but this advantage is somewhat offset by the long slotted opening which can give rise to feeding problems.

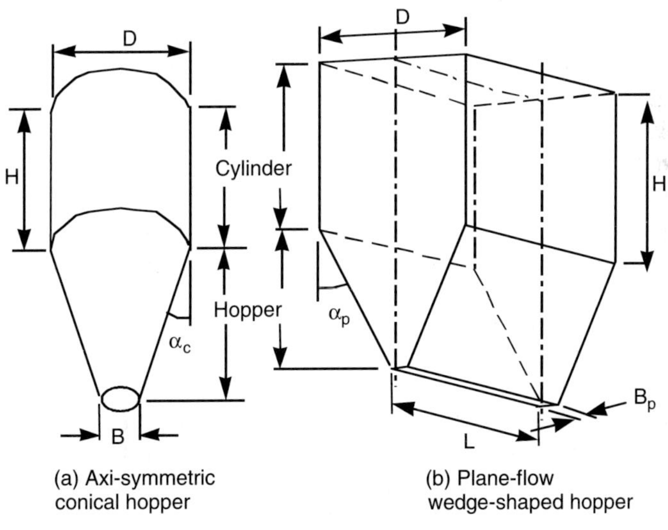

Figure 3.2 Basic shapes of mass-flow bins.

Other types of mass-flow bins are illustrated in Figure 3.3. The transition hopper (Figure 3.3(a)), which has plane-flow sides and conical ends, offers a more acceptable opening slot length. The chisel-shaped hopper (Figure 3.3(b)) is a variation of plane-flow when the bin is of circular cross-section. Pyramidal hoppers (Figure 3.3(c)), while simple to manufacture, have the potential for buildup of material that is likely to occur in the

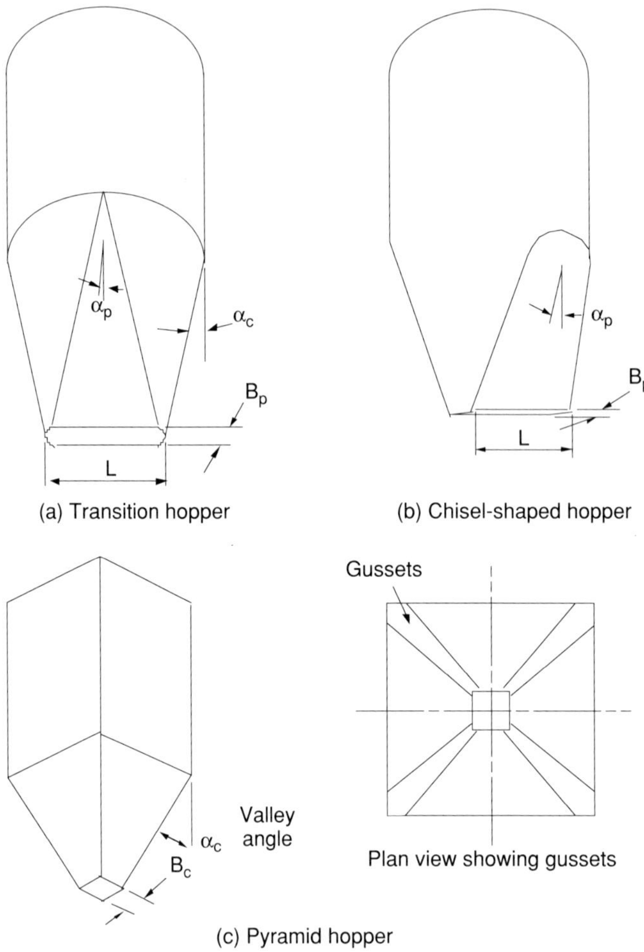

Figure 3.3 Variations of mass-flow bins.

sharp corners of in-flowing valleys. Such problems may be overcome, preferably, by radiusing the corners of the hopper or by fitting suitably shaped gusset plates in the valleys, as illustrated.

3.2.3 *Funnel-flow bins*

Typical funnel-flow bins are illustrated in Figure 3.4. Such bins may have flat or tapered bottoms, but in the case of the latter, the hopper half-angles are larger than for mass-flow bins. Since funnel-flow bins are prone to 'ratholing' with reduced live capacity, they often require the use of flow promotion aids to initiate and maintain flow. In the case of the funnel-flow bin with the slotted opening of Figure 3.4(b), the width of the slot B should be large enough to prevent a stable arch from forming across the opening.

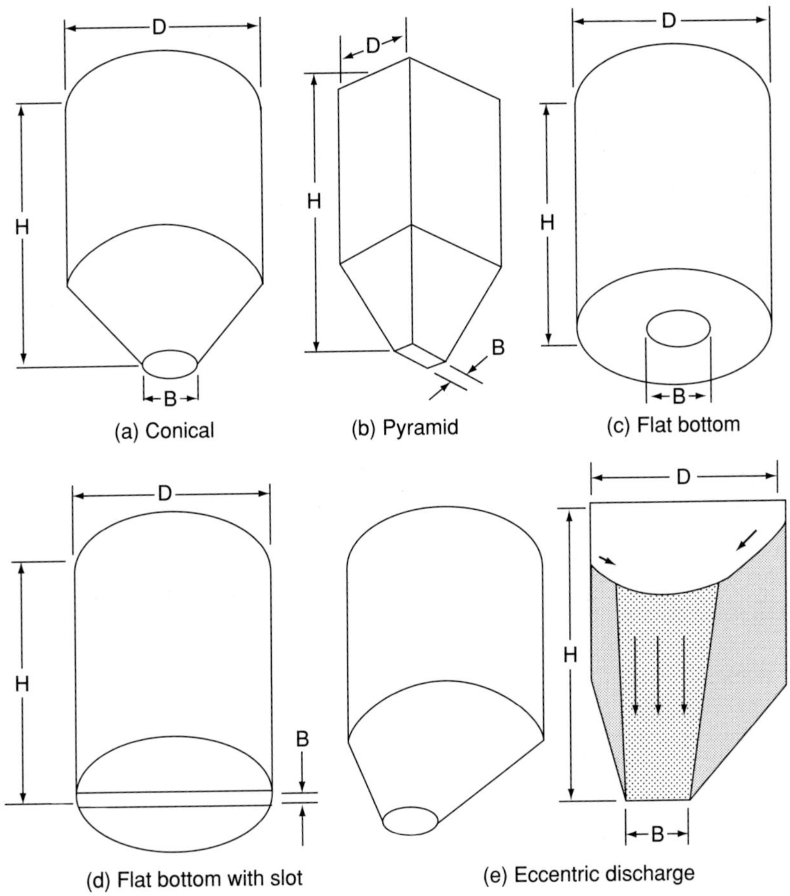

Figure 3.4 Typical funnel-flow bins.

3.2.4 *Expanded flow*

The disadvantages of funnel-flow are overcome by the use of expanded flow, as illustrated in Figure 3.5. This combines the wall protection of funnel-flow with the reliable discharge of mass-flow. For complete discharge, the dimension at the transition of the funnel-flow and mass-flow sections must be at least equal to the critical rathole dimension, D_f, at that level. Expanded-flow bins are particularly suitable for storing large quantities of bulk solids while maintaining acceptable head heights. The concept of expanded-flow may be used to advantage in the case of bins or bunkers with multiple outlets.

3.2.5 *Bin symmetry*

Generally speaking, symmetric bin shapes provide the best performance. Asymmetric shapes often lead to segregation problems with free-flowing materials of different particle sizes and make the prediction of wall loads much more difficult.

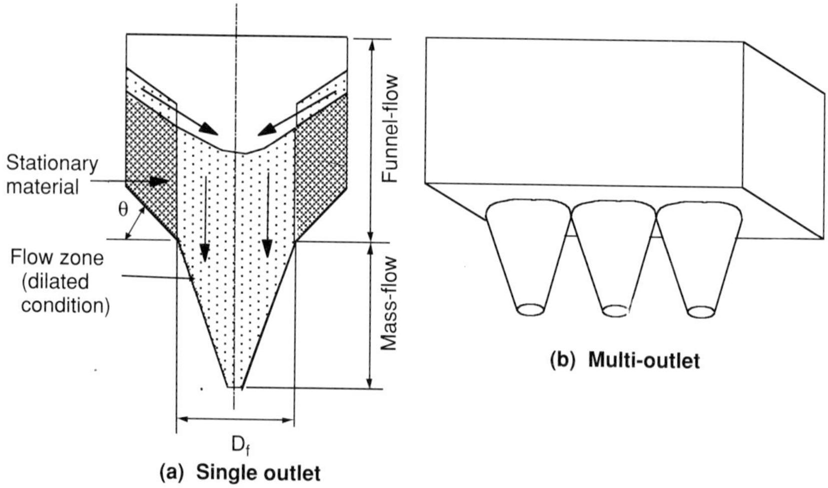

Figure 3.5 Expanded flow.

3.3 Bulk solid parameters

For bin, stockpile and chute design, knowledge of the stress and flow characteristics of bulk solids is important. The necessary bulk solid parameters must be related to the desired flow modes developed in the particular flow channels formed by hoppers and chutes.

3.3.1 *Stress fields in mass-flow bins*

The stress fields generated in symmetrical mass-flow storage bins are shown in Figure 3.6. The major and minor principal stresses are σ_1 and σ_2, respectively, σ_1 being referred to as the major consolidating stress or pressure. When the bin is filled from the empty condition, a peaked or active stress field is developed, as shown in Figure 3.6(a). In this case, σ_1 acts almost vertically, the variation of σ_1 being depicted in Figure 3.6(c). Once flow has been initiated, the vertical support provided by the gate is removed, and the load is transmitted to the hopper walls. The stress field in the hopper changes from a peaked stress field to an arched or passive stress field, as indicated in Figure 3.6(b).

The change or 'switch' in the stress field is initiated at the outlet and travels up the hopper during a short time interval until the transition of the cylinder and hopper is reached. Once steady-state flow conditions are established, the switch remains 'locked in' at the transition as shown, with a peaked stress field being retained in the cylinder. The stress field of Figure 3.6(b) for the flow condition is a stable one and is retained even if the flow is stopped after some discharge has occurred. The only way the stress field of Figure 3.6(a) can be re-established is to empty the bin completely and fill it again from the empty condition.

The characteristics of the arched stress field are illustrated in Figure 3.6(c) for the flow case. Consolidating pressures are high in the region of the cylinder and hopper transition but decrease towards the outlet as indicated. As Jenike has shown, a radial stress field is developed in the lower portion of the hopper (Jenike 1961, 1964). This implies that the

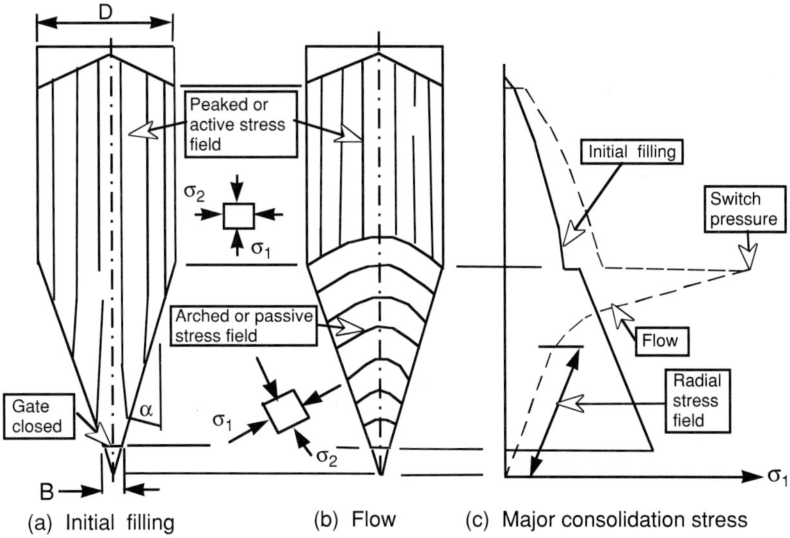

Figure 3.6 Stress fields for initial filling and flow in mass-flow bins.

consolidation stresses or pressures are proportional in magnitude to the radial distance measured from the point of intersection of the sides. In the arched stress field, the major principal stress or pressure, σ_1, acts in a direction tangential to the arches as indicated. A major proportion of the surcharge load due to the parallel or cylindrical section of the bin is carried by the upper region of the hopper where the pressures are highest; the stresses near the outlet are quite low, and this means that the loads on the gate when flow is stopped are significantly lower than for the initial filling case.

The stress fields have a major influence on both the magnitude and distribution of the pressures acting on the bin walls and the magnitude of the load's action on discharge gates and feeders.

3.3.2 *Yield loci defining flow of bulk solid*

When a bulk solid flows, it does so by internal shear or yielding. Within the mass of the solid itself, there is internal shear, whereas at the boundary wall of a hopper or chute, there is boundary shear. Consider an element of bulk solid flowing by gravity in a symmetrical hopper, as depicted in Figure 3.7. Since an arched stress field exists, the major and minor principal stresses, σ_1 and σ_2, respectively, are as shown.

The relationship between the normal and shear stresses during flow is defined by the yield locus, YL, or, more specifically, the instantaneous yield locus, IYL, which is tangential to the two Mohr semicircles as indicated. The major Mohr semicircle is defined by the principal stresses or pressures, σ_1 and σ_2, while the Mohr semicircle through the origin tangential to the yield locus defines the unconfined yield strength, σ_c. The extrapolation of the yield locus to the shear stress axis allows the cohesive stress, τ_o, to be determined; τ_o is the shear stress at zero normal stress.

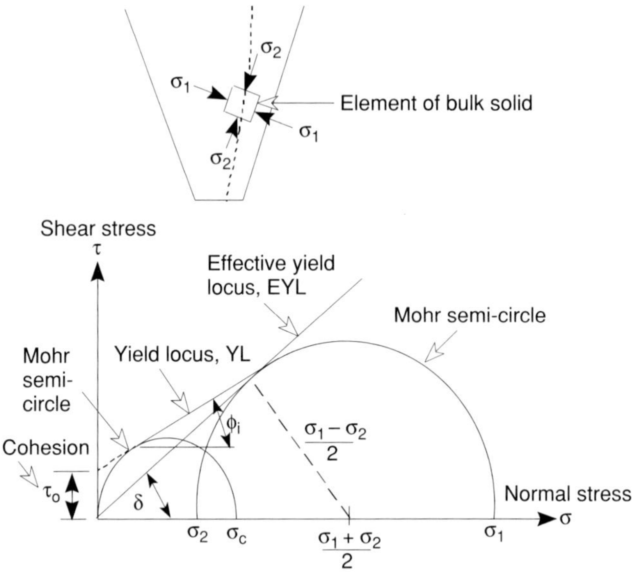

Figure 3.7 Yield loci for consolidated bulk solid.

The straight line through the origin and tangential to the major Mohr semicircle is the effective yield locus, EYL. The angle, δ, defining the slope EYL is called the effective angle of internal friction. For a free-flowing non-cohesive bulk solid, such as dry sand or grain, both τ_o and σ_c are zero. Hence, the internal friction angles δ and ϕ_i are the same. For a cohesive bulk solid, the EYL may be thought of as the yield locus for an equivalent free-flowing bulk solid having similar properties to that of the actual bulk solid. In bulk solids analysis, the effective angle of internal friction, δ, is a more convenient parameter to use than the cohesive stress, τ_o. The angle δ defines the major and minor principal stresses, σ_1 and σ_2, as follows:

$$\sin\delta = \frac{\sigma_1 - \sigma_2}{\sigma_1 + \sigma_2}, \tag{3.1}$$

$$\text{hence } \frac{\sigma_2}{\sigma_1} = \frac{1 - \sin\delta}{1 + \sin\delta}. \tag{3.2}$$

3.3.3 *Instantaneous yield loci (IYL) and flow function (FF)*

Figure 3.8 shows the instantaneous yield loci, IYL, for a bulk solid for three consolidation conditions defined by the major consolidation pressures $(\sigma_1)_1$, $(\sigma_1)_2$ and $(\sigma_1)_3$. The instantaneous condition is that corresponding to the instant when the consolidation pressure acts. In the case of a storage bin, it is the instant that flow occurs just after the initial filling of the bin. The three consolidation conditions represent three positions of an element of flowing bulk solid in the lower hopper during discharge.

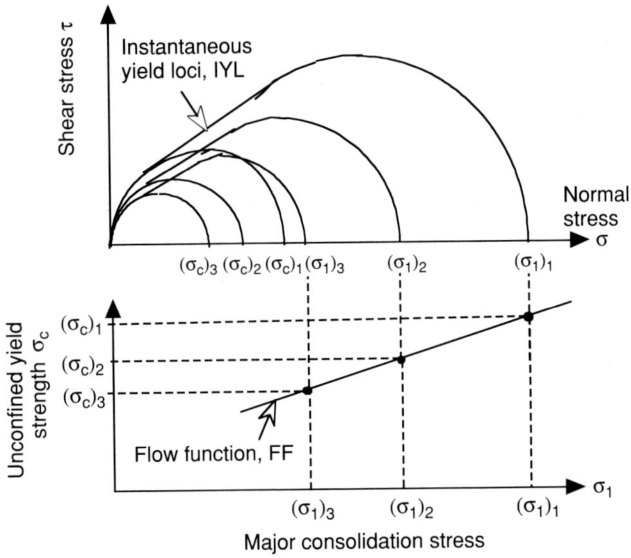

Figure 3.8 Yield loci and flow function, FF.

Referring to Figure 3.8, the Mohr stress semicircles through $(\sigma_1)_1$, $(\sigma_1)_2$ and $(\sigma_1)_3$ are tangential to the respective yield loci, as shown. The Mohr semicircles through the origin and tangential to the yield loci define the unconfined yield strength or stress corresponding to each consolidation condition. The unconfined yield strength is denoted by σ_c. The three strength values $(\sigma_c)_1$, $(\sigma_c)_2$ and $(\sigma_c)_3$ correspond, respectively, to the three consolidation pressures $(\sigma_1)_1$, $(\sigma_1)_2$ and $(\sigma_1)_3$. The bulk strength, represented by the flow function, FF, is the unconfined yield strength, σ_c, expressed as a function of the major consolidation stress, σ_1. It is obtained from the IYL, as shown in Figure 3.8.

3.3.4 *Time yield loci (TYL) and flow function (FF_t)*

Many cohesive bulk solids will gain strength during undisturbed storage. Referring to Figure 3.9, the application of the major consolidation pressure σ_1 over an extended period of time will cause the unconfined yield strength to increase from σ_c to σ_{ct}. The time yield locus, TYL, defines the yielding condition following undisturbed time storage. The angle ϕ_t is called the static angle of internal friction. It is the angle defined by the slope of the TYL at the point of tangency with the Mohr semicircle through σ_{ct}.

3.3.5 *Types of flow functions for bulk solids*

The various forms of flow functions for bulk solids are illustrated in Figure 3.10. Most cohesive bulk solids tend towards a limiting strength as the consolidation pressure increases; their flow functions are convex upward in shape, as indicated by curve (a) of Figure 3.10. Simple bulk solids have a linear flow function over the range of consolidation pressures

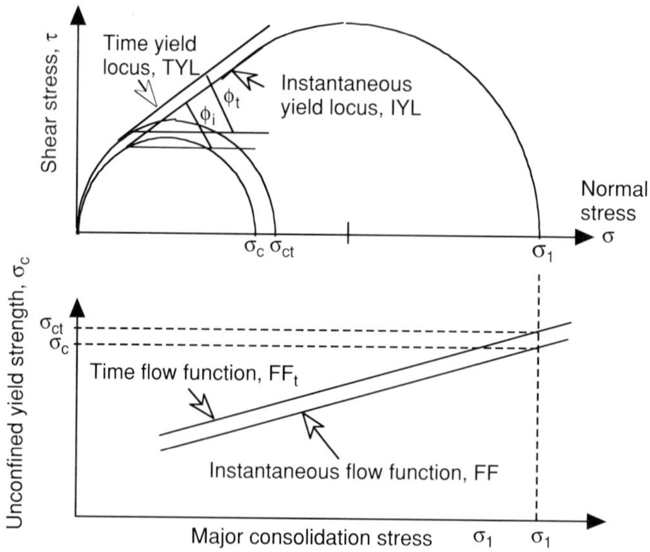

Figure 3.9 Time yield locus, TYL, and time flow function, FF_t.

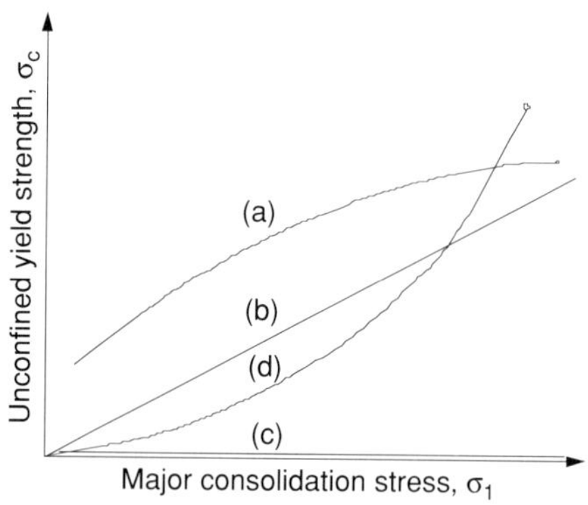

Figure 3.10 Various forms of flow functions for bulk solids.

considered as in curve (b). Free-flowing bulk solids have zero strength, and their flow function coincides with the horizontal axis, as depicted by curve (c). Occasionally, bulk solids are encountered in which the bulk strength increases non-linearly with increase in consolidation pressures as in curve (d). Such bulk solids are free-flowing under low consolidation pressures but tend to cake or bond together as the consolidating pressures increase. A typical material exhibiting these characteristics is ammonium nitrate prill (Roberts *et al.* 1979). Bulk solids with flow functions of the form given by curve (d) are very difficult to handle. When stored in mass-flow bins, they give rise to arching problems high up in the hopper where the consolidation stresses are high. They should be stored in mass-flow bins consisting of hoppers only without upper, cylindrical sections.

3.3.6 *Angles of internal friction*

The three angles of internal friction, ϕ_i, ϕ_t and δ, previously defined, are normally plotted against the major consolidation pressure, σ_1. The typical form of the graphs is as illustrated in Figure 3.11.

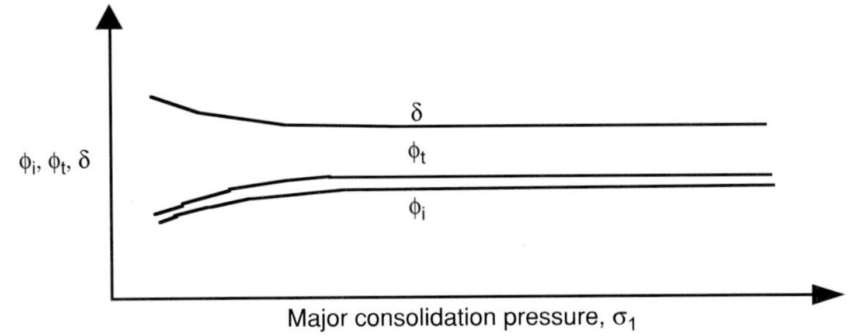

Figure 3.11 Angles of internal friction.

3.3.7 *Wall yield loci and wall or boundary friction*

At a boundary surface, such as that of a hopper or chute, the flow is characterised by the wall yield locus (WYL), which relates the shear stress at the wall to the corresponding normal stress, or pressure. The WYL and stress conditions at the boundaries of a hopper and chute are illustrated in Figure 3.12. Figures 3.12(a) and 3.12(b) show the flow channels in hoppers and chutes, respectively. Figure 3.12(c) shows the bulk solid yield functions and Mohr semicircle, which relates the normal and shear stresses at a hopper wall to the principal stresses of the bulk solid.

Typically, the WYL for most bulk solids is slightly convex, as shown in Figure 3.12(c). In some cases, the extrapolated WYL does not pass through the origin, indicating adhesion forces or forces of attraction between the bulk solid and the wall surface. This is highlighted in Figure 3.13, which shows, diagrammatically, the low pressure or stress end of the WYL. For illustration purposes, the WYL is shown extended into the tensile normal stress region where the tensile stress for zero shear stress is given by σ_o. This stress is referred to as the adhesion. The cohesive stress, τ_o, is the shear stress corresponding to zero normal stress; sometimes it is also referred to as the adhesive stress.

The wall friction angle, ϕ, is defined by

$$\phi = \tan^{-1}\left[\frac{\tau_W}{\sigma_W}\right], \tag{3.3}$$

where τ_W = shear stress at the wall; σ_W = pressure acting normal to the wall.

Since the WYL for cohesive bulk solids is often convex upward in shape and, when extrapolated, intersects the shear stress axis at τ_o, then the wall friction angle, ϕ, will decrease with an increase in normal pressure. This is illustrated in Figure 3.14. It should be noted that the wall friction angle cannot be larger than the effective angle of internal friction, δ. The angle, δ, is, therefore, an upper bound limit for ϕ. This means that at low normal pressures, the bulk solid will flow by internal shear rather than by sliding against the wall.

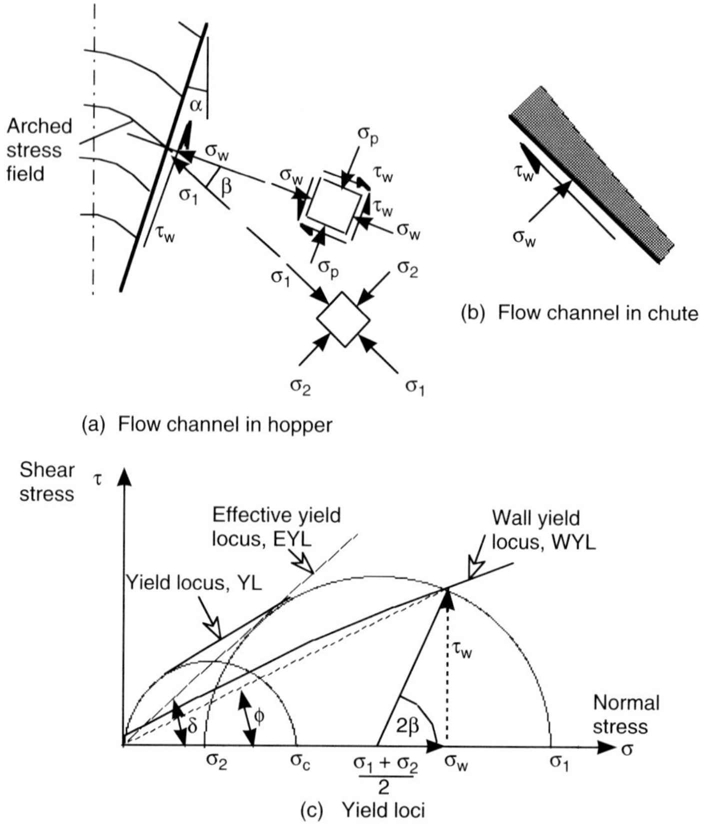

Figure 3.12 Bulk solid yield and flow channels.

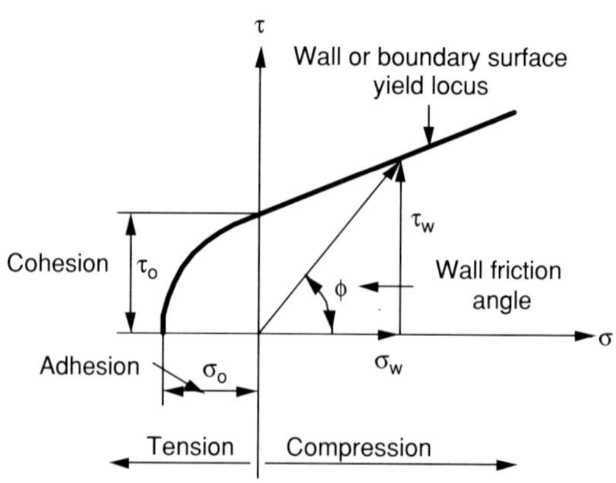

Figure 3.13 Wall yield locus, cohesion and adhesion.

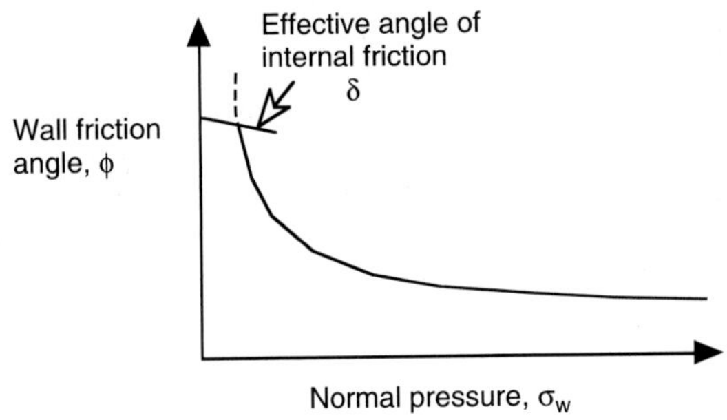

Figure 3.14 Variation of wall-friction angle with normal pressure.

3.4 Determination of bulk solids flow properties

3.4.1 *Introductory remarks*

The design of bulk solids handling plant requires a knowledge of the strength and flow properties of the bulk solids under operating conditions. The latter conditions include loading and consolidation for instantaneous and extended time storage as well as environmental factors such as temperature, moisture and humidity. There are now well-established laboratory test procedures for determining the necessary flow properties. Further details on flow property measurement can be found in the literature. The paper by Schwedes (2002) provides a comprehensive overview of the various test methods.

Focusing specifically on bin, hopper and stockpile design, the laboratory tests aim to duplicate field conditions and provide the designer with such parameters as:

(1) Yield loci and flow functions, FF, for instantaneous and time storage conditions for the range of moisture contents and, as relevant, temperatures occurring in practice. The flow functions represent the variation of unconfined yield strength with major consolidation stress as occurs during storage and flow.
(2) Effective angle of internal friction, δ, as a function of major consolidation stress.
(3) Static angle of internal friction, ϕ_t, as a function of major consolidation stress.
(4) Wall friction angles, ϕ, as a function of normal stress for different bin and chute wall materials and finishes.
(5) Bulk density, ρ, as a function of major consolidation stress.

For fine powders, flow rate predictions may be critical. For the determination of flow rates, it is necessary to measure:

(1) solids density
(2) permeability of the solids as a function of major consolidation stress.

Additional tests may be conducted where the influence of vibrations on the consolidation of bulk solids needs to be examined. Of particular interest from the point of view of flow promotion by vibration is the reduction in both unconfined yield strength and wall friction. This aspect of bin design has been examined in some detail by Roberts (1997). In general, best results are achieved when vibration frequencies equal to or greater than 100 Hz and small amplitudes, 1 mm or less, are employed.

During the flow of a bulk solid with mixed particle sizes, the large particles move bodily while the material shears across the fines. Therefore, the yield strength of the bulk solid, which may cause a flow blockage to occur, is dependent on the yield strength of the fines component. For this reason, the flow properties of a bulk solid are determined by testing the fines component, which is normally the −4 mm size fraction.

3.4.2 *Shear test apparatus*

The strength and flow characteristics of bulk solids are based on the various yield loci obtained from laboratory tests. Such tests involve a consolidation phase, in which the sample is brought to the critical state condition in terms of voidage, followed by a shear phase. In general, the shear vs. consolidation characteristics may be obtained by direct shear, using either the linear or rotational type shear testers, or indirectly using the biaxial or triaxial cells. The shear test methods are indicated schematically in Figure 3.15. These tests are applicable to bulk solids over the range of moisture conditions up to, but not beyond, the saturated moisture content.

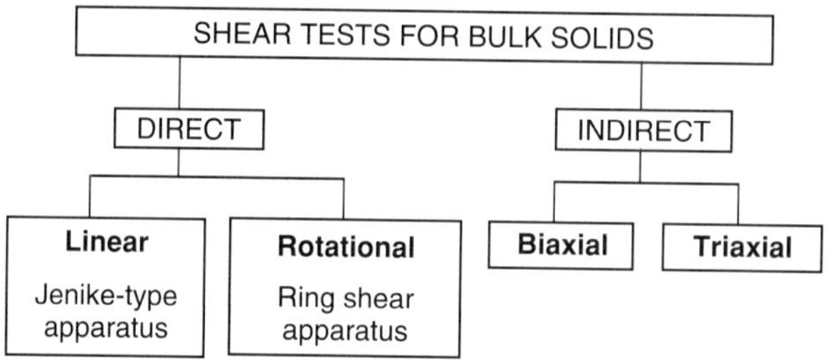

Figure 3.15 Shear tests for bulk solids.

The biaxial and triaxial tests are suitable for the determination of bulk strength but are not relevant for wall or boundary friction determination. In particular, the biaxial testers that have been developed are sophisticated test machines used for the fundamental characterisation of powders (Harder & Schwedes 1985; Schwedes & Harder 1986; Maltby 1993; Janssen 2001; Schwedes 2002; Janssen & Zetzener 2003). Hence, for design work, the direct shear testers are the more suitable (Jenike 1961, 1964; Arnold *et al.* 1982; Australian Standard AS 3880-1991; Standard Shear Testing Procedure for Particulate Solids Using the Jenike Shear Cell; Schulze 1994; Roberts 1998; D6128-00 Standard Method for Shear Testing Using the Jenike Shear Cell; Schulze & Wittmaier 2003). Other testers have been developed for flow property and testing, and quality-control analysis. These include the Johanson Indicisers (Johanson 1992; Bell *et al.* 1994) and uniaxial testers (Maltby 1993; Maltby & Enstad 1993; Kwade *et al.* 1994; Goelema *et al.* 1995; Schwedes 2002; Wiche *et al.* 2003), the latter being, in effect, unconfined compression tests.

3.4.3 *Jenike-type direct shear tester*

Of the above methods, the linear direct shear tester of the type developed by Jenike (Jenike 1961, 1964; Arnold 1982; Australian Standard AS 3880-1991; D6128-00 Standard Method for Shear Testing Using the Jenike Shear Cell; Roberts 1998; Schwedes 2002; Standard Shear Testing Procedure for Particulate Solids Using the Jenike Shear Cell) is widely used particularly for industrial design applications. It allows measurement of the relevant bulk solid parameters, including wall or boundary friction. With the aid of consolidation benches, the Jenike test is very amenable to time consolidation tests.

The Jenike-type direct shear test apparatus employs a shear cell of circular cross-section, as shown in Figure 3.16. The test involves two phases, the filling of the cell and pre-consolidation phase, as in Figure 3.16(a), and the shear consolidation followed by shear phase as in Figure 3.16(b). The normal load is applied to the top cover by means of a gravity load frame, while the shearing action is applied by means of an electro-mechanically driven loading stem, which is driven at a rate of 2.5 mm/min. The shear force is measured using a strain-gauge load cell and displayed on either a chart recorder or a computer via a data logger.

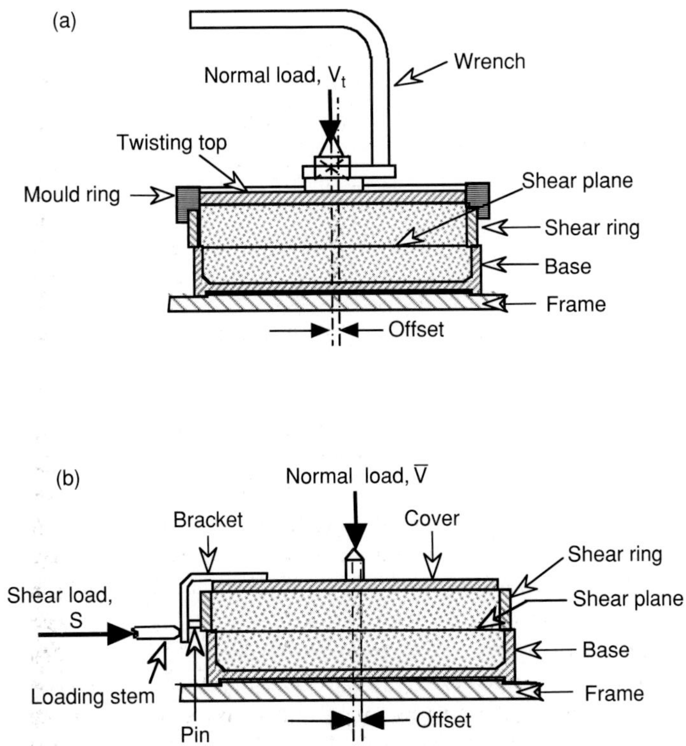

Figure 3.16 Jenike type direct shear test: (a) pre-consolidation of sample; (b) shear consolidation followed by shear to failure.

The shear consolidation phase involves shear deformation under the nominated load, V, and, hence, normal stress $\sigma = V/A_C$, the aim being to bring the sample to the critical state condition such that the voidage reaches, asymptotically, its limiting value corresponding to the stress condition. As indicated in Figure 3.16(b), the consolidation phase commences with the shear ring offset, the objective being for the sample to reach the critical state condition by the time the shear ring has moved to a position of concentricity with the base. Because of the limited travel for this condition to be reached, it is necessary to pre-consolidate the sample by applying the nominated normal force V_t and giving the twisting top a predetermined number of oscillatory twists through an amplitude of, approximately, $\pm 10°$ to $\pm 15°$. Normally, the load V_t equals the load V selected for the shear consolidation phase of the test. However, on some occasions, particularly when testing very cohesive bulk solids, it is necessary to select values of $V_t > V$. On completion of pre-consolidation, the mould ring is removed and the sample screeded level with the top of the shear ring to remove excess material. The cell is then set up, as indicated in Figure 3.16(b) for the shear consolidation to be completed.

Following consolidation to the critical state condition, the sample is then sheared to failure under a nominated load $\overline{V}$ smaller than the consolidation load, V. The process is repeated by consolidating samples to the same critical state condition and each time choosing a different value of load $\overline{V}$ for the shear to failure test. The shear failure stresses corresponding to the normal stresses due to the applied loads, $\overline{V}$, together with the consolidation shear stress corresponding to the load, V, are plotted to obtain the instantaneous yield locus (IYL), as illustrated in Figure 3.17. While Figure 3.17 shows the normal and shear loads, it is usual to plot the yield loci using stress units, as illustrated in Figure 3.17.

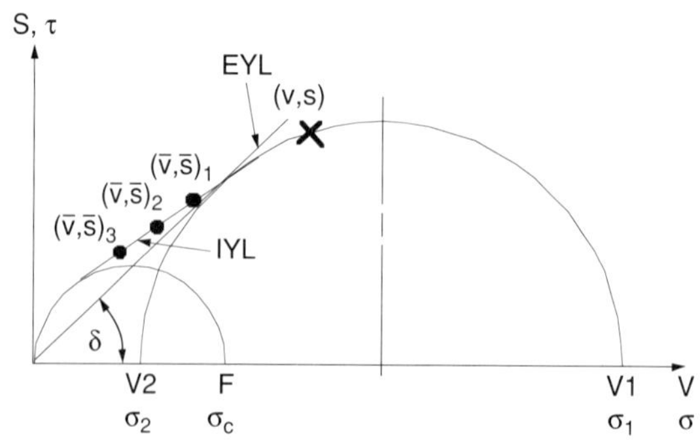

Figure 3.17 Determination of yield loci using direct shear tester.

For the time yield loci tests, each sample is consolidated as previously described and then removed from the shear tester and placed in a consolidation bench. The load, $V1$, is applied, and the sample remains in the consolidation bench for the prescribed storage time. The sample is then removed from the consolidating bench, placed back in the shear tester and sheared under the nominated applied load, $\overline{V}$. The time yield locus, illustrated in Figure 3.9, is then obtained.

3.4.4 *Torsional or ring shear apparatus*

A disadvantage of the Jenike-type direct shear tester is the limited travel it offers during the consolidation phase, as previously indicated. As is evident from Figure 3.7, travel is limited, effectively, by the thickness of the shear ring. The torsional or ring shear testers have unlimited travel, which allows the consolidation of the sample to be performed without the need of the pre-consolidation phase as in the Jenike-type shear test. However, ring or torsional shear testers have the following disadvantages:

(1) Owing to the rotational motion, shear strain is not linear; it is a function of the radius and hence is not uniform for all regions of the bulk solid sample.
(2) Time consolidation tests are difficult to perform.
(3) Wall friction tests may be performed, but the rotary motion will mask out the unidirectional surface properties of wall lining materials.

While there have been several types of annular or ring shear testers developed, the Schulze tester (Schulze 1994; Schulze & Wittmaier 2003) is gaining increasing acceptance.

3.4.5 *Wall yield locus (WYL) and wall friction angle ϕ*

Of the various parameters influencing the gravity flow performance of bulk solids in hoppers and chutes, the most significant is the friction characteristic at the contact zone of the bulk solid and wall or boundary surface. For hopper design, the importance of low wall friction in permitting an acceptable hopper half angle for mass-flow in hopper design is emphasised. Low wall or boundary friction is also required for efficient design and performance. Primarily, wall friction depends on the interaction between three groups of variables, those relating to the bulk solid, those relating to the wall or boundary surface and those which arise from the loading and environmental conditions. This interaction is illustrated in Figure 3.18.

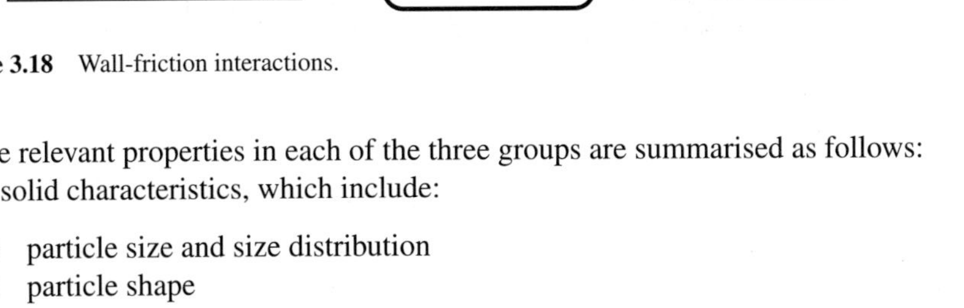

Figure 3.18 Wall-friction interactions.

The relevant properties in each of the three groups are summarised as follows:
Bulk solid characteristics, which include:

(1) particle size and size distribution
(2) particle shape
(3) particle hardness
(4) moisture content
(5) particle density

(6) bulk density
(7) surface chemistry characteristics
(8) temperature
(9) undisturbed storage time.

Wall surface characteristics, which include:

(1) surface roughness
(2) hardness
(3) chemical composition.

Loading and environmental factors, which include:

(1) normal pressure between bulk solid and wall surface
(2) relative rubbing or sliding velocity
(3) temperature and humidity or moisture conditions
(4) wall vibrations.

The importance of wall friction and the associated properties of wear of lining materials in bins, hoppers and chutes is discussed in various literature (Johanson & Royal 1982; Roberts *et al.* 1984, 1988, 1993; Ooms & Roberts 1985; Keys & Wiche 1993; Roberts & Wiche 1993).

3.4.6 *Wall yield loci (WYL) and wall friction angle ϕ*

The WYL is obtained using the direct shear apparatus shown in Figure 3.19. A sample of the wall lining material is clamped to the base of the shear tester, and the bulk solid is sheared across the surface. It is customary to progressively reduce the normal load in steps from a maximum value to a minimum value by removing weights as the shear strength is recorded. This reproduces the same action as in the flow in an actual hopper or chute where the pressure normally reduces in the direction of flow at that location.

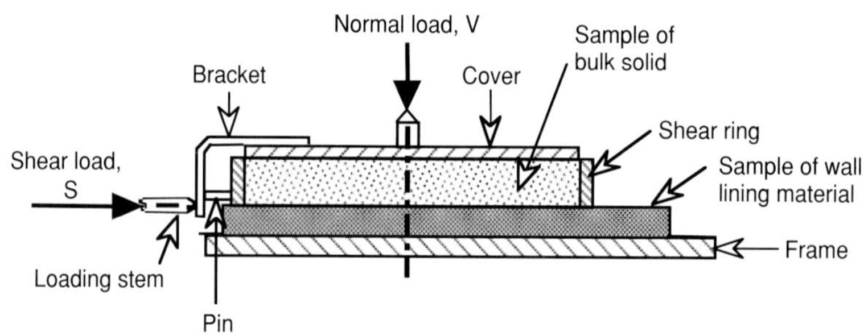

Figure 3.19 Test arrangement for the wall-yield locus and wall-friction determination.

For chute design, the wall friction angles at low consolidation stress as well as the cohesion, τ_o, and adhesion, σ_o (Figure 3.13), are required. The difficulty associated with the Jenike direct shear test of Figure 3.8 is that the lowest pressure that can be generated is that due to the weight of the ring and the contained bulk solid plus the weight of the lid.

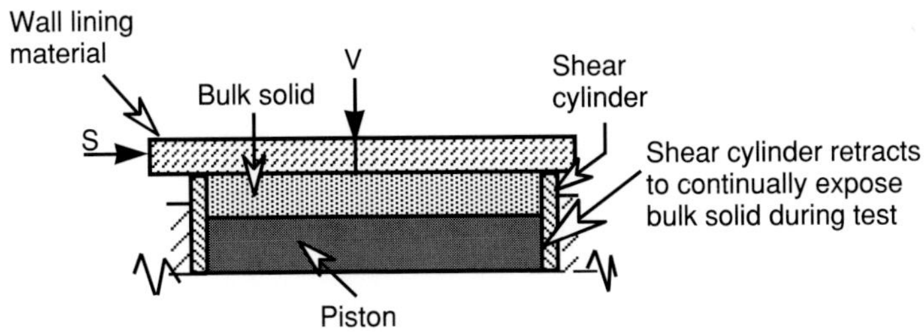

Figure 3.20 Inverted shear tester for determination of wall-yield loci.

The inverted shear tester of Figure 3.20 overcomes this problem. Two such testers have been developed at the University of Newcastle, Australia, one with a 95-mm-diameter shear cylinder and another with a 300-mm-diameter shear cylinder. The latter allows a much larger size range of bulk solids to be tested, which is more in keeping with the requirements of chute design. Using these devices, it is possible to measure the WYL into the tensile stress zone.

3.4.7 *Bulk density*

This is determined using the compressibility tester, which allows the variation of bulk density with major consolidation pressure to be measured. Typically, this variation has the characteristic shape illustrated in Figure 3.21. The bulk density may be expressed by

$$\rho = \rho_o + C(\sigma_1)^n, \tag{3.4}$$

where C, ρ_o and n are constants.

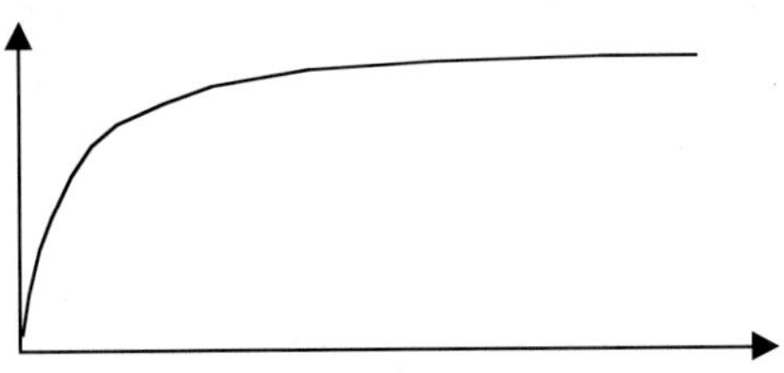

Figure 3.21 Bulk density vs. major consolidation pressure.

3.4.8 *Permeability and particle solids density*

In the case of fine powders, permeability and particle solids density are properties that need to be measured. The settling and consolidation characteristics of fine powder type bulk solids, when placed in storage, and the gravity discharge characteristics during flow from storage are governed, to a large extent, by the ability of air to permeate through the stored mass. Permeability is determined as a function of consolidation pressure by measuring the pressure drop across a sample as a function of pressure drop. The relationship is given by Darcy's equation:

$$v = c\ \mathrm{d}p/\mathrm{d}x \approx c\ \Delta p/h, \tag{3.5}$$

where v = air velocity; c = permeability factory; Δp = pressure drop across sample; h = sample height. The permeability factor is given by:

$$c = \frac{c_o}{\sigma_1^a}, \tag{3.6}$$

where c_o = constant; σ_1 = major consolidation pressure; a = permeability constant for the bulk solid.

In addition to measurements of bulk density, information on the particle solids density is sometimes required. In particular, the estimation of flow rates of fine powders requires a knowledge of the particle solids density. A simple method for measuring the solids density is to use an air comparison pycnometer, which measures the relative density of the solid particles in air. In this way, the true particle volume of a powder sample of known mass is obtained. The solids density is then

$$\rho_S = \frac{\text{mass}}{\text{particle volume}}. \tag{3.7}$$

3.5 Mass-flow and funnel-flow limits for symmetrical bins

3.5.1 *Established theory due to Jenike*

The mass-flow and funnel-flow limits have been defined by Jenike on the assumption that a radial stress field exists in the hopper (Jenike 1961, 1964). These limits are well known and have been used extensively and successfully in bin design. The limits for axi-symmetric or conical hoppers and hoppers of plane-symmetry depend on the hopper half-angle, α, the effective angle of internal friction, δ, and the wall friction angle, ϕ. Once the wall friction angle and effective angle of internal friction have been determined by laboratory tests, the hopper half-angle may be determined. In functional form,

$$\alpha = f(\phi, \delta). \tag{3.8}$$

In the derivation of the theory of flow in converging channels, Jenike showed that the bounds between mass-flow and funnel-flow in conical or axi-symmetric channels are given by:

$$\alpha = \frac{\pi}{2} - \frac{1}{2}\cos^{-1}\frac{(1-\sin\delta)}{2\sin\delta} - \beta, \tag{3.9}$$

$$\text{where } \beta = \frac{1}{2}\left[\phi + \sin^{-1}\left(\frac{\sin\phi}{\sin\delta}\right)\right]. \tag{3.10}$$

Equation (3.10) follows from the geometry of Figure 3.12(c). Since the flow in the hopper is based on radial stress field theory, no account is taken of the influence of the head in the cylinder on the flow pattern developed, particularly in the region of the transition. For plane-flow hoppers, since the discharge is by planar converging motion, the bounds between mass-flow and funnel-flow are much less severe than for conical hoppers where the discharge flow converges radially inward. This means that much larger hopper half-angles are possible in the case of plane-flow bins than in bins with conical hoppers. As a consequence, for bins with plane-flow hoppers, the discharging bulk solid will undergo a significant change in direction as it moves from the cylinder to the hopper with the potential for build up to occur at the cylinder and hopper transition. To minimise this problem by controlling the slope of the hopper, Jenike (1964) has adopted recommended design limits to decide on the hopper half-angle. These design limits may be represented by the empirical equation,

$$\alpha = \frac{1}{0.725(\tan\delta)^{0.2}}\left\{\exp\left[3.75(1.01)^{(0.1\delta-3)}\right] - \phi\right\}. \tag{3.11}$$

Equation (3.11) applies for $\phi < (\delta - 3°)$; δ and ϕ are in degrees.

The bounds for conical and plane-flow hoppers for three values of δ are plotted in Figures 3.22 and 3.23, respectively. In the case of conical hoppers, it is recommended that, as a safety precaution, the hopper half angle be generally 3° less than the limiting value. For plane-flow, the design limit may be selected; if the transition of the cylinder and hopper is sufficiently radiused or chamfered by means of a gusset plate, as illustrated in Figure 3.24, then it is even possible to select a hopper half angle 3–4° larger than the design limit.

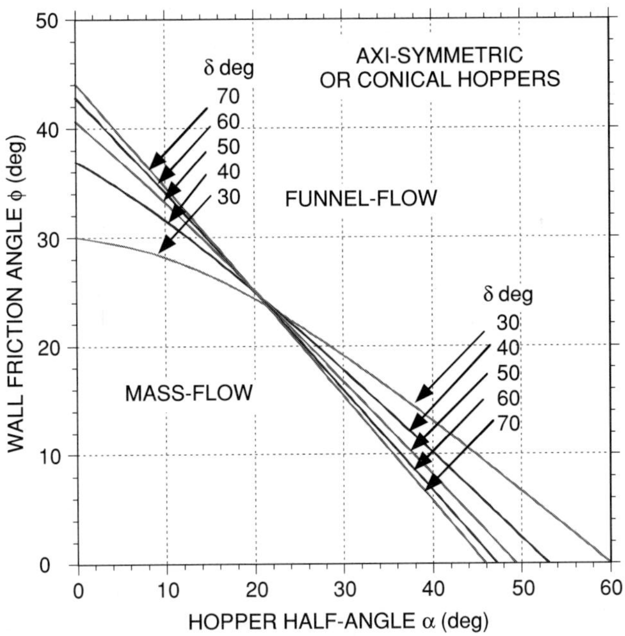

Figure 3.22 Limits for mass flow for conical or axi-symmetric hoppers.

Consider, for example, a conical hopper handling coal with $\delta = 50°$. If the hopper is of mild steel with a mill scale surface and some possible corrosion, flow property tests indicate that the angle, ϕ, is likely to be approximately 30°. On this basis, $\alpha = 14 - 3° = 11°$, which makes for a very steep hopper. If the hopper is lined with stainless steel type 304 with 2B finish, flow-property tests indicate that the friction angle, ϕ, is likely to be 20° or even lower.

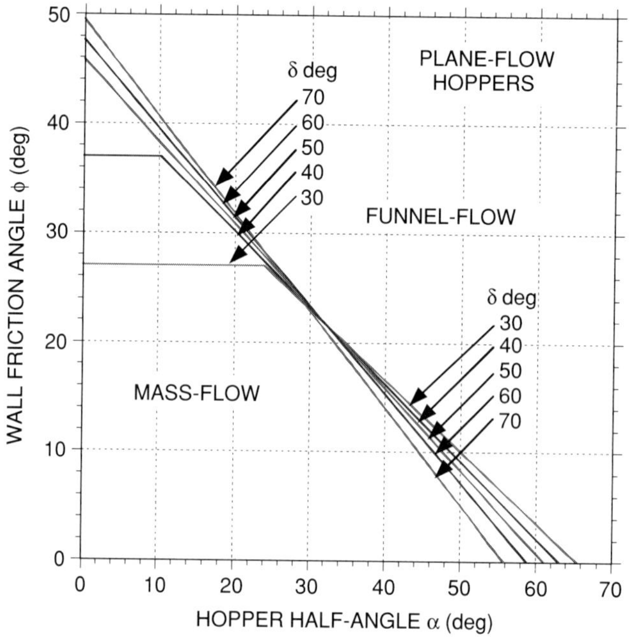

Figure 3.23 Limits for mass flow for plane-flow hoppers.

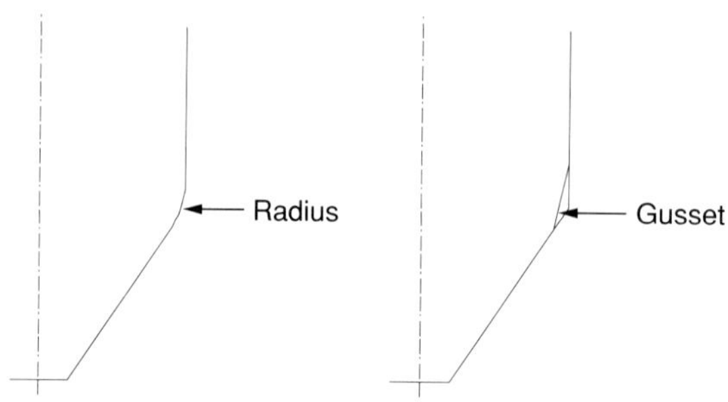

Figure 3.24 Relieving transition angle in plane-flow hoppers.

On this basis, $\alpha = 25 - 3^\circ = 22^\circ$. The corresponding angles for plane-flow are $\alpha = 22^\circ$ for $\phi = 30^\circ$ and $\alpha = 34^\circ$ for $\phi = 20^\circ$. This shows the advantage of plane-flow over axi-symmetric or conical flow, since there is a substantial increase in the hopper half-angle. However, as previously discussed, the advantage is offset, to some extent, by the fact that plane-flow or wedge-shaped hoppers have long, slotted openings, which makes it difficult to obtain uniform drawdown when feeding along the slot.

The importance of the wall friction angle in determining the mode of flow is clearly demonstrated. For a given hopper geometry, a small increase in ϕ can change a conical hopper operating under mass-flow into a hopper operating under funnel-flow.

While mass-flow is very sensitive to small variations in the wall friction angle, the mass-flow, funnel-flow limits are much less sensitive to variations in the effective angle of internal friction, δ, particularly in the region of the crossover points of Figures 3.22 and 3.23, where the limits are the same for δ values ranging from 30 to 70°.

3.5.2 *Modification to mass-flow limits – more recent research*

Since, in the work of Jenike, flow in a hopper is based on the radial stress field theory, no account is taken of the influence of the surcharge head due to the cylinder on the flow pattern developed, particularly in the region of the transition. It has been known for some time that complete mass-flow in a hopper is influenced by the cylinder surcharge head. For instance, there is a minimum level, H_{cr}, which is required to enforce mass-flow in the hopper (Thomson 1997). For the mass-flow bin of Figure 3.1(a), this height ranges from approximately 0.75 to 1.0 D.

More recent research has shown that the mass-flow and funnel-flow limits require further explanation and refinement. For instance, Jenike (1987) published a new theory to improve the prediction of funnel-flow; this led to new limits for funnel-flow which give rise to larger values of the hopper half-angle than previously predicted, particularly for high values of the wall friction angle. In the earlier theory, the boundary between mass-flow and funnel-flow was based on the condition that the stresses along the centre line of the hopper became zero. In the revised theory, the flow boundary is based on the condition that the velocity becomes zero at the wall.

Whereas the radial stress theory ignores the surcharge head, Benink (1989) has shown that the surcharge head has a significant influence on the flow pattern generated. He identified three flow regimes, mass-flow, funnel-flow and intermediate flow, as illustrated in Figure 3.25. Benink developed a new theory, namely the 'arc theory', to quantify the boundaries for the three flow regimes. This theory predicts the critical height, H_{cr}, at which the flow changes. Benink derived a fundamental relationship for H_{cr} in terms of the various bulk solid and hopper geometrical parameters, notably the H/D ratio of the cylinder and the effective angle of internal friction δ.

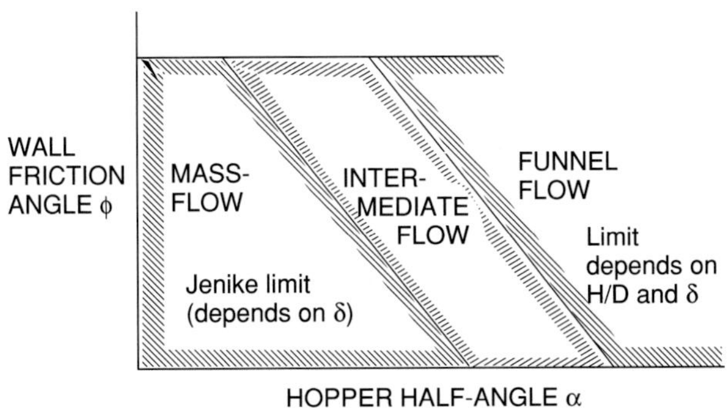

Figure 3.25 Flow regimes for plane-flow hopper defined by Benink (1989).

Intermediate flow occurs in mass-flow type bins with low surcharge head. The flow pattern, which is illustrated in Figure 3.26, is characterised by the bulk solid moving more

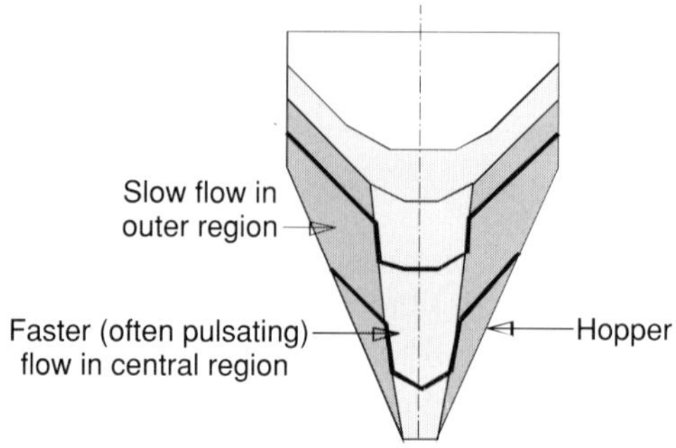

Figure 3.26 Intermediate flow.

quickly in the central flow channel, often with a pulsating motion. This may give rise to dynamic loads on the structure.

3.6 Determination of mass-flow hopper geometry

3.6.1 *Critical arching dimension*

The opening dimension for mass-flow must be large enough to prevent a cohesive arch from forming. A cohesive arch is one which forms as a result of the bulk strength due to the consolidation of the bulk solid. At the same time, the opening dimension should also be large enough to prevent the formation of a mechanical arch which occurs as a result of the interlocking of solid particles. Mechanical and cohesive arches are illustrated in Figure 3.27. To prevent a mechanical arch from forming, the opening dimension B should be at least four times the maximum particle or lump size.

Figure 3.27 Mechanical and cohesive arches.

The aim in mass-flow design is to determine the hopper half-angle, α, for flow along the walls and the minimum hopper opening dimension B to prevent a cohesive arch from forming. Normally, the actual opening dimension will be chosen to be larger than the minimum value so that the required flow rate can be obtained, while also ensuring that a mechanical arch cannot form.

Consider a bulk solid flowing in the mass-flow bin of Figure 3.28. The solid was unconsolidated when it was first deposited in the bin, but as it flows downward, it becomes consolidated under the action of the major consolidating pressure, σ_1. The variation of σ_1 is as indicated in Figure 3.28.

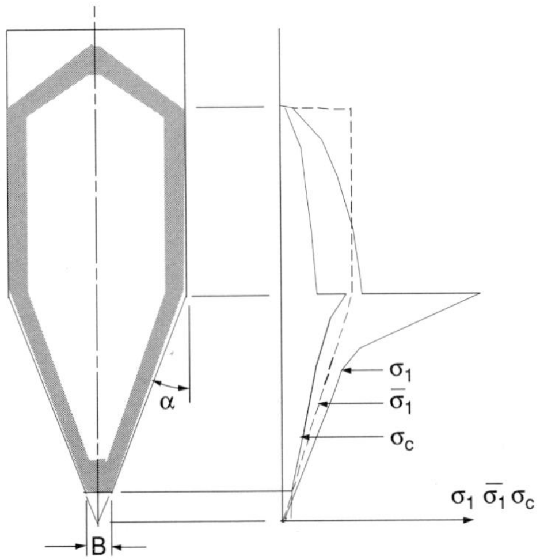

Figure 3.28 Critical arch determination for mass flow.

As a result of this consolidation, the strength, σ_c, of the bulk solid which is related to σ_1 by the flow function, FF, also varies as shown. The stress, $\bar{\sigma}_1$, set up in a stable arch is assumed to be proportional to the width, B, of the arch. The variation of $\bar{\sigma}_1$ is also shown in Figure 3.28. Whenever $\bar{\sigma}_1 > \sigma_c$, the bulk solid has insufficient strength to support a cohesive arch so that failure and flow can occur. When $\bar{\sigma}_1 < \sigma_c$, the stress in the arch is too small to cause the arch to collapse. Hence, the critical or minimum opening dimension is determined for the critical condition $\bar{\sigma}_1 = \sigma_c$.

For design purposes, the foregoing is translated into the diagram of Figure 3.29. In this figure, the relationship between the unconfined yield strength, σ_c, and the major consolidating pressure, σ_1, is given by the flow function, FF, which is a bulk solid parameter. The stress, $\bar{\sigma}_1$, acting in an arch is related to the major consolidating pressure, σ_1, by the 'flow factor', ff, as defined by Jenike (1961, 1964). The flow factor is a flow channel parameter and describes the stress condition in the hopper during flow. It provides an indication of 'flowability'; the lower the value of ff, the better the 'flowability' of the channel. The flow factor is given by

$$\mathrm{ff} = \frac{\sigma_1}{\bar{\sigma}_1}. \tag{3.12}$$

Referring to Figure 3.29, the critical opening dimension, B_{cr}, is obtained for the condition when the stress in the arch, $\bar{\sigma}_1$, equals the unconfined yield strength, σ_c. This condition

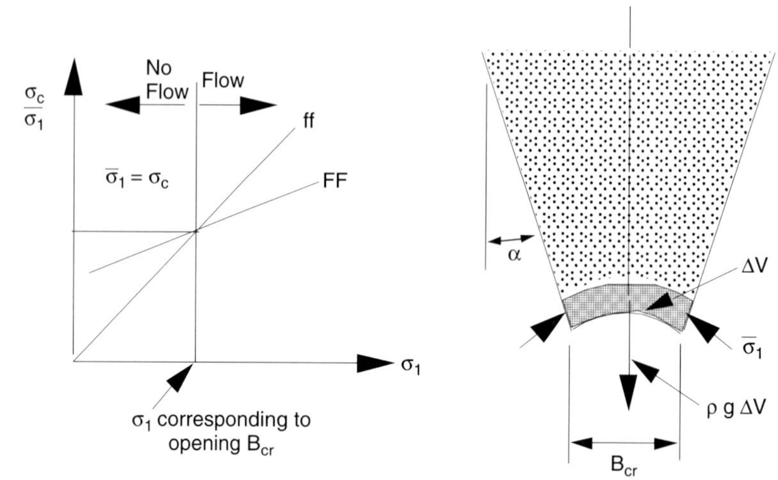

Figure 3.29 Flow/no flow condition for mass-flow design.

occurs at the intersection point of the flow function, FF, and flow factor, ff. Following the work of Jenike, the minimum or critical opening dimension, B_{cr}, for flow to occur is given by

$$B_{cr} = \frac{\bar{\sigma}_1 H(\alpha)}{\rho g}. \quad (3.13)$$

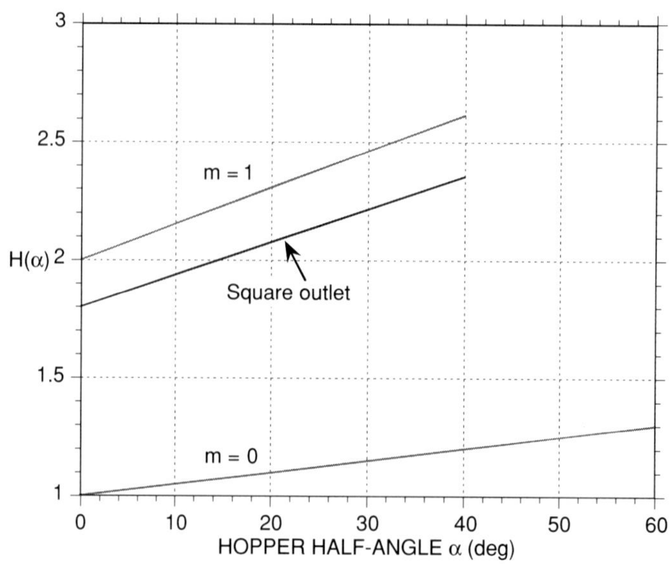

Figure 3.30 Function $H(\alpha)$.

It is often more convenient to express Equation (3.13) as

$$B_{\mathrm{cr}} = \frac{\sigma_1 H(\alpha)}{\mathrm{ff}\,\rho g}, \tag{3.14}$$

where g = acceleration due to gravity; ρ = bulk density.

The function $H(\alpha)$ from Jenike and Leser (1963) is an arch thickness parameter. It is not known explicitly but may be represented by the following equation:

$$H(\alpha) = \left[\frac{(130+\alpha)}{65}\right]^m \left[\frac{(200+\alpha)}{200}\right]^{1-m}. \tag{3.15}$$

$H(\alpha)$ calculated using Equation (3.15) is plotted in Figure 3.30; it is given in Jenike (1961, 1964) as well as Arnold *et al.* (1982) and Roberts (1998). For a square opening, $H(\alpha)$ is obtained by multiplying the $H(\alpha)$ for a conical hopper (that is $m = 1$) by 0.9.

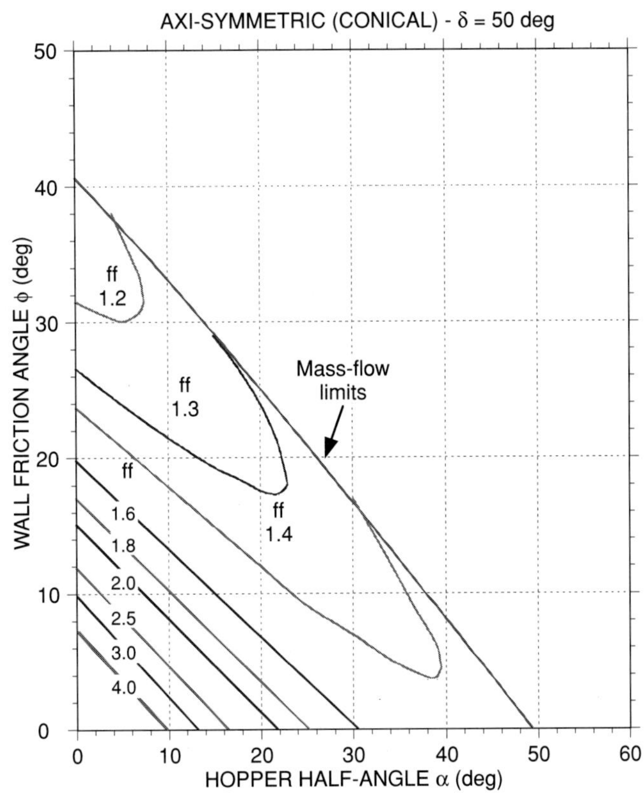

Figure 3.31 Flow factor chart for conical hopper, $\delta = 50°$.

3.6.2 *Flow factor and design charts*

The flow factor, ff, given by Equation (3.12), is a function of the effective angle of internal friction, δ, hopper half-angle, α, and wall friction angle, ϕ. That is:

$$\text{ff} = f(\delta, \alpha, \phi). \tag{3.16}$$

In the work of Jenike (1961, 1964), the flow factor, ff, is derived from the equilibrium equations for the flow stress fields in the hopper. To facilitate hopper design, Jenike presented his results in graphical form for both axi-symmetric and plane-flow hoppers for values of δ ranging from 30 to 70° in 10° steps. Arnold *et al.* (1982) have also solved the flow equations and presented the design charts over the same range of δ values, but in 5° steps. By way of illustration, two flow factor charts for $\delta = 50°$ are presented in Figures 3.31 and 3.32, for the axi-symmetric and plane-flow cases, respectively.

In order to simplify the mass-flow design procedures, Moore and Arnold (1985) developed two design charts which present, in a more concise way, the essential information for hopper

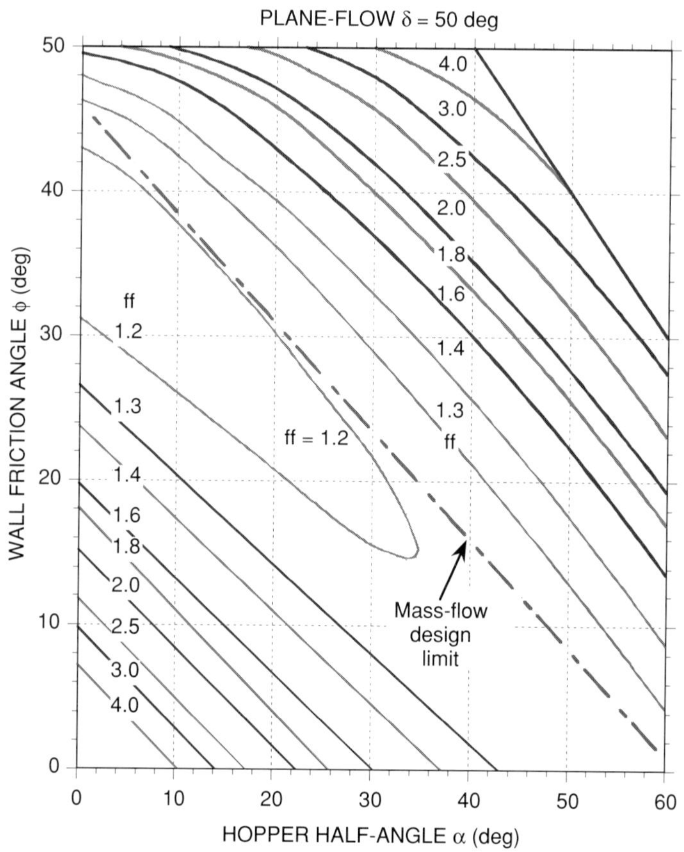

Figure 3.32 Flow factor chart for plane-flow hopper, $\delta = 50°$.

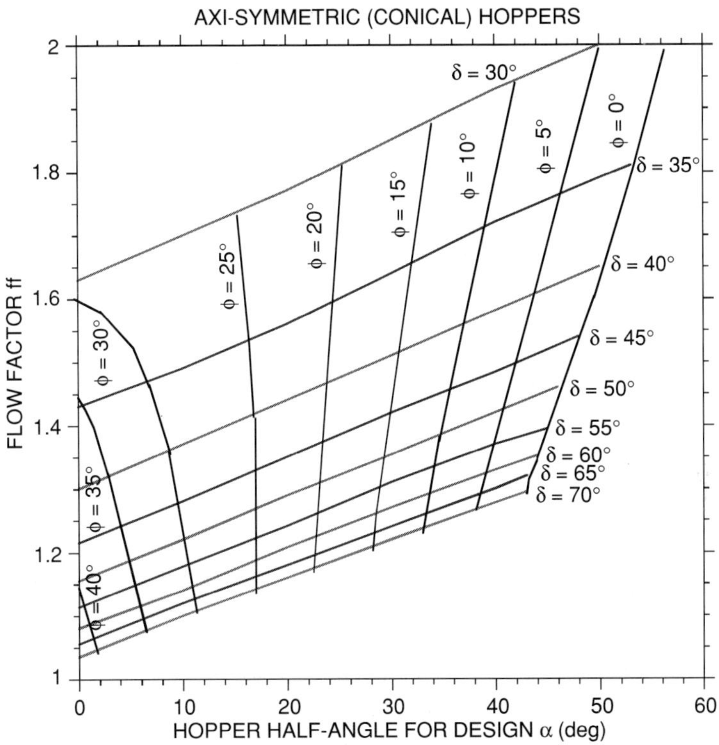

Figure 3.33 Design chart for conical hoppers based on $\alpha = \alpha_{(critical)} - 3^{\circ}$.

design purposes. Alternative presentations to those of Moore and Arnold of the mass-flow design data for axi-symmetric and plane flow hoppers for a range of δ and ϕ values are presented in Figures 3.33 and 3.34, respectively. The charts display the critical design values near the limits for mass-flow and funnel-flow and considerably simplify the selection of hopper half-angles and corresponding flow factors.

For axi-symmetric hoppers (Figure 3.33), the hopper half angle, α, has been reduced by 3° from the limiting values at the mass-flow/funnel-flow boundaries; this follows recommended design practice to ensure that the design solutions lie within the mass-flow zones. For plane-flow (Figure 3.34), the hopper half-angles, α, are based on the recommended design limits.

3.6.3 *Approximate analytical solutions to the determination of flow factors*

While the flow factor charts provide the basis for hopper design, it is noted that their derivation involves the numerical solutions to the equilibrium equations, a somewhat tedious process. To facilitate hopper design, Arnold *et al.* (Arnold *et al.* 1982; Arnold & McLean 1976) derived equations which allow flow factors to be calculated more conveniently. The equations are based on the arched stress field proposed by Enstad (1975) for plane-flow hoppers, the theory being extended to also include axi-symmetric hoppers.

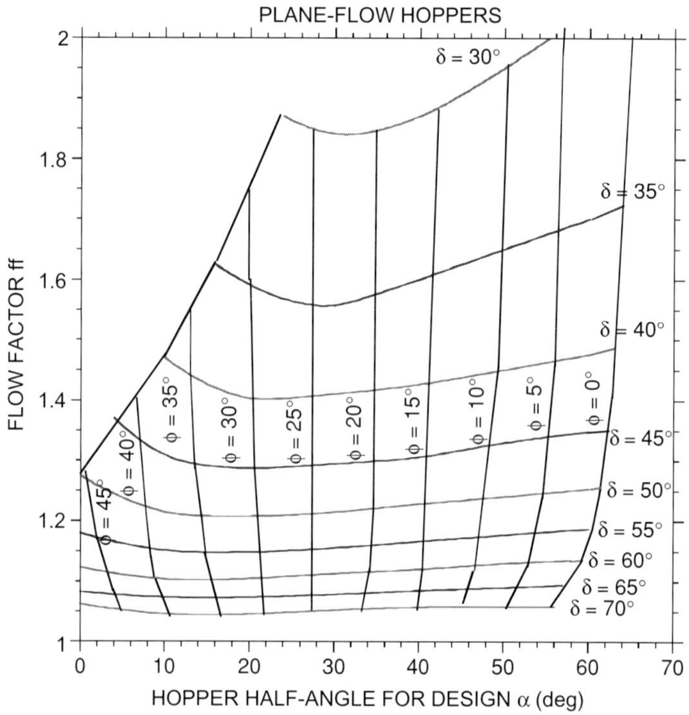

Figure 3.34 Design chart for plane-flow hoppers based on α for mass-flow limits.

From Equation (3.13) the flow factor, ff, can be expressed as

$$\text{ff} = \frac{\sigma_1 H(\alpha)}{\rho g B}. \tag{3.17}$$

From Arnold *et al.* (1982), the non-dimensional stress function has been determined as

$$\frac{\sigma_1}{\rho g B} = \frac{Y(1+\sin\delta)}{2(X-1)\sin\alpha}.$$

$$\text{Hence, ff} = \frac{Y(1+\sin\delta)H(\alpha)}{2(X-1)\sin\alpha}, \tag{3.18}$$

$$\text{where } X = \frac{2^m \sin\delta}{1-\sin\delta}\left[\frac{\sin(2\beta+\alpha)}{\sin\alpha}+1\right] \tag{3.19}$$

$$Y = \frac{[2(1-\cos(\beta+\alpha))]^m\,(\beta+\alpha)^{1-m}\sin\alpha + \sin\beta\sin^{1+m}(\beta+\alpha)}{(1-\sin\delta)\sin^{2+m}(\beta+\alpha)}, \tag{3.20}$$

where $m = 0$ for plane-flow channels; $m = 1$ for axi-symmetric channels. Note that in the numerator term $(\beta+\alpha)^{1-m}$ of Equation (3.20), β and α are in radians.

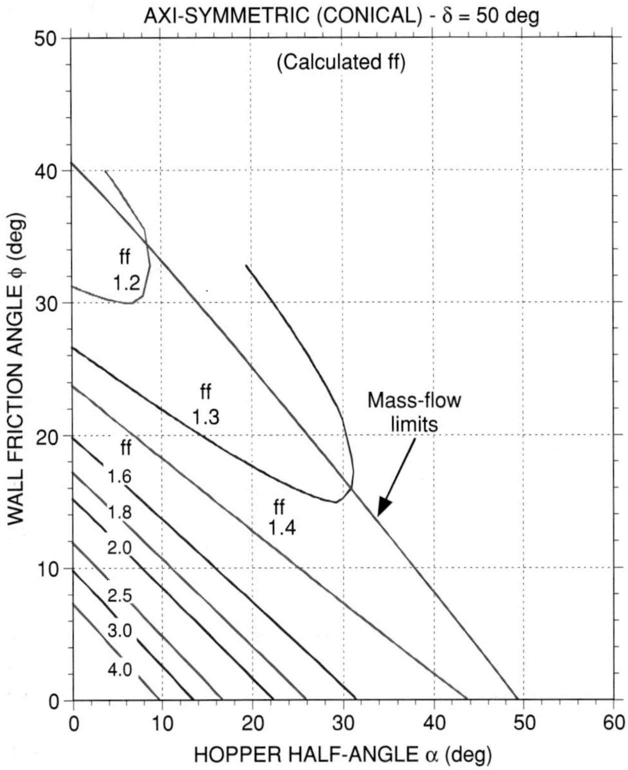

Figure 3.35 Approximated flow factors for conical hoppers for $\delta = 50°$.

As shown in Figure 3.12, β is the angle between the normal stress at the wall and the major principal stress and is given by

$$\beta = \frac{1}{2}\left[\phi + \sin^{-1}\left(\frac{\sin\phi}{\sin\delta}\right)\right]. \tag{3.21}$$

As an example, using the procedure outlined above, the calculated flow-factors for $\delta = 50°$ are shown plotted in Figure 3.35 for axi-symmetric or conical hoppers. Comparing Figure 3.31 with Figure 3.35, the agreement is generally quite good. There is some variation in the flow factors in the region of the mass-flow and funnel-flow limits. In general, the approximated flow factors in the region of the design limits lead to slightly less conservative design than the Jenike flow factors for δ values above 40° and slightly more conservative design for δ values less than 40°.

3.7 Mass-flow design example

To illustrate the procedures for mass-flow hopper design, the case of a 'Run-of-Mine' (ROM) coal at 13% moisture content (wet basis) is considered. The flow properties are given in Figures 3.36 and 3.37. The lower graph of Figure 3.36 shows the instantaneous and

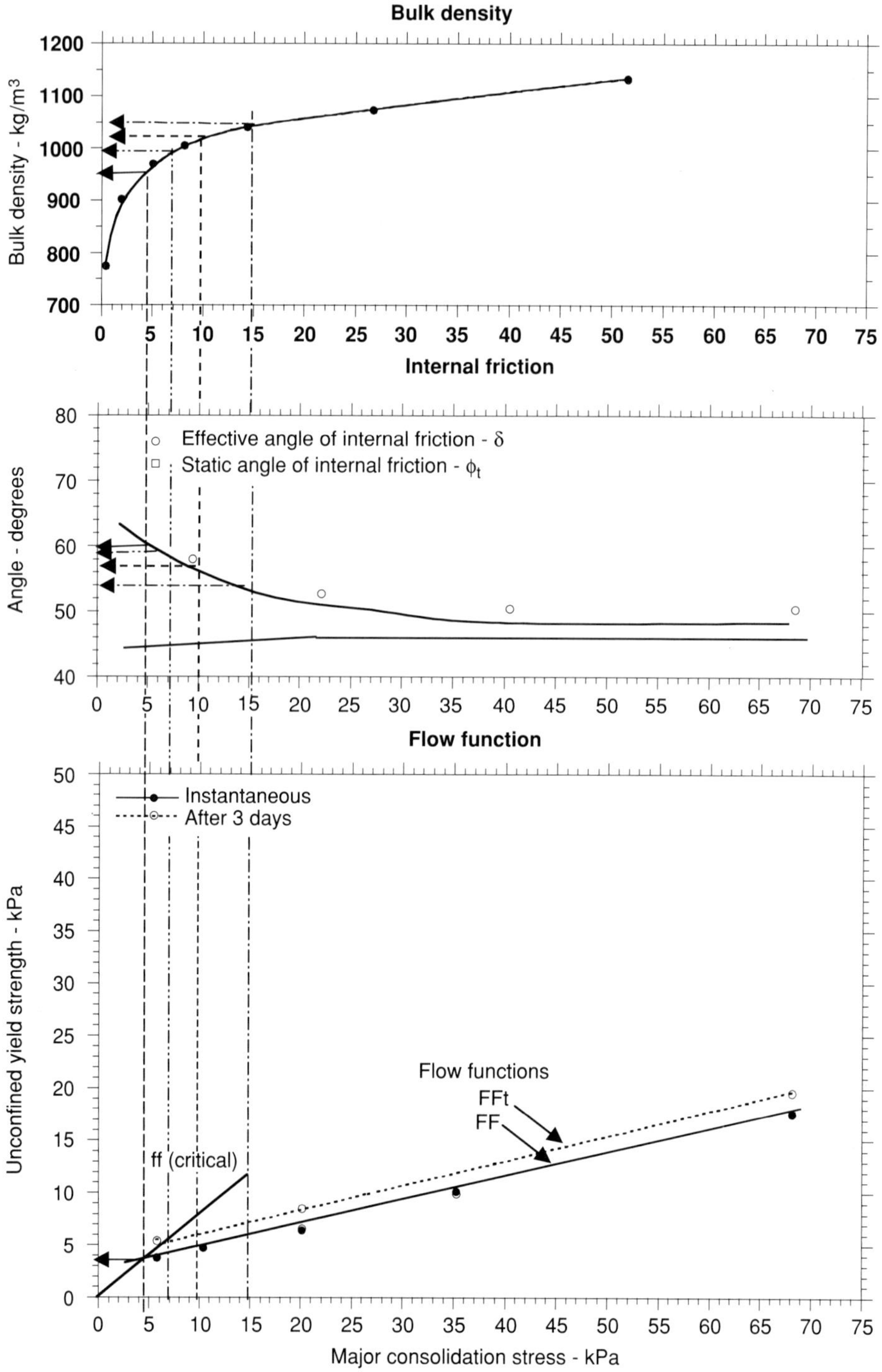

Figure 3.36 Flow properties for ROM coal at 13% moisture content.

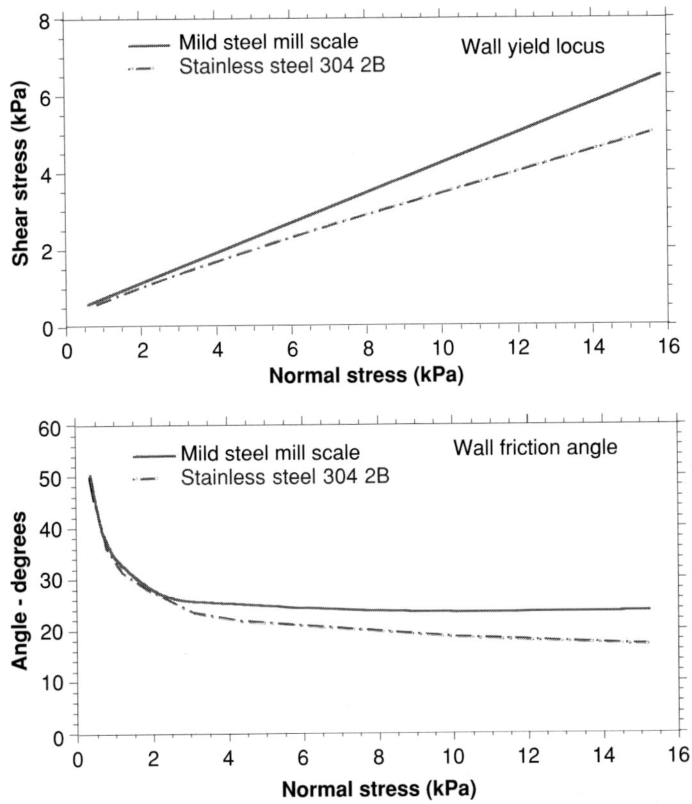

Figure 3.37 Wall friction for ROM coal at 13% moisture content.

three daytime flow functions; the middle graph shows the static, ϕ_t, and effective δ angles of internal friction; and the upper graph shows the bulk density. In all cases, the independent variable is the major consolidation stress, σ_1.

Figure 3.37 presents the wall friction data for the ROM coal in contact with mild steel with mill scale finish and stainless steel Type 304 with 2B finish. The upper graph depicts the wall yield loci (WYL), while the lower graph shows the wall friction angles (ϕ), both graphs being plotted against the normal pressure at the boundary surface.

The aim is to determine the required geometry of a conical or axi-symmetric mass-flow hopper. The procedure is as follows:

3.7.1 *Limiting or critical geometry*

(1) Estimate δ and ϕ at the outlet: Assume $\delta = 60°$, $\phi = 23°$; From Figure 3.33, $\alpha = 20°$ and ff $= 1.21$.

(2) Draw ff $= 1.21$ line on the FF graph of Figure 3.36: The intersection point is $\sigma_1 =$ 4.5 kPa and $\bar{\sigma}_1 = 3.5$ kPa; checking $\delta = 60°$ and, based on Figure 3.12(c), $\phi = 23°$; hence, it is not necessary to perform any iterations.

(3) Determine minimum opening dimension B: From Figure 3.36, $\rho = 0.95$ t/m^3; from Figure 3.30, $H(\alpha) = 2.32$; substitution into Equation (3.12) gives $B = 0.87$ m.

3.7.2 *α vs. B and flow-rate curves*

For practical reasons, the opening dimension needs to be larger than the value given above. Since the wall friction angle decreases with increase in normal pressure, as indicated by Figure 3.37, the relationship between the hopper half-angle, α, and opening dimension, B, provides important information for hopper design. The application of Equation (3.13) in conjunction with the flow property graphs provides the means for determining the α vs. B graph, which is shown in Figure 3.38.

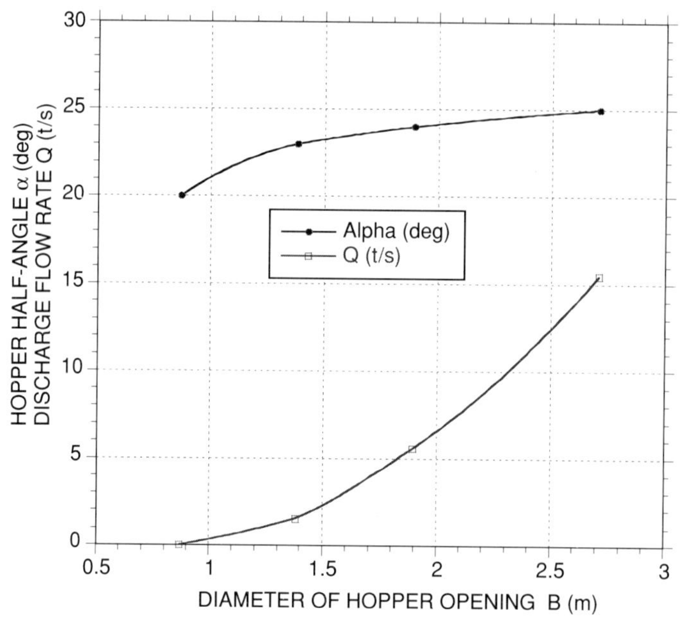

Figure 3.38 Hopper half-angle and flow rate vs. opening diameter. Conical hopper stainless steel line – ROM coal at 13% MC.

Also shown in Figure 3.38 is the plot of estimated flow rate computed using the method for coarse, highly permeable, cohesive bulk solids proposed by Johanson (1965). The steady state flow rate is given by

$$Q = \rho B^{(1+m)} L^{(1-m)} \left(\frac{\pi}{4}\right)^m \sqrt{\frac{Bg}{2(m+1)\tan\alpha}\left[1 - \frac{\mathrm{ff}}{\mathrm{ff}_a}\right]}, \tag{3.22}$$

where $m = 0$ for plane-flow hopper; $m = 1$ for axi-symmetric or conical hopper; B = width of slot or diameter of circular opening; L = length of slot in case of plane-flow hopper; α = hopper half-angle; ff = critical flow factor; $\mathrm{ff_a} = \sigma_1/\sigma_\mathrm{c}$ = actual flow factor.

Table 3.1 Mass-flow design example.

σ_1 (kPa)	$\delta(°)$	σ_w (kPa)	$\phi(°)$	ff	ff_a	ρ (t/m^3)	$\alpha(°)$	$H(\alpha)$	B (m)	Q (t/s)
4.5	60	3.75	23	1.21	1.21	0.95	20	2.32	0.87	0.00
7.0	59	6.06	20	1.23	1.56	1.00	23	2.35	1.37	1.51
10.0	57	8.80	19	1.25	2.0	1.02	24	2.37	1.90	5.58
15.0	54	13.34	18	1.28	2.58	1.05	25	2.38	2.71	15.46

In the majority of cases, the flow rate determined above for unimpeded discharge will be well in excess of the plant requirements. For this reason, feeders are used to control the discharge rate to the required value. The calculated values for the foregoing example are given in Table 3.1. Flow rate analysis is also discussed by other researchers, such as those listed in McLean (1979), Arnold and Gu (1990) and Arnold *et al.* (1990).

3.8 Funnel-flow and expanded flow bin design

3.8.1 *General remarks*

As discussed in Section 3.2, funnel-flow is characterised by material sloughing off the top surface and flowing down the central flow channel which forms above the opening. Flow will continue until the level of the bulk solid in the bin drops an amount, H_D, equal to the drawdown. At this level, the bulk strength of the contained material is sufficient to sustain a stable pipe or 'rathole' of diameter $D_F = B_f$, as illustrated in Figure 3.39. B_f will be slightly larger than the hopper opening, B. Once the level defined by H_D is reached, there is no further flow, and the material below this level represents 'dead' storage. For complete drawdown, B_f should be equal to D_{fm}, which is the rathole diameter calculated for the base of the bin. However, for cohesive bulk solids, D_{fm} is often several metres

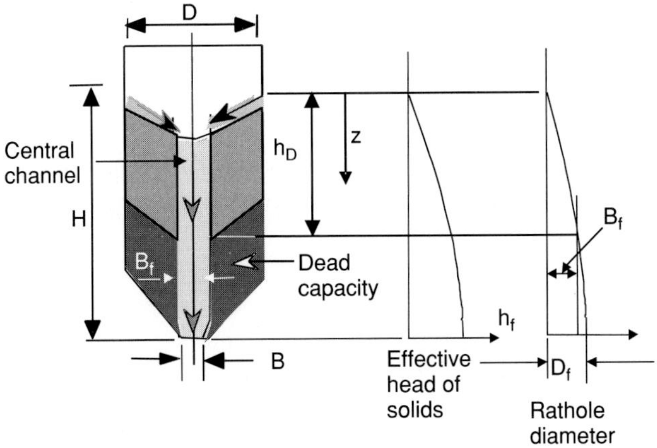

Figure 3.39 Funnel flow showing effective drawdown.

rendering funnel-flow impracticable. Expanded flow, as shown in Figure 3.5, where D_{fm} defines the minimum transition diameter, is then a better proposition.

Based on previous literature (Roberts & Teo 1990; McBride 2000, 2001; Roberts 2000a–c; Roberts & McBride 2000), the computation of drawdown, h_D, and live capacity involves the following steps:

(1) Determination of the relationship between the rathole diameter, D_f, vs. effective head of solids, h_f. This determination is derived entirely from the flow properties of the bulk solid without reference to a particular bin or stockpile geometry. Usually, D_f vs. h_f is expressed in graphical form.

(2) Using the information in (a), the relationship between the rathole dimension, D_f, and drawdown, h_D, can be obtained for a particular bin or stockpile geometry. Again, graphical representation is usually used. With this information, the dimensions of the outlet or, in the case of expanded flow, the dimensions of the lower mass-flow hopper at the transition with the funnel-flow section can be selected.

(3) For a particular geometry for the bin outlet or mass-flow hopper transition, the shape of the rathole is estimated, and this information is used in conjunction with the D_f vs. h_f relationship of (b) to determine the actual drawdown.

(4) The live capacity is then estimated by calculating the volume of the craters formed after drawdown has occurred.

3.8.2 *Flow properties and D_f vs. h_f graph*

The relevant flow property graphs and design lines are shown in Figure 3.40. The first step is to draw the critical rathole diameter, D_f, as a function of the effective head of solids, h_f. Three possible solutions are considered, the upper bound, the 'average' and the lower bound.

In all cases, the relevant value of the unconfined yield strength is determined, and the corresponding critical 'rathole' diameter is calculated in accordance with the following equation due to Jenike (1961, 1964):

$$D_f = \frac{\sigma_c G(\phi_t)}{\gamma}, \tag{3.23}$$

where $\gamma = \rho g$ (bulk specific weight).

The function $G(\phi_t)$ has been given by Jenike as a design graph, as shown in Figure 3.41. For the range of ϕ_t values plotted in Figure 3.41, $G(\phi_t)$ may be represented by the empirical equation

$$G(\phi_t) = 4.3 \tan(\phi_t). \tag{3.24}$$

For the chosen value of σ_1, the effective head of solids is

$$h_f = \frac{\sigma_1}{\gamma}. \tag{3.25}$$

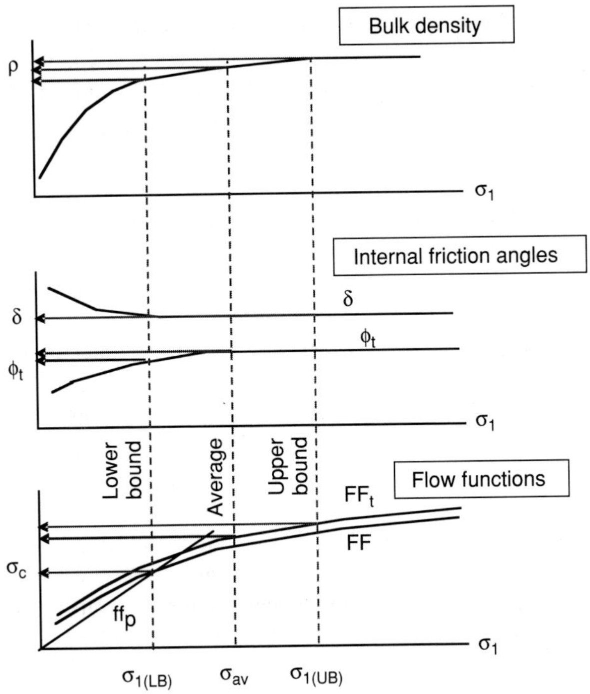

Figure 3.40 Flow properties and design lines.

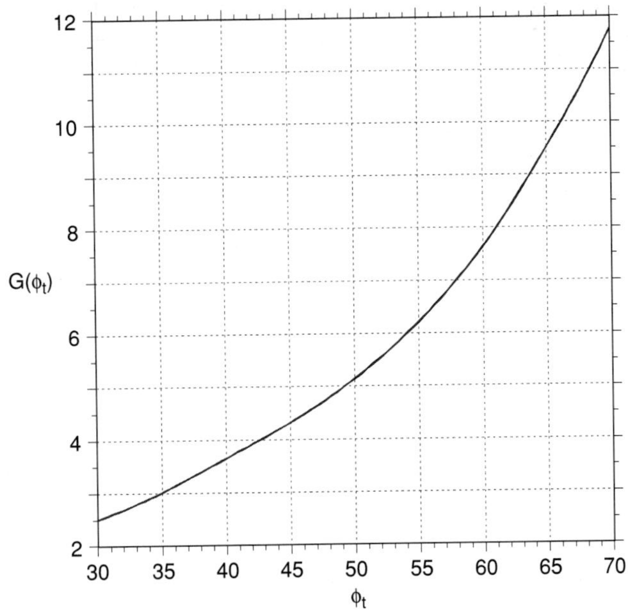

Figure 3.41 Function $G(\phi_t)$.

3.8.2.1 *Upper bound solution* For the upper bound case, a value of the major consolidation stress, σ_1, is selected and the corresponding values of the unconfined yield strength, σ_c, the static angle of internal friction, ϕ_t, and bulk density, ρ, read from the graphs. The process is repeated above for a range of σ_1 values enabling the D_f vs. h_f graph to be plotted.

3.8.2.2 *Lower bound solution* The lower bound critical rathole diameter is based on the instantaneous flow function and represents the condition when the bin is being filled and emptied at the same time. The lower bound value usually underestimates the rathole diameter. Based on the original work of Jenike, the lower bound critical rathole dimension is determined using the piping flow factor, ff_p, which is given in graphical form as a function of ϕ_t and δ in Jenike (1961, 1964) and Arnold *et al.* (1982). The ff_p line is drawn on the instantaneous flow function as in Figure 3.40 to define the critical value of σ_c. For the corresponding value of σ_1, the values of ϕ_t and ρ are also read from the flow properties graph. D_f is determined from Equation (3.23).

3.8.2.3 *'Average' solution* Experience has indicated that the upper bound solution to the rathole size determination is often too conservative, particularly in the case of drawdown in stockpiles. For this reason, the 'average' solution is proposed. While somewhat empirical, the method is based on the hoop strength of the rathole, which is estimated from the average of the major and minor principal stresses occurring in axi-symmetry:

$$\sigma_{av} = \frac{\sigma_1}{1 + \sin\delta}. \tag{3.26}$$

For the chosen value of σ_1 and, hence, δ, the corresponding value of σ_{av} is determined using Equation (3.26). The value of σ_c is read from the flow property graphs and the value of D_f calculated using Equation (3.23). The graphs of D_f vs. h_f are plotted as illustrated in Figure 3.42.

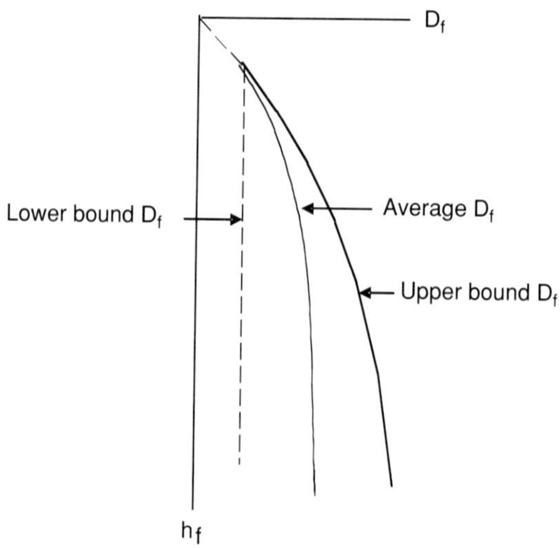

Figure 3.42 Graph of D_f vs. h_f.

3.8.3 *Rathole dimension, D_f, vs. drawdown, h_d, graph*

The information in Figure 3.42 may now be converted into a graph depicting rathole dimension vs. drawdown for a funnel flow bin or gravity reclaim stockpile. The objective here is to determine the relationship between the actual head, z, and drawdown, h_D, as a function of the effective head, h_f.

3.8.3.1 *Funnel-flow bin* Referring to Figure 3.39, the effective head of solids h_f as a function of depth, h, is given by Janssen's equation as follows:

$$h_f = \frac{R}{K_j \tan\phi}\left(1 - e^{-K_j \tan\phi\, z/R}\right), \tag{3.27}$$

where R = area of bin cross-section/perimeter of bin cross-section; K_j = ratio of horizontal to vertical pressure in the bin (assumed to be 0.4); ϕ = wall friction angle in degrees; z = actual head of solids. For a circular bin, $R = D/4$.

Referring to Figure 3.39 and substituting h_D for z in Equation (3.27) and solving for h_D gives

$$h_D = \frac{D}{4\,K_j \tan\phi} \ln\left(\frac{D}{D - 4K_j h_f \tan\phi}\right). \tag{3.28}$$

3.8.3.2 *Gravity reclaim stockpile* Drawdown in a gravity reclaim stockpile is illustrated in Figure 3.43. In this case, the consolidation vs. bulk strength conditions are difficult to predict precisely, since they depend on the loading history, variations in the properties of the bulk solid and environmental influences such as temperature and moisture content. The most conservative case is when the effective head, h_f, equals the actual head, z. That is,

$$h_f = z. \tag{3.29}$$

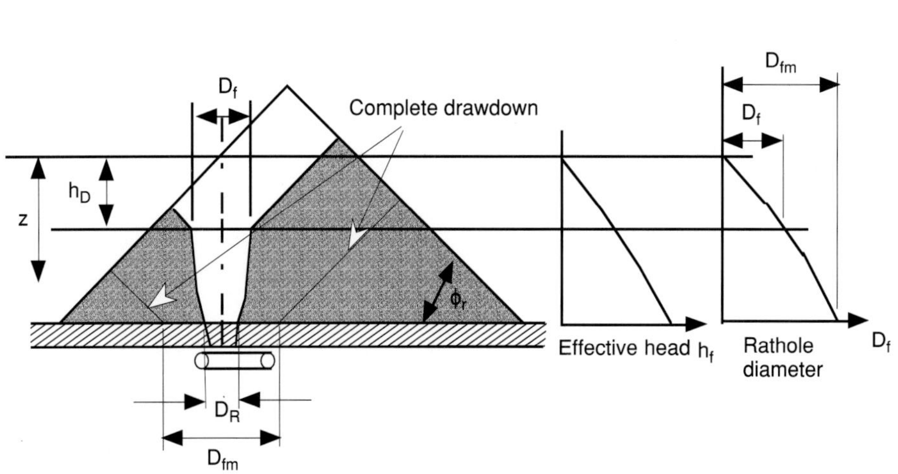

Figure 3.43 Drawdown in gravity reclaim stockpile.

Experience has shown that this is too conservative. A less conservative and, perhaps, more realistic estimate of h_f is

$$h_f = z \cos \phi_r, \tag{3.30}$$

where ϕ_r = angle of repose.

Substituting h_D for z in Equation (3.30) gives

$$h_D = \frac{h_f}{\cos\phi_r}. \tag{3.31}$$

In the case of mineral ores, the height of stockpiles may be in the order of 30–40 m, and the consolidation pressures at the base are often in the order of a megapascal. Consequently, complete drawdown, as depicted in Figure 3.43, is not practical, since the required opening dimension which is directly related to the maximum rathole dimension, D_{fm}, would be too large. Consequently, a smaller opening dimension, D_R, is necessary leading to the rathole geometry illustrated in Figure 3.43. To meet live capacity requirements, it is usual to employ multi-outlets, as illustrated by the twin outlet stockpile of Figure 3.44. The drawdown and live capacity may be optimised by selecting the appropriate spacing distance 'X'.

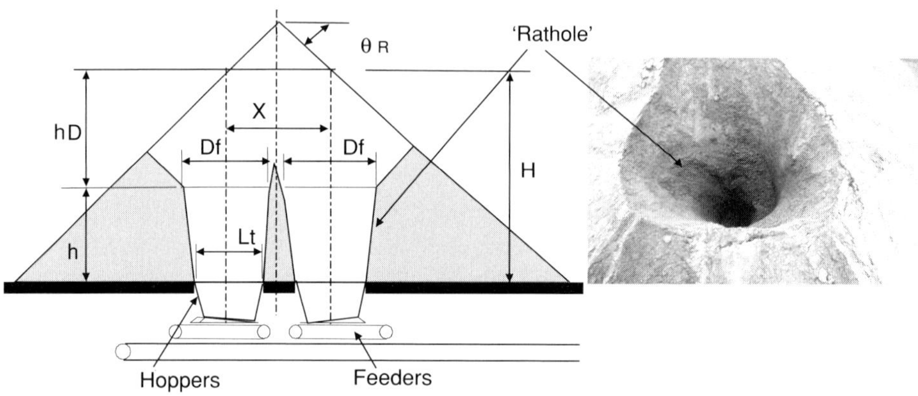

Figure 3.44 Gravity reclaim stockpile with twin outlets.

3.9 Rathole geometry

3.9.1 *Rathole shape*

Following the work of Roberts (2000a), the general case of a rectangular opening associated with a transition hopper in a gravity reclaim stockpile is considered as illustrated in Figure 3.45. The rathole forms at the top of the hopper across the diagonal D_R of the

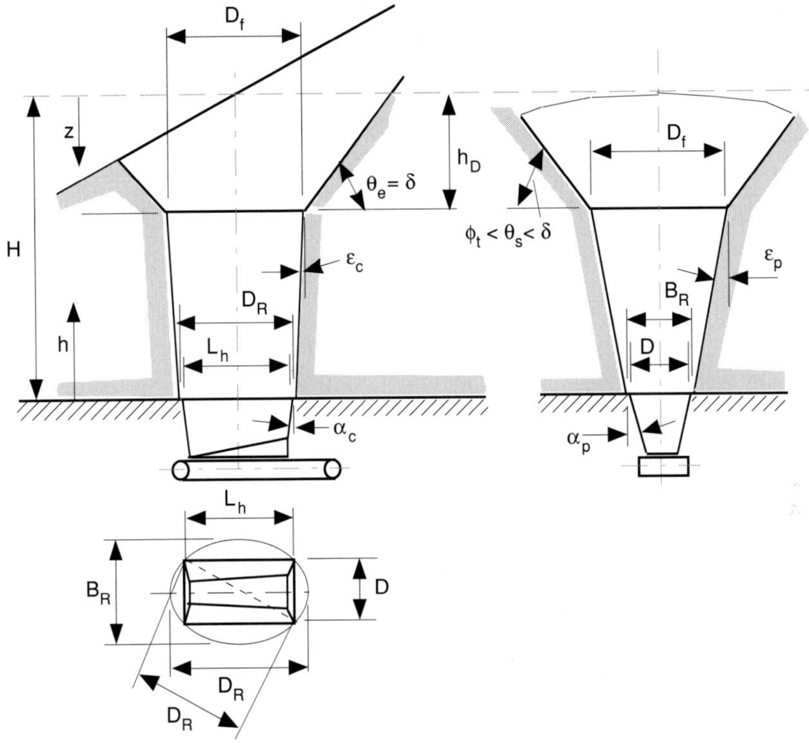

Figure 3.45 Rathole geometry.

rectangle, such that

$$D_{\mathrm{R}} = \sqrt{L_{\mathrm{h}}^2 + D^2}, \tag{3.32}$$

where L_{h} = length of hopper transition; D = width of hopper transition.

While the actual shape of the rathole is quite complex with the walls slightly curved rather than straight, a close approximation to the shape may be made using the following assumptions based on the limiting rathole wall slope half-angles plotted in Figure 3.46:

(1) Immediately above the rectangular opening corresponding to the hopper transition, the rathole is approximately elliptical in shape with D_{R} the major axis and B_{R} the minor axis. The determination of B_{R} is based on the following assumptions:
   (a) For rectangular opening in which $L_{\mathrm{h}} \gg D$ (Figure 3.45), the flow channel opens out at a half-angle of 30° along the sides, and at an average angle of 5° at the end.
   (b) For a square opening, the flow channel opens out at an angle of 7° along the sides.
   (c) For $D < L_{\mathrm{h}}$, transitional arrangements apply.

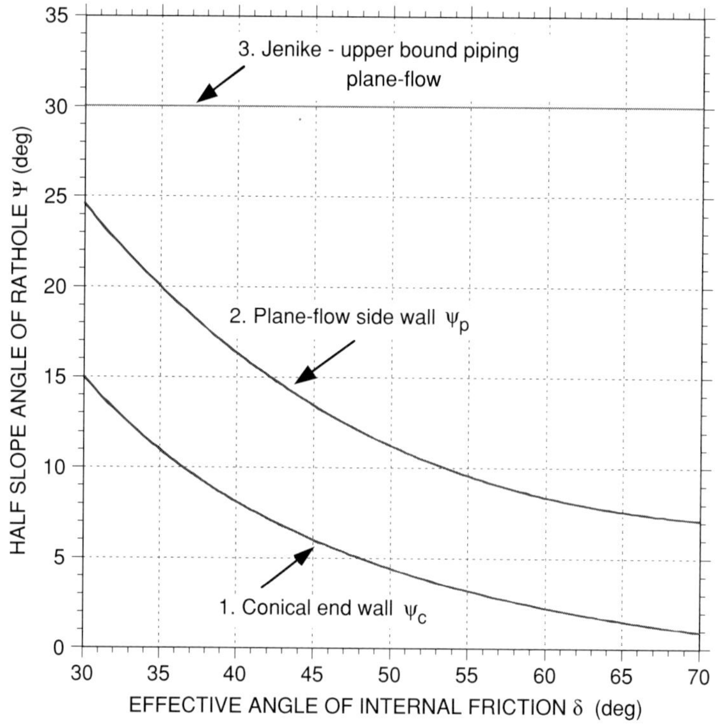

Figure 3.46 Limiting half slope angles for ratholes.

On this basis, the following relationship is proposed:

$$\frac{B_R}{D} = \frac{L_h}{D}\left\{1 + C_R\left[1 - \exp\left(-\frac{\frac{L_h}{D} - 1}{C_L}\right)\right]\right\}\left\{\sqrt{1 + \left[\frac{D}{L_h}\right]^2} - 1\right\} + 1, \quad (3.33)$$

where $C_R = 5.6$ and $C_L = 8$.

(2) The ratio of the area of the approximated ellipse to the area of the rectangular opening at the hopper transition is

$$A_r = \frac{\frac{\pi}{4} D_R B_R}{L_h D} = \frac{\pi}{4}\frac{B_R}{D}\sqrt{1 + \left[\frac{D}{L_h}\right]^2} \geq 1.0 \quad (3.34)$$

(3) The rathole opens out from the elliptical shape at the bottom to a circular shape of diameter D_F (the critical rathole diameter) at the top corresponding to the drawdown h_D.

(4) The end walls of the rathole are sloped at an angle, ε_c, which is equal to or less than the limiting conical piping angle, ψ_c.

(5) The side walls of the rathole are sloped at an angle, ε_p, which is equal to or less than the limiting plane-flow piping angle, θ_p. Normally, $\varepsilon_p < \psi_p$, since the slope of the side walls is controlled by the rathole becoming circular at the drawdown transition.

(6) Above the drawdown level, the crater sloughs off at an angle, θ, which normally ranges between the static angle of internal friction, ϕ_t, and the effective angle of internal friction, δ. Referring to Figure 3.45, the limiting slope will normally be $\theta_e = \delta$ for the end slope. The slope of the sides is such that $\phi_t \leq \theta_s \leq \delta$.

3.9.2 *Slope of rathole walls*

Jenike (1961) determined the flow factors, ff_p, for no piping and the corresponding limiting slope angles for the funnel-flow channel. Since, in the case of funnel-flow, the slope of the flow channel is not defined, the Jenike limits are taken as upper bound values for plane-flow and conical surfaces, respectively. Jenike assumed that for a rectangular slot, the side walls of the rathole can slope out to allow the rathole to expand upward to a circular shape. He recommends that in the design for rectangular outlets, a limiting $ff_p = 1.7$ be adopted. This corresponds to a half slope angle for plane-flow, $\psi_p = 30°$. This applies only to the transitional region immediately above the side walls of the opening, and not to the side walls of the rathole that forms above the lower transition.

A more realistic, yet somewhat empirical, approach is to use the mass-flow limits as a guide. For instance, it is assumed that, as Jenike did, at the surface of the rathole, the friction angle is given by

$$\phi_S = \tan^{-1}(\sin\delta). \tag{3.35}$$

The assumption embodied in Equation (3.35) is the failure by internal shear occurs at the maximum shear stress. Using the values of ϕ_s given by Equation (3.35), the limiting half slope angles for plane-flow and conical channels are plotted as curves 1 and 2, respectively, in Figure 3.46. The 30° limit line for plane-flow proposed by Jenike as previously discussed is also shown in Figure 3.46.

3.9.3 *Arching under funnel flow*

In the case of bins with circular or square openings operating under funnel-flow, arching over the opening will not normally occur. However, for a rectangular opening, while the rathole or flow channel forms across the diagonal, the width of the opening must be large enough to prevent a cohesive arch from forming. Following the discussion in the previous section, the procedure involves the application of the flow factor, $ff_p = 1.7$, based on a half-slope-angle of 30° for the sides of the rectangle. The critical arching dimension, B_r, is computed as follows:

(1) Draw the $ff_p = 1.7$ line on the flow function graph (instantaneous or time, whichever is relevant to the design). The value of σ_c at the intersection of ff_p and FF is read from the graph.

(2) Compute B_r using Equation (3.13) with $\bar{\sigma}_1 = \sigma_c$ and $H(\alpha) = 1.15$, the latter being obtained from Figure 3.30 with $\alpha = 30^\circ$ for plane flow. That is

$$B_r = \frac{1.15\sigma_c}{\gamma}. \tag{3.36}$$

3.9.4 *Drawdown and live capacity*

Assuming the rathole geometry, as illustrated in Figure 3.45, and choosing the slope of the end walls of the flow channel such that $\varepsilon_c \le \psi_c$ and the side walls $\varepsilon_p \le \psi_p$, as given by Figure 3.46, the rathole diameter, D_f, and drawdown, h_D, may be determined as indicated in Figure 3.47. The D_f vs. h_D graph is drawn, and the expansion lines for the rathole side and end walls are drawn as indicated. The upper intersection point, in this case the end wall expansion, defines the critical condition for drawdown.

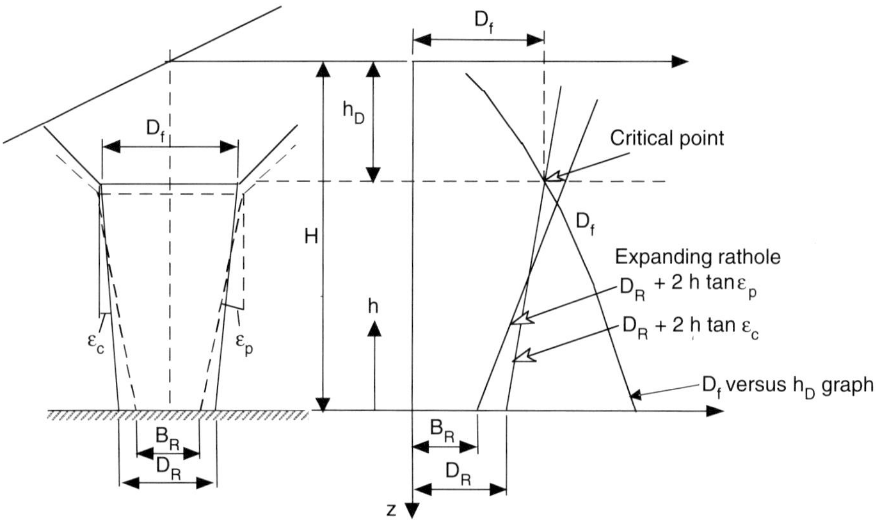

Figure 3.47 Rathole diameter and drawdown determination.

With the various geometrical parameters defining the rathole geometry, the reclaimed volume can be determined. The procedures outlined in McBride (2000, 2001), Roberts (2000a–c) and Roberts and McBride (2000) may be used for this purpose. One option is to use scale modeling, as illustrated in Roberts (2000c). However, it is now well established that the application of CAD geometrical modeling packages, coupled with flow property testing and rathole geometry analysis, as illustrated in Figure 3.47, provides a ready means for simulating stockpile drawdown performance.

As an example, Figure 3.48 shows the predicted and actual drawdown performance of a large, kidney-shaped, iron ore stockpile with twin outlets. Figure 3.48(a) shows the simulated, CAD model of the stockpile which was generated using the measured flow properties of the iron ore. The model was produced in advance of the plant construction to predict the drawdown and live capacity of the proposed stockpile. The photograph of Figure 3.48(b)

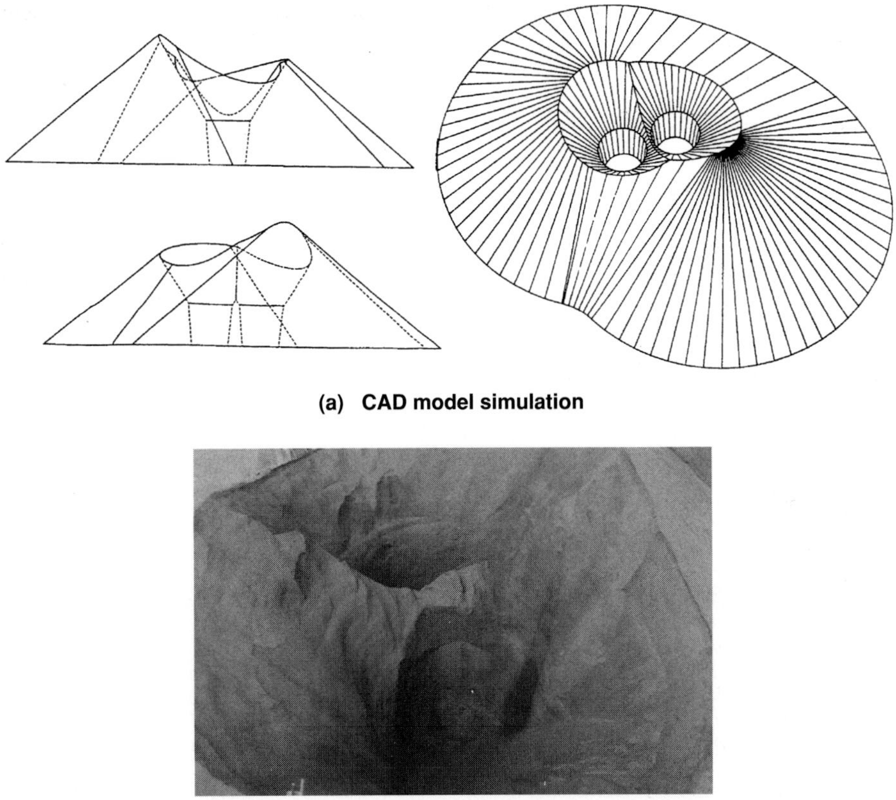

Figure 3.48 Simulation and drawdown performance of iron ore stockpile: (a) CAD model simulation; (b) view of actual stockpile showing drawdown and craters.

shows the drawdown of the stockpile after it had been in service. The comparison between the predicted and actual performance is shown to be reasonable, confirming the validity of the design procedures outlined in this section.

3.10 Closing remarks

The material presented in this chapter has outlined the methods for determining the relevant flow properties of bulk solids and how this information is integrated into the design procedures for bins and gravity reclaim stockpiles. The procedures, which have their foundation on the original research of Dr Andrew Jenike in the 1950s and 1960s, have been developed over the subsequent years by a number of researchers in various countries of the world. At the same time, the practical application of this research to the solution of actual industrial design problems has been widespread and well proven. The interaction between researchers, engineers and practitioners in industry has been invaluable in identifying the many complex problem areas where the research effort should be focused. It is important that this process be continued into the future.

References

Arnold, P.C. & Gu, Z.H. (1990) The effect of permeability on the flow rate of bulk solids from mass-flow bins. *Int. J. Powder Handling Process.*, **2**, 229–38.

Arnold, P.C., Gu, Z.H. & McLean, A.G. (1990) On the flow rate of bulk solids from mass-flow bins. In: *Proceedings of 2nd World Congress Particle Technology*, September 19–22, Kyoto, Japan, 2, pp. 2–9.

Arnold, P.C. & McLean, A.G. (1976) Improved analytical flow factors for mass-flow hoppers. *Powder Technol.*, **15**, 279–81.

Arnold, P.C., McLean, A.G. and Roberts, A.W. (1982) *Bulk Solids: Storage, Flow and Handling*. The University of Newcastle Research Associates (TUNRA), Australia.

Australian Standard AS 3880-1991 Bin Flow Properties of Coal.

Bell, T.A., Ennis, B.J., Grygo, R.J., Scholten, W.J.F. & Schenkel, M.M. (1994) Practical evaluation of the Johanson hang-up indicizer. *Bulk Solids Handling*, **14**, 117–25.

Benink, E.J. (1989) *Flow and stress analysis of cohesionless bulk materials in silos related to codes*. Doctoral thesis, The University of Twente, Enschede, The Netherlands.

D6128-00 Standard Method for Shear Testing Using the Jenike Shear Cell (2003) ASTM International, USA.

Enstad, G.G. (1975) On the theory of arching in mass-flow hoppers. *Chem. Eng. Sci.*, **30**, 1273–83.

Enstad, G.G. & Ose, S. (2003) Uniaxial testing and performance of a pallet press. *Chem. Eng. Technol.*, **26**, 171–76.

Goelema, G.G., Maltby, L.P. & Enstad, G.G. (1993) *Use of a Uniaxial Tester for the Determination of Instantaneous and Time Consolidated Flow Properties of Powders*. REPLPOWFLO II, EFChE Publication Series 96, Oslo, Norway, pp. 139–52.

Harder, J. & Schwedes, J. (1985) The development of a true biaxial tester. *Particle Charact.*, **2**, 149–53.

Janssen, R.J.M. (2001) *Structure and shear in a cohesive powder*. Ph.D. thesis, Delft University of Technology, The Netherlands.

Janssen, R.J.M. & Zetzener, H. (2003) Measurements on cohesive powder with two biaxial shear testers. *Chem. Eng. Technol.*, 26, 147–51.

Jenike, A.W. (1961) *Gravity Flow of Bulk Solids*. Bulletin 108, The University of Utah, Engineering Experimental Station, USA.

Jenike, A.W. (1964) *Storage and Flow of Solids*. Bulletin 123, The University of Utah, Engineering Experimental Station, USA.

Jenike, A.W. (1987) A theory of flow of particulate solids in converging and diverging channels based on a conical yield function. *Powder Tech.*, **50**, 229–36.

Jenike, A.W. & Leser, T. (1963) A flow no-flow criterion in the gravity flow of powders in converging channels. In: *Proceedings of 4th International Congress on Rheology*, Pt. 3, pp. 125–41.

Johanson, J.R. (1964) Stress and velocity fields in the gravity flow of bulk solids. *Trans. ASME, J. Appl. Mech. E*, **31**, 499–605.

Johanson, J.R. (1965) Method of calculating rate of discharge from hoppers and bins. *Trans. Min. Eng. AIME*, **232**, 69–80.

Johanson, J.R. (1992) The Johanson Indicizer™ System vs. the Jenike Shear Tester. *Bulk Solids Handling*, **12**, 141–44.

Johanson, J.R. & Jenike, A.W. (1962) *Stress and Velocity Fields in the Gravity Flow of Bulk Solids*. Bulletin 116, The University of Utah Engineering Experimental Station, USA.

Johanson, J.R. & Royal, T. A. (1982) Measuring and use of wear properties for predicting life of bulk materials handling equipment. *Bulk Solids Handling*, **2**, 517–23.

Keys, S. & Wiche, S.J. (2001) The development of an improved wall material wear tester. In: *Proceedings of the 7th International Bulk Materials Storage, Handling and Transportation Conference*, Australian Society for Bulk Solids Handling, Newcastle, Australia, October 2001, pp. 487–98.

Kwade, A., Schulze, D. & Schwedes, J. (1994) Determination of the stress ratio in uniaxial compression tests. Part 2. *Powder Handling Process.*, **6**, 199–203.

Maltby, L.P. (1993) *Investigation of the behaviour of powders under and after consolidation*. Ph.D. thesis, Telemark Institute of Technology, Porsgrunn, Norway.

Maltby, L.P. & Enstad, G.G. (1993) Uniaxial tester for quality control and flow property characterisation of powders. *Bulk Solids Handling*, **13**, 135–39.

McBride, W. (2000) Effective gravity reclaim stockpile design. *Bulk Solids Handling (Systems and Design) Notes*. Centre for Bulk Solids and Particulate Technologies, The University of Newcastle, Australia, July, pp. 635–44.

McBride, W. (2001) *Mechanics of bulk solids stockpiles*. Ph.D. thesis, The University of Newcastle, Australia.

McLean, A.G. (1979) *Flow rates of simple bulk solids from mass-flow bins*. Ph.D. thesis, The University of Wollongong, Australia.

Moore, B. & Arnold P.C. (1985) An alternative presentation of the design parameters for mass-flow hoppers. *Powder Technol.*, **42**, 79–89.

Ooms, M. & Roberts, A.W. (1985) Significant influence of wall friction in the gravity flow of bulk solids. *Int. J. Bulk Solids Handling*, **5**, 1271–77.

Roberts, A.W. (1991) Bulk solids handling, recent developments and future directions. *Bulk Solids Handling*, **11**, 1–19.

Roberts, A.W. (2000a) *Rathole Geometry and Draw-Down During Funnel-Flow in Bins and Gravity Reclaim Stockpiles. Bulk Solids Handling (Systems and Design) Notes*. Centre for Bulk Solids and Particulate Technologies, The University of Newcastle, Australia, July, pp. 609–18.

Roberts, A.W. (2000b) *Design Considerations for Maximum Reclaim Capacity of Gravity Reclaim Stockpiles. Bulk Solids Handling (Systems and Design) Notes*. Centre for Bulk Solids and Particulate Technologies, The University of Newcastle, Australia, July, pp. 653–75.

Roberts, A.W. (2000c) Dynamic modelling of bulk solids storage, discharge and handling systems. *Bulk Solids Handling*, **20**, 183–89.

Roberts, A.W. (1997) Vibration of fine powders and its application. In: *Handbook of Powder Science and Technology* (eds M.E. Fayed and L. Otten), 2nd edn, pp. 146–201. Chapman & Hall, London.

Roberts, A.W. (1998) *Modern Basic Principles of Bulk Solids, Storage, Flow and Handling*. Centre for Bulk Solids and Particulate Technologies, The University of Newcastle, Australia.

Roberts, A.W. & McBride, W. (2000) *Predicting the Performance of Gravity Reclaim Stockpiles. Bulk Solids Handling (Systems and Design) Notes*. Centre for Bulk Solids and Particulate Technologies, The University of Newcastle, Australia, July, pp. 645–52.

Roberts, A.W., Ooms, M. & Scott, O.J. (1984) Surface friction and wear in the storage, gravity flow and handling of bulk solids. In: *Proceedings of Conference on War on Wear. Wear in the Mining and Mineral Extraction Industry*. Institute of Mechanical Engineers, Nottingham, UK, pp. 123–34.

Roberts, A.W., Ooms, M. & Wiche, S.J. (1990) Concepts of boundary friction, adhesion and wear in bulk solids handling operations. *Bulk Solids Handling*, **10**, pp. 189–98.

Roberts, A.W., Scott, O.J. & Tiedemann, W. (1979) Bulk handling of ammonium nitrate: flow properties and storage bin design criteria. In: *International Powder and Bulk Solids Conference*. Philadelphia, USA, May.

Roberts, A.W., Sollie, L.A. & de Silva, S.R. (1993) The interaction of bulk solid characteristics and surface parameters in surface or boundary friction measurements. *Tribol. Int.*, **26**, 335–43.

Roberts, A.W. & Teo, L.H. (1990) Design considerations for maximum reclaim capacity of conical stockpiles. *Bulk Solids Handling*, **10**, 9–15.

Roberts, A.W. & Wiche, S.J. (1993) Prediction of lining wear life of bins and chutes in bulk solids handling operations. *Tribol. Int.*, **26**, 345–51.

Schulze, D. (1994) New ring shear tester for flowability and time consolidation measurements. In: *Proceedings of the 1st International Particle Technology Forum*, Part III, AIChE, Denver, CO, pp. 11–16.

Schulze, D. & Wittmaier, A. (2003) Flow properties of highly dispersed powders at very small consolidation stresses. *Chem. Eng. Technol.*, **26**, 133–37.

Schwedes, J. (2002) Consolidation and flow of cohesive bulk solids. *Chem. Eng. Sci.*, **57**, 287–94.

Schwedes, J. & Harder, J. (1986) Influence of stress history on the yield limit of bulk solids. In: *Proceedings of Second International Conference on Bulk Materials Storage, Handling and Transportation*. Institute of Engineers, Australia, pp. 94–97.

Standard Shear Testing Procedure for Particulate Solids Using the Jenike Shear Cell (1989) Institution of Chemical Engineers, Rugby, UK.

Thomson, F.M. (1997) Storage of particulate solids. In: *Handbook of Powder Science & Technology* (eds M.E. Fayed & L. Otten), 2nd edn. Chapman & Hall, London, pp. 389–486.

Wiche, S.J., Roberts, A.W. & McBride, W. (2003) A flowability tester for bulk solids flow measurement. In: *Proceedings of the 4th International Conference for Conveying and Handling of Particulate Solids*, Budapest, Hungary, May, pp. 5.7–5.12.

4 The characterisation of bulk powders

DEREK GELDART

4.1 Introduction

Characterisation is a means of describing something or someone by distinctive qualities. If a person is described as *large,* we want to know if the term relates to weight or height or both, and these two qualities can be measured easily. It was Lord Kelvin, writing in general of scientific matters, who stated:

> *'When you can measure what you are talking about,*
> *and express it in numbers, you know something of your subject.*
> *But if you cannot measure it, when you cannot express it in numbers,*
> *your knowledge is of a meagre and unsatisfactory kind.'*

In the case of human beings, there are many qualities that are impossible to quantify, such as kindness, sense of humour, wit and compassion. And these are often more important than those other characteristics that can be quantified.

With powders, some qualities can be measured and expressed in numbers, with various degrees of difficulty or precision (bulk density, mean particle size, particle-size distribution), while others which may be much more important for the smooth and efficient operation of a production unit cannot. Will the powder start to flow out of a storage hopper in a controllable manner? Will it suddenly stop flowing, or will it continue to flow when it is desired to shut off the flow, the phenomenon known as *flooding*?

Generally speaking, these properties are not always directly related to the properties of the single particles that comprise the powder (though attempts to seek quantitative connections exist) because local conditions such as the gas in the spaces between the particles, the size and geometry of the equipment may have an influence, too. Consequently, specific tests on the bulk powder may have to be made under well-defined conditions and interpreted by experienced technologists to predict the behaviour of the powder in process plant.

The purpose of this chapter is to explain the present state of knowledge regarding the relationship between particle and bulk powder properties, and to outline some of the tests that are available to measure bulk powder properties of interest in operations such as storage for reliable flow, operation of fluidised beds, powder transfer systems, mixers, etc.

4.2 Characterisation in the absence of an imposed gas flow through the powder

Carr (1965) was one of the first to propose making a set of simple tests on bulk powders with the aim of combining them into a score corresponding to descriptions of relative flowability ranging from cohesive to free-flowing. The particle properties that favour free-flowing behaviour are large mean size (small specific surface area); spherical shape; a smooth particle surface; hard materials that are not easily deformed; a minimum of inducible electrostatic activity; non-hygroscopicity; and a high particle density. In addition, there are several other properties that are not dependent on the powder, such as the relative humidity and temperature of the gas surrounding the particles, and their time in storage. Carr measured four basic parameters: angle of repose, compressibility, angle of spatula and coefficient of uniformity of powders having a wide size distribution. A score of up to 25 points was assigned to each of these properties from 25 for an ideal material down to zero for one that is very poor. The scores were added up to give a global score, which was then used to put any given powder into one of seven categories, ranging from excellent to very, very poor, from the point of view of flowability. These ideas and the separate tests were incorporated into one machine by the Hosakawa powder company and their Powder Tester has proved useful for monitoring changes in flowability in the process plant on a day-to-day basis, though the individual tests do not always represent the best possible way of obtaining data for design purposes.

4.2.1 *Angle of repose*

It is a matter of almost universal experience – by children playing with sand on a beach, by adults using flour in the kitchen or handling grain in a barn, gravel on a road, or mixing concrete – that when poured into a heap, granular materials form a conical pile that makes an angle, α, with the flat surface on which it is supported. This angle of repose is empirically observed to be small for coarse, rounded particles and large for fine and/or angular materials. The first truly scientific studies on angles of repose were carried out by Train (1958), who identified at least four methods to measure angle of repose. Subsequently, extensive experiments were done by Brown (1961), and these authors agreed in so far as they found that the four methods all gave different values for α, and that when using method 1 (poured heap), powders with an angle less than 40° flowed easily, while those with angles greater than 50° flowed poorly. This apparently simple method of judging flowability was neglected for many years, probably because there was no consensus on the best technique to use to yield consistent, reproducible results. It was only in 1990, in cooperation with a UK soda ash producer, that an improved test method was developed (Geldart *et al.* 1990) and subsequently further improved (Wouters & Geldart 1996) so that it could be used as a means of process control. Powder is poured on to an upper vibrating chute, which directs it into the conical hopper, from which it flows on to the lower chute and against the vertical wall, where it forms a semi-cone with the Perspex base. The graduations on the vertical and horizontal surfaces and simple geometry allow the angle, α, to be calculated. The latest version of the device mark 4, shown in Figure 4.1, is available commercially. (A rig and further information are available from Powder Research Ltd, 7 Westminster Gate, Harrogate HG3 1LU, UK.)

4.2.2 *Bulk density*

We now consider methods for measuring the bulk densities (BDs) of powders in a standardised manner. BDs can be measured conveniently in the devices shown in Figures 4.2 and 4.3. These features are available on the Hosakawa powder characteristics tester. Particular care is taken to ensure that the centre of the stationary chute is in alignment with the centre of the preweighed 100 cm^3 cup. The most loosely packed bulk density, ρ_{BLP}, is obtained by

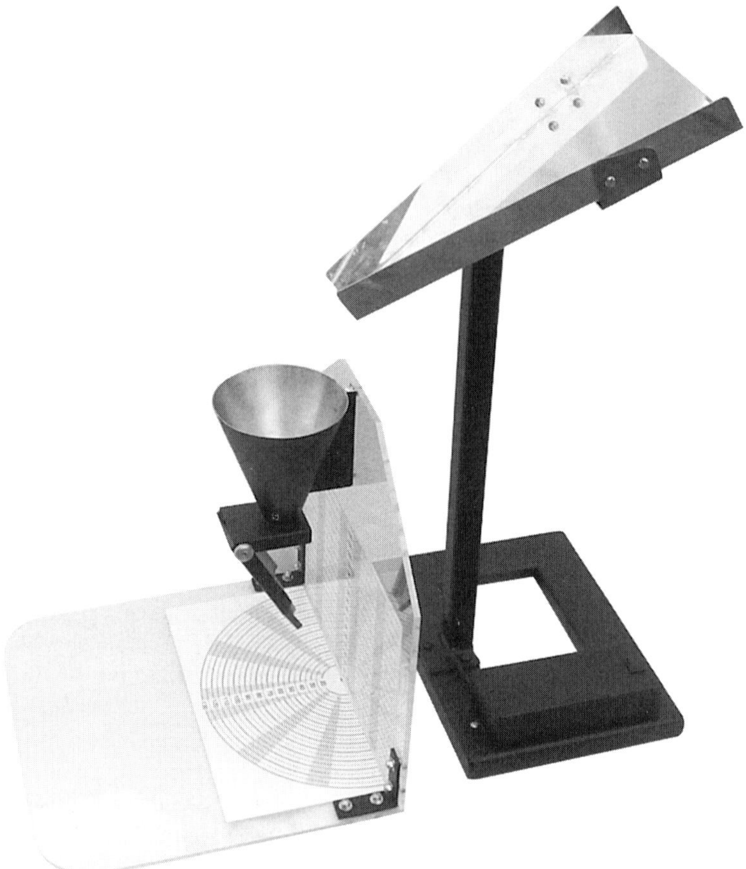

Figure 4.1 Equipment for measuring the angle of repose, α.

pouring the powder through a vibrating sieve and allowing it to fall a fixed height (about 25 cm) into a cylindrical cup. The amplitude of the vibration is set so that even cohesive powders will flow and fill the cup in 20–30 s. The excess powder is removed from the cup by drawing the flat edge of a steel rule across the top. Care should be taken to ensure that the cup is not jarred, as this causes the powder to compact and gives a false reading. The powder and cup are weighed, and an extension piece is then fixed to the cup. The cup is tapped 180 times (with a fall from a height of about 3 cm), while extra powder is added to ensure that the powder level does not drop below the rim of the cup. When tapping ceases, the extension is carefully removed, the excess powder scraped off, and the powder and the cup reweighed.

Strong interparticle forces prevent the particles from rolling over each other when poured slowly into the cup, giving an open powder structure and low values of ρ_{BLP} for cohesive powders. However, this open structure is disrupted when work is done on the system, so that the powder is made to settle and so produce a closer matrix under mechanical vibration. This phenomenon gives a high ratio of ρ_{BT} to ρ_{BLP} in cohesive solids and a low ratio in

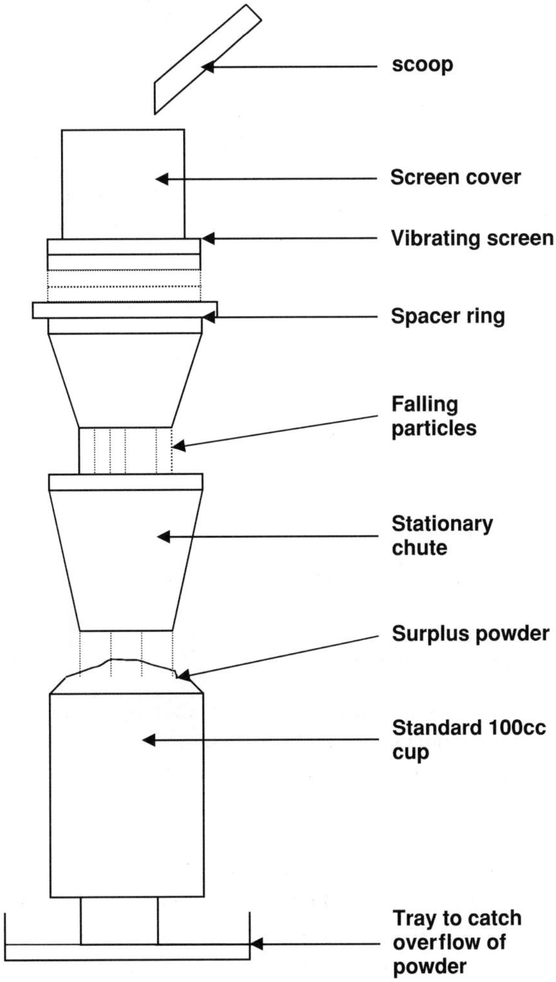

Figure 4.2 Equipment for measuring aerated bulk density of powder.

free-flowing powders. The ratio ρ_{BT}/ρ_{BLP} is called the Hausner ratio (HR) (Grey & Beddow 1968/1969). If the ratio HR $>$ 1.4, the powder is considered to be cohesive and classified as being in Group C (see below and Geldart 1973); if 1.25 $<$ HR $<$ 1.4, the powder is in the transition AC Group; and if HR $<$ 1.25, the powder is in Groups A, B or D.

The 'particle' or apparent density is especially important, since it appears in all the equations needed to characterise or correlate fluidisation properties of powders. For non-porous particles, this is the same as the specific gravity of the material that forms the particle, but for porous particles, the 'particle density', which is defined as [mass of particle/(volume of particle including all pores open and closed)], may be very different. The particle density can be estimated by a number of techniques, and the two most useful are described below. Commercial gas pycnometers measure the absolute density of the material that forms the particle because gas penetrates all the open pores and may even diffuse into some closed pores.

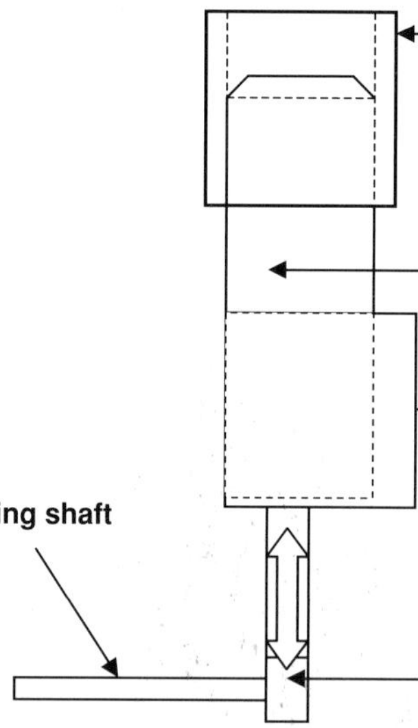

Figure 4.3 Equipment for measuring tapped bulk density of powder.

4.2.3 *Method 1: Comparative method*

The *comparative method* is based on the assumption that the minimum packed bed voidage is virtually the same for powders comprising particles with a similar shape and within the same narrow size range. A non-porous powder of known particle density, ρ_{pc}, is used as a control powder. It is put into a cylinder with a volume of at least 100 cm^3, height $\approx$ diameter, fitted with an open-ended plastic extension piece (or sleeve). This is overfilled and then tapped mechanically (preferably) at least 180 times. If a mechanical tapper is not available, tapping may be done manually by tapping the container on a lab bench. When tapping ceases, the sleeve is carefully removed, the excess powder is scraped off, and the container + powder is weighed so that the bulk density, ρ_{BTC}, can be calculated (Abrahamsen & Geldart 1980). The procedure is repeated with the unknown porous powder X to give ρ_{BTX}. Then:

$$\rho_{pX} = k\frac{\rho_{BTX}}{\rho_{BTC}}\rho_{pc}. \tag{4.1}$$

The empirical factor, k, is introduced because in practice it is not always possible to find a control powder having the same particle shape as that of the unknown powder. $k = 1$ if x and c are approximately the same shape; $k \approx 0.82$ if x is rounded or spherical, and c is angular; $k \approx 1/0.82$ if c is rounded or spherical, and x is angular.

4.2.4 *Method 2: Gas flow technique*

A modification (Geldart 1990) of the gas flow technique proposed by Ergun (1952) is particularly suitable for powders in which there is a significant difference between their aerated and tapped densities; that is, the powder volume changes at least 5% when tapped from a loosely packed condition. However, because the method depends on measuring accurately the pressure drop across a fixed bed, it is not suitable for cohesive powders in which channelling is likely to occur. A bed consisting of a known weight of the porous powder is fluidised to mix it well, and the gas is turned off so that the powder settles gently and thus assumes its maximum voidage. The bed height should be at least 400 mm. The bed height, L_A, is measured, and the pressure drop, Δp, is measured for at least four gas velocities. The plastic column is then tapped (with the hand on the side of the tube to avoid damage) so that it reaches as low a voidage as possible, the new bed height L_T is measured, and the pressure drops are measured again at the same gas flow rates. It is important to ensure that the bed does not become fluidised. Plots of Δp against U, the superficial velocity, are made, and their slope, S, is measured. Now, the first term on the right-hand side of the Ergun equation can be written as:

$$\frac{\Delta p}{L} = \frac{K(1-\varepsilon)^2}{\varepsilon^3} \frac{\mu}{d_{sv}^2}. \tag{4.2}$$

Rearranged, with subscript A used for the first (aerated) experiment, the slope of the graph

$$S_A = \left[\frac{\Delta p}{U}\right]_A = \frac{K\mu}{d_{sv}^2} L_A \frac{(1-\varepsilon_A)^2}{\varepsilon_A^3}. \tag{4.3}$$

Similarly, for the tapped condition,

$$S_T = \left[\frac{\Delta p}{U}\right]_T = \frac{K\mu}{d_{sv}^2} L_T \frac{(1-\varepsilon_T)^2}{\varepsilon_T^3}. \tag{4.4}$$

Now:

$$\frac{L_A}{L_T} = \frac{\rho_{BT}}{\rho_{BA}} \tag{4.5}$$

and

$$1-\varepsilon_A = \rho_{BA}/\rho_p \tag{4.6}$$

$$1-\varepsilon_T = \rho_{BT}/\rho_p. \tag{4.7}$$

Dividing Equation (4.2) by Equation (4.3) and substituting for L_A, L_T, ε_A and ε_T

$$\frac{S_A}{S_T} = \frac{\rho_{BA}}{\rho_{BT}} \left(\frac{\rho_p - \rho_{BT}}{\rho_p - \rho_{BA}}\right)^3 \tag{4.8}$$

Substituting Y for $\left(\frac{S_A}{S_T}\frac{\rho_{BT}}{\rho_{BA}}\right)$ (4.9)

$$\rho_p = \frac{\rho_{BT} - Y^{1/3}\rho_{BA}}{1 - Y^{1/3}}. \tag{4.10}$$

4.3 Characterisation in the presence of an imposed gas flow through the powder

4.3.1 *Equipment*

A small column is attractive because it requires only small amounts of powder. However, interparticle forces and powder/wall effects are of some importance, especially if the powders have mean sizes below about 30 μm, and these forces start to dominate in columns with a diameter smaller than about 100 mm.

A Perspex column 100 or 150 mm in diameter may be used, and the pressure drop across the bed alone should be measured by inserting into it from the top a long, small-diameter (~4 mm) pipe. This is open at the top and connected to a water manometer or a pressure transducer. The lower end of the probe is closed but has a small (~1 mm diameter) hole drilled in the side close to the bottom. This is covered with a piece of filter paper to prevent powder getting in. This tube is inserted into the bed while it is fluidised so that it rests as near to the distributor as possible. Alternatively, the pressure-measuring device may be connected to a small pressure tapping just above the gas distributor. When the bed is fluidised, the recorded pressure drop should be within a few per cent of the theoretical value, that is, the weight of the bed divided by the cross-sectional area.

4.4 Flow and pressure drop through packed beds

Through the work of Darcy and Poiseuille, it has been known for more than 120 years that the average velocity through a packed bed, or through a pipe, is proportional to the pressure gradient:

$$U \propto \Delta p/H. \tag{4.11}$$

A packed bed may be thought of as a large number of smaller tortuous pipes of varying cross-section, and a number of workers developed this approach, notably Kozeny (1927), Carman (1937) and Ergun (1952). The reasoning behind these equations is summarised in Coulson *et al.* (2002). At Reynolds numbers less than about 1 (laminar flow), the Carman–Kozeny equation applies:

$$\frac{\Delta p}{H} = \frac{K\mu}{d_{\mathrm{sv}}^2}\frac{(1-\varepsilon)^2}{\varepsilon^3}U. \tag{4.12}$$

U is the superficial or empty tube velocity, and the Reynolds number is defined as:

$$\mathrm{Re} = \frac{\rho_{\mathrm{g}} U d_{\mathrm{sv}}}{\mu}. \tag{4.13}$$

K is generally assumed to be 180, but there are indications that this may be correct only for narrow cuts, voidages between 0.4 and 0.5, and Reynolds numbers between 0.1 and 1. Abrahamsen and Geldart (1980) found that for fine powders of mean size 30–80 μm, K had average values of 263 ($\pm$35%) for spherical or near-spherical particles, and about 291

(±26%) for other shapes. For Reynolds numbers greater than about 1, the Ergun equation has proved satisfactory:

$$\frac{\Delta p}{H} = \frac{P_1}{\bar{p}}\left[150\frac{(1-\varepsilon)^2}{\varepsilon^3}\cdot\frac{\mu U_1}{\bar{d}_{sv}^2} + 1.75\frac{(1-\varepsilon)}{\varepsilon^3}\frac{\rho_{g1}U_1^2}{\bar{d}_{sv}}\right] \tag{4.14}$$

The term $P_1/\bar{p}$ is a correction factor for compressibility where $\bar{p}$ is the average absolute pressure in the bed. Subscript 1 denotes conditions at the inlet to the bed. Under laminar flow conditions (Re < 1), the first term on the right-hand side dominates:

$$\frac{\Delta p}{H} \propto \frac{\mu U}{d_{sv}^2}. \tag{4.15}$$

In fully turbulent flow (Re > 1000), the second term dominates and:

$$\frac{\Delta p}{H} \propto \frac{\rho_g U^2}{d_{sv}}. \tag{4.16}$$

Note that the surface/volume size, d_{sv}, is used: if only sieve sizes are available, depending on the particle shape, an approximation can be used for non-spherical particles:

$$d_{sv} \approx 0.87 d_p, \tag{4.17}$$

where d_p is the mean sieve size calculated from Equation (4.18):

$$\bar{d}_p = 1\Big/\sum\frac{X_i}{d_{pi}}, \tag{4.18}$$

where X_i is the mass fraction trapped between two adjacent sieves with the average mesh size of the two sieves being d_{pi}

$$d_V \approx 1.13 d_p. \tag{4.19}$$

4.5 Minimum fluidisation velocity

When gas is passed upwards through a packed bed unrestrained at its upper surface (Figure 4.4), the pressure drop increases with gas velocity (Figure 4.5) according to Equations (4.12) or (4.14) until, on the microscopic scale, the drag on an individual particle exceeds the force exerted by gravity or, on the macroscopic scale, the pressure drop across the bed equals the weight of the bed per unit area, Mg/A. If the bed has been compacted or is composed of interlocked, very angular or cohesive particles, then an excess pressure is required to free them from each other or from the wall, and they adopt a higher voidage configuration causing a fall back to the theoretical pressure drop. Further increases in velocity cause the formation of small bubbles whose size increases with gas velocity. The bed pressure drop begins to fluctuate, and if the bed is deep enough ($H \gg 2D$), the bubbles occupy a substantial proportion of the cross-sectional area. These large bubbles are called *slugs*, and they cause regular piston-like movements of the upper surface of the bed. The

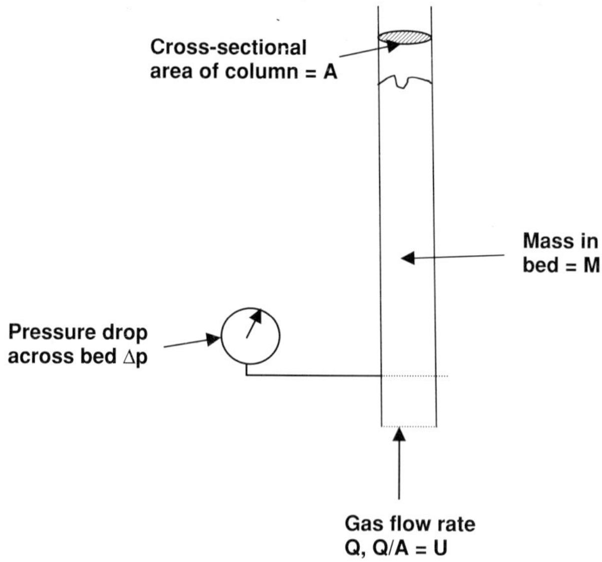

Figure 4.4 Diagrammatic sketch of equipment for determining minimum velocity of fluidisation.

average bed pressure drop then remains about constant, as shown in Figure 4.5. If the gas velocity is now reduced, depending on the size distribution of the powder, the pressure drop declines along curve ZXO (Figure 4.5). If the powder has a narrow size range line, ZYO is followed; if it has a wide size range line, ZXO is followed as particles settle out progressively on the distributor. Point Z represents the minimum velocity required to fully support the solids (though not necessarily in a well-mixed state) and is called the minimum velocity

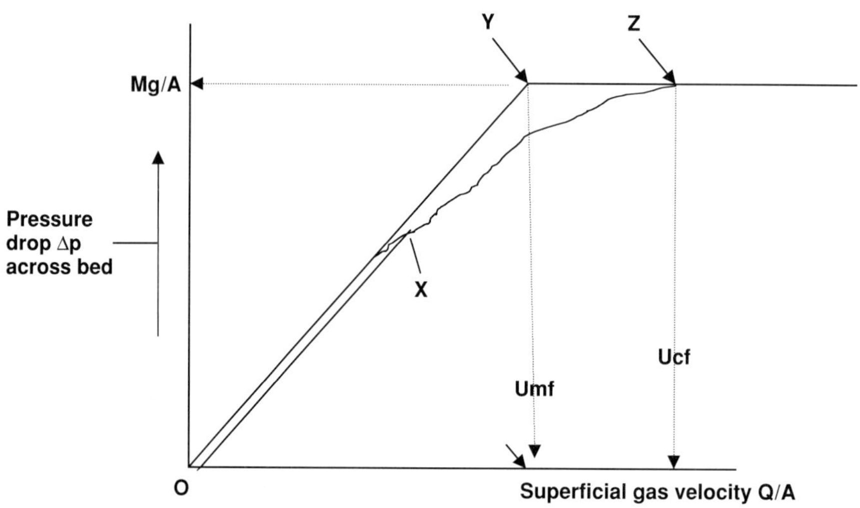

Figure 4.5 Plot of pressure drop across a bed of powder as a function of gas velocity.

of complete fluidisation, U_{cf}. U_{mf} is usually defined as the intersection of the horizontal fluidised bed line, ZY, and the sloping packed bed line, OX. The velocity corresponding to point Z is of practical interest, but relatively little work has been done to predict it. For the present, the best recourse is to make measurements and visual observations.

The pressure drop across a fluidised bed is the only parameter which can be accurately predicted:

$$\Delta p_F = \frac{Mg}{A}\ \text{N/m}^2 \tag{4.20}$$

$$\text{or } \Delta p_F \text{ cm w.g.} = \frac{0.1M\ \text{kg}}{A\ \text{m}^2} \tag{4.21}$$

$$\frac{\Delta p_F}{H} = (\rho_p - \rho_g)(1 - \varepsilon_{mf})g. \tag{4.22}$$

ε_{mf} is the bed voidage at U_{mf}, and a close approximation to it can be obtained by measuring the aerated or most loosely packed bulk density, ρ_{BLP}.

If Equations (4.14) and (4.22) are combined (with $p_1/\bar{p} = 1$), $\Delta p/H$ is eliminated, and U and ε set equal to U_{mf} and ε_{mf}, respectively, then:

$$\frac{\rho_g d_{sv}^3(\rho_p - \rho_g)g}{\mu^2} = \frac{150(1-\varepsilon_{mf})}{\varepsilon_{mf}^3}\frac{\rho_g \bar{d}_{sv}}{\mu}U_{mf} + \frac{1.75}{\varepsilon_{mf}^3}\frac{\rho_g^2 \bar{d}_{sv}^2}{\mu^2}U_{mf}^2. \tag{4.23}$$

The first group, on the left-hand side, is dimensionless and known as the Archimedes number, Ar. Wen and Yu (1966) correlated many data in this form. They combined the numerical constants with the voidage terms and, using the volume diameter, d_v, instead of d_{sv}, proposed:

$$Ar = 1650\,\text{Re}_{mf} + 24.5\,\text{Re}_{mf}^2, \tag{4.24}$$

where

$$\text{Re}_{mf} = \frac{\rho_g U_{mf} d_v}{\mu}.$$

There have been several studies on the independent effects of temperature and pressure on U_{mf} (see review paper by Yates 1996). In general, the qualitative effects are as predicted by Equations (4.22) and (4.24): in fine powders, U_{mf} decreases with increasing temperature and is hardly affected by pressure, while in coarse powders, increased temperature causes an increase in U_{mf}, and an increase in pressure causes a decrease. It should be noted that several strange effects can occur in beds of coarse particles at high temperatures (Botterill & Teoman 1980). Also, in any powder, if softening or sintering occurs so as to produce agglomeration, none of the equations is valid. Rearranged, Equation (4.23) becomes:

$$U_{mf} = \frac{\mu}{\rho_g \bar{d}_v}[(1135.7 + 0.0408Ar)^{1/2} - 33.7], \tag{4.25}$$

where

$$Ar = \frac{\rho_g \bar{d}_v^3(\rho_p - \rho_g)g}{\mu^2}.$$

Equation (4.29) should be used for particles larger than 100 μm (i.e. Groups B and D solids) in conjunction with Equations (4.1), (4.2) and (4.30). For small particles, $d_p < 100\,\mu$m,

Baeyens' equation gives the best agreement with experiments (Geldart & Abrahamsen 1981).

$$U_{mf} = \frac{(\rho_p - \rho_g)^{0.934} g^{0.934} \bar{d}_p^{1.8}}{1111 \mu^{0.87} \rho_g^{0.066}}. \tag{4.26}$$

4.6 Minimum bubbling velocity, U_{mb}

Fine powders (cracking catalyst is a typical example) exhibit a type of behaviour not found in coarse solids, namely the ability to be fluidised at velocities higher than U_{mf} without the formation of bubbles. The bed expands, apparently smoothly and homogeneously, until a velocity is reached at which bubbles appear at the surface. These must not be confused with the continuous channels or spouts which often appear and resemble miniature volcanoes that disappear when the side of the column is gently tapped. Further increases in velocity produce, at first, a further slight increase in bed height (see Figure 4.6) followed by a reduction. Relatively large bubbles burst through the bed surface periodically, causing the bed to collapse rapidly; it then 're-inflates' slowly, only to collapse again as another swarm of bubbles bursts through. Reduction in gas velocity produces a retracing of the height–velocity graph, and finally the last bubble disappears, giving once again a quiescent bed. The average of the velocities at which the bubbles appear and disappear is called the minimum bubbling velocity or bubble point, and it generally coincides with the maximum bed height for deep beds; premature bubbling can be caused by non-uniform distributors or protuberances in the bed, but the maximum values of U_{mb} have been correlated by Abrahamsen and Geldart (1980) and found to depend on the gas and particle properties:

$$U_{mb} = 2.07 \exp(0.716F) d_p \rho_g^{0.06} / \mu^{0.347}, \tag{4.27}$$

where F is the mass fraction of the powder less than 45 μm. The numerical constant is dimensional, and SI units must be used. If $F \approx 0.1$, and the powder is fluidised by air under ambient conditions,

$$U_{mb} \approx 100 d_p. \tag{4.28}$$

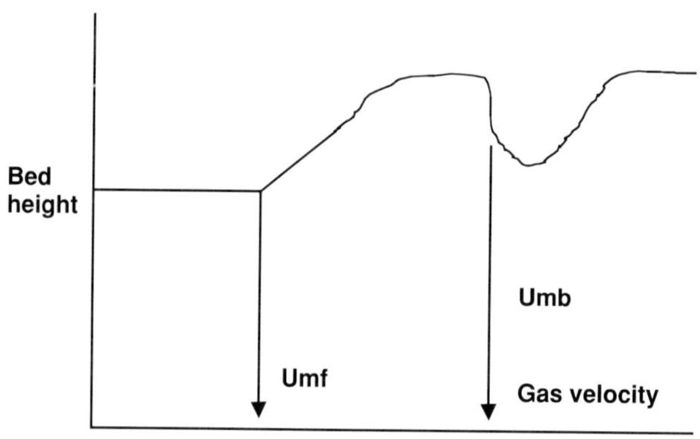

Figure 4.6 Bed expansion of Group A powder as a function of gas velocity.

Note that Equation (4.31) implies that U_{mb} is not dependent on particle density, a finding confirmed independently by other workers. It must be stressed that equations (4.26) and (4.27) are valid only for powders which are fine enough (in the main, a mean size less than 100 μm) to have values of U_{mf} less than U_{mb}. If calculations show that U_{mf} (from Equation (4.26)) is larger than U_{mb}, the powder will start to bubble at or very slightly above incipient fluidisation, and Equation (4.27) should not be used. This is discussed further later.

4.7 Effect on U_{mb} of temperature, pressure and type of gas

This is still a relatively unknown area, but see the review by Yates (1996). Equation (4.27) indicates that U_{mb} increases with pressure, and this trend is confirmed by the work of other researchers. Rietema and Piepers (1990) provide data which show that the power depends on the type of gas: up to $p = 15$ bar, U_{mb} for a catalyst fluidised in H_2 is affected little; in N_2, the dependency is $\rho_g^{0.13}$; and in argon, the dependency is $\rho_g^{0.18}$. The reason for this influence of the chemistry is not clear, but adsorption of the gas on the surface of the catalyst may have played a part, and it may be that not all fine powders are affected in the same way. Omitting any sintering effects, increasing the temperature reduces U_{mb} according to $1/\mu^{0.347}$.

4.8 Velocity for complete fluidisation, U_{cf}

In materials with a wide size range, particularly those having a large mean size, segregation by size tends to occur at velocities close to U_{mf} for the mixture. Although a value of U_{mf} can be defined for the mixture based on d_p, its usefulness is limited. Knowlton (1974) defined a velocity, U_{cf}, at which all the particles are fully supported (even though there may be segregation) and suggested that:

$$U_{cf} = \sum_{o}^{n} X_i U_{mfi}, \tag{4.29}$$

where U_{mfi} and X_i refer to a fraction of size d_{pi}. Alternatively, an approximation to U_{cf} can be found by calculating the average size of the largest 10% of the distribution and then calculating its U_{mf}. Both these methods appear to give a reasonable agreement with experimental values, even at high pressures.

4.9 Voidages in fluidised beds

Almost all gas-fluidised beds operate in the bubbling regime and consist of two phases: bubbles and the emulsion (or dense) phase. Conditions in the dense phase (gas velocity, U_D, and the corresponding voidage, ε_D) are the subject of much discussion and speculation, largely because it is believed that (1) most of the chemical conversion occurs there, and (2) in fine powders, the equilibrium size of the bubbles may be controlled by the voidage, ε_D. The dense phase also figures in modes of flow in standpipes or downcomers. U_D and ε_D are extremely difficult to measure directly in bubbling beds, though they can be predicted for quiescent fluidisation of fine powders.

4.9.1 *Voidage in non-bubbling (quiescent) beds*

The earliest attempts to measure and correlate the expansion of quiescent (or particulate) fluidised beds in gas/solid systems were made by Godard and Richardson (1968), who adapted the approach used by Richardson for liquid/solid sedimentation and fluidisation:

$$\frac{U}{v't} = \varepsilon^n. \tag{4.30}$$

In liquid fluidisation, n is a function of d_p/D and the terminal velocity Reynolds number, and varies between 4.65 and 2.4. The form of the relationship appears to hold for gas/solid systems, but experimental values of n between 3.84 and 19.7 have been reported (Godard & Richardson 1968). Geldart and Wong (1985) have shown that both n and v'_t increase with decreasing particle size below about 60 μm. v'_t is the intercept of the log ε–log U plot at $\varepsilon = 1$, and v_t is the Stokes terminal velocity.

An alternative approach based on the Carman–Kozeny equation was used by Abrahamsen and Geldart (1980). Combining Equations (4.20) and (4.31) (with $\varepsilon_{mf} = \varepsilon$ and $d_p = d_{sv}$) gives:

$$\frac{(\rho_p - \rho_g)g\bar{d}p^2}{\mu}\frac{\varepsilon^3}{1-\varepsilon} = K_f U. \tag{4.31}$$

Plots of all their data using 48 gas/solid systems gave a general expression to predict the non-bubbling expansion of a bed of fine powder, namely:

$$\frac{\varepsilon^3}{1-\varepsilon}\frac{(\rho_p - \rho_g)g\bar{d}_p^2}{\mu} = 210(U - U_{mf}) + \frac{\varepsilon_{mf}^3}{1-\varepsilon_{mf}}\frac{(\rho_p - \rho_g)gd_p^2}{\mu}. \tag{4.32}$$

The standard deviation on the numerical constant is $\pm 22\%$. ε_{mf} can be found from the simple experiment to determine the aerated voidage described earlier; U_{mf} is obtained from Equation (4.29), and U_{mb} from Equation (4.31).

4.9.2 *Voidage in bubbling beds*

The average overall voidage of a bubbling bed, that is the bed expansion, is largely caused by the bubble hold-up and is outside the scope of this chapter. The voidage of the dense phase (the portion of the powder where bubbles are absent) is generally taken as being ε_{mf} for Group B and D systems but can be significantly higher for Group A powders. Rietema (1967), Abrahamsen and Geldart (1980) and Yang *et al.* (1985), using the collapse technique and certain assumptions, concluded that $\varepsilon_{mf} < \varepsilon_D < \varepsilon_{mb}$ and hence $U_{mf} < U_D < U_{mb}$.

The *bed collapse technique* (Figure 4.7) is a good method for characterising Group A powders (see below). The powder is fluidised at a chosen velocity (say 20 cm/s), and when the gas is suddenly shut off, the bed height deceases linearly and is recorded as a function of time (Figure 4.8); it falls rapidly in the first few seconds as the bubbles rise to the surface and then much more slowly as the interstitial gas flows out, and the dense phase collapses. The straight-line portion of the curve is extrapolated back to time zero, and it is assumed that the intercept gives the height H_D which the dense phase would occupy in the bubbling

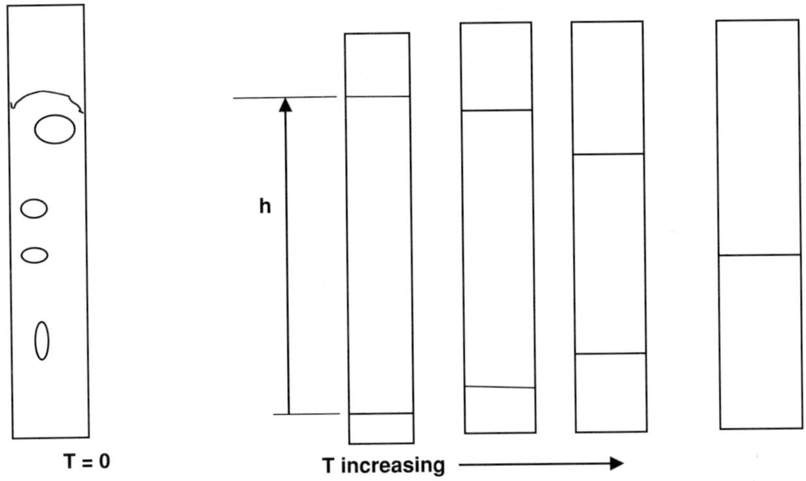

Figure 4.7 Bed collapse experiment.

bed. By further assuming that Darcy's law holds within the dense phase in a bubbling bed, U_D can be calculated. Abrahamsen and Geldart (1980) give:

$$\frac{H_D}{H_{mf}} = \frac{2.54\rho_g^{0.016}\mu^{0.066}\exp(0.09F)}{\bar{d}_p^{0.1}g^{0.118}(\rho_p - \rho_g)^{0.118}H_{mf}^{0.043}} \tag{4.33}$$

and

$$\frac{U_D}{U_{mf}} = \frac{188\rho_g^{0.089}\mu^{0.371}\exp(0.508F)}{\bar{d}_p^{0.568}g^{0.663}(\rho_p - \rho_g)^{0.663}h^{0.244}}. \tag{4.34}$$

The constants are dimensional, and SI units must be used. It should be noted that these empirical equations indicate that the dense phase 'opens up' as the mean particle size and particle density decrease and as gas density (pressure) and viscosity (temperature) increase.

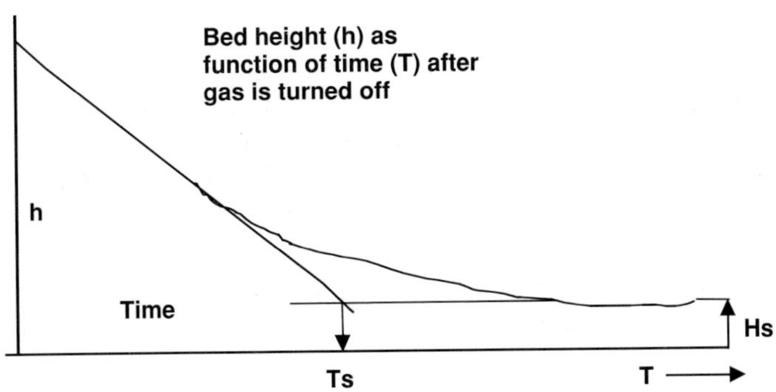

Figure 4.8 Bed heights from Figure 4.7 vs. time.

Table 4.1 Summary of fluidization properties of powder groups.

Property \ Group (Size increasing →)	C	A	B	D
Bed Expansion	very small	high, bubble-free initially	moderate	low
Collapse Time	long	long	short	very short
Bubble Shape	no bubbles–channels			
Solids Mixing	none	very high	moderate	poor
Gas Backmixing	none	very high	moderate	little
Slugging Mode	solid slugs	axi-symmetric	axi-symmetric	solid slugs
Spouting	none	none	only in small shallow beds	yes
Effect of DSV Within Group	large	large	some	some
Effect of PSD	?	some	little	?
Size Segregation	none	little	moderate	high

Increasing the fines fraction, F, also increases the dense phase voidage and gas velocity relative to conditions at incipient fluidisation.

4.10 Powder groups

Geldart (1972, 1973) postulated that the fluidized behaviour of powders could be classified into four Groups A, B, C, D, each having its own set of characteristics, summarised in Table 4.1. The groups are broadly characterised by their range of particle density and surface volume mean size, as shown in Figure 4.9.

4.10.1 *Group A*

Most commercial fluidised bed catalytic reactors such as units producing acrylonitrile, vinyl chloride, phthalic anhydride and large units for the catalytic cracking of petroleum use powders comprising low-density particles in the size range 20–100 μm, giving a mean size of about 75 μm. Consequently, a considerable amount of research work has been devoted to trying to understand the behaviour of such powders, and major contributions have been made by Donsi *et al.* (1975) and Rietema (1991) and their co-workers. It has been shown experimentally that interparticle forces are present in those powders, of which the cracking catalyst is the most typical example, and theoretical studies (Ye *et al.* 2004) using discrete particle modelling confirm that weak van der Waals forces are responsible for the non-bubbling expansion between U_{mf} and U_{mb}. Powders in this group expand considerably between minimum fluidisation and minimum bubbling velocities, and it is believed that in this non-bubbling condition, the particles in the bed are held in contact with each other by

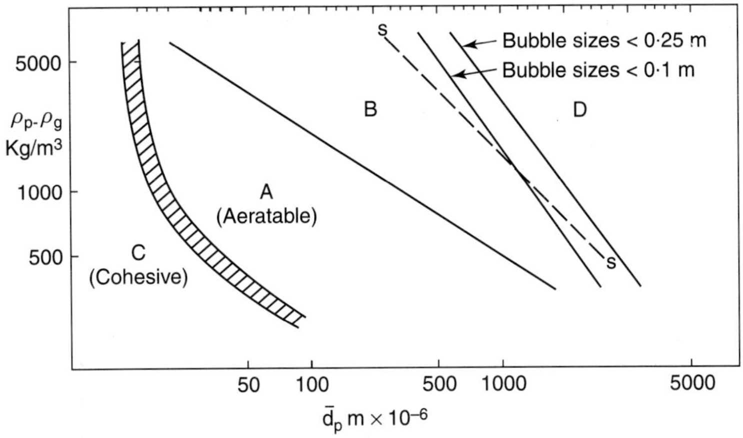

Figure 4.9 Classification of fluidised powders according to Geldart (1973).

weak interparticle forces to give a porous network resembling a three-dimensional porous structure. As the gas velocity is increased above the minimum bubbling condition, the bubbles that form rise up through the bed, disrupting the weak unstable structure of the expanded dense phase, sucking in the gas from the voids between the particles and causing the structure to assume a lower voidage condition. Gross circulation of the powder in the bed is induced by the growing bubbles, akin to convection currents in liquids, thereby producing rapid mixing and making the bed surface resemble a boiling liquid. There is continuous coalescence and splitting of the bubbles; this produces good gas–solid contacting and gives rise to an equilibrium bubble size that is correlated with the free fall or terminal velocity of the particles.

The criterion for identifying whether or not a powder is in Group A is given by the ratio of U_{mb}/U_{mf}. If

$$(U_{mb}/U_{mf}) > 1, \tag{4.35}$$

then the powder belongs to Group A.

This is a good way of characterising Group A powders, and the time, T_s, taken for the bed height to fall to the settled bed height, H_s, is standardised as T_s/H_s. This was called the standardised collapse time, SCT, by Geldart and Wong (1985) and was found to be a power function of the ratio (U_{mb}/U_{mf}). For any powder, the larger the SCT, the easier it is to fluidise, and the less likely it is to deaerate or settle out in transfer lines.

4.10.2 *Group B*

Powders in the size range 60–500 μm with particle densities of 1000–4000 kg/m^3 are in Group B and are not subject to interparticle forces, provided there are no capilliary forces, due to the presence of large amounts of liquids, or cohesive forces caused by triboelectric charging. Dry beach sand is a typical Group B powder. In all Group B powders, gas bubbles start to form as soon as minimum fluidisation velocity is exceeded. These bubbles grow rapidly by coalescence as they rise through the bed and as the gas velocity is increased. Most of the gas fed to the bed appears as bubbles which rise typically at velocities of 0.5–1.5 m/s, much faster than the superficial gas velocity, and this gives much poorer gas solid contact than in beds of Group A powders. If the bubbles grow to a size exceeding about 30–50% of the column diameter, the rise velocity depends on the tube diameter giving rise to a pulsating mode of operation, called *slugging*.

4.10.3 *Group D*

Large and/or dense particles have such high minimum fluidisation velocities that the gas bubbles that form as soon as U_{mf} is exceeded rise more slowly than the superficial gas velocity. This allows gas to bypass the solids by taking short cuts through the bubbles. These coarse powders are more prone to segregation by size in the bed because the solids mixing is poorer unless very high gas velocities are used. If the gas is admitted to the bed through only one central nozzle, a phenomenon known as *spouting* occurs. Particles are transported upwards in the central jet that resembles dilute phase pneumatic conveying and flow back to the base in an annular region bounded by the walls and the central jet. However, the principal difference between Groups B and D powders is that the latter can be made to form spouted beds in large-diameter columns, and the broken line S–S on Figure 4.9 divides the two groups using this criterion (Baeyens & Geldart 1973). The theory and applications relating to spouted beds are described well in most books on fluidisation and in Mathur and Epstein's (1974) book.

4.10.4 *Group C*

Powders that are cohesive and difficult to fluidise are in this group, typically powders with mean sizes less than about 15 μm and which are very soft or have sticky surfaces so that interparticle forces are much larger than the hydrodynamic forces exerted by the gas or by the force of gravity. Powders belonging to Group C are often very fine and light. Normal fluidisation with gas alone is difficult because the fluidising gas tends to bypass the powder bed through vertical channels. Materials showing this behaviour are said to be *cohesive*. These problems can be overcome by using mechanical stirrers to break up the channels or by adding coarser Group B particles to make a so-called powder-particle fluidised bed.

Van der Waals forces are attractive forces which act between particles if they are close enough. They arise because of local polarisations, induced by randomly excited electrons, which generate an electric field with solids and, in the case of facing bodies, within the dielectric medium between them. Lifshitz (see Rumpf 1990) suggested the following equation for the van der Waals force between a sphere of radius, R, and an infinite half-space at a distance, z_0, from the sphere :

$$F_{vdW} = \frac{\hbar\omega}{8\pi z_0^2} R. \qquad (4.36)$$

To calculate the van der Waals force between two spheres of radii R_1 and R_2, Equation (4.36) is adjusted to:

$$F_{vdW} = \frac{\hbar\omega}{8\pi z_0^2} \frac{R_1 R_2}{R_1 + R_2}. \qquad (4.37)$$

A comprehensive survey of interparticle forces was published by Krupp (1967) and by Visser (1989). Perhaps the most important point to make about this classification of powders into four groups is that phenomena observed when using a powder in one group may not be valid if changes to the powder during processing then affect its size and/or density so that it moves to another group. Although this classification was developed for gas–solid fluidised systems, it is used far beyond fluidisation purposes. For example, it has been found that Group A powders are much more likely to be subject to flooding when flowing out of hoppers because they are easily aerated and deaerate only very slowly (Geldart & Williams 1985). When studying the flowability of materials, our main interest lies with Group A and C powders: two powders may have similar values of particle size and density but show a completely different flow behaviour caused by their different particle shapes or surface properties. There is no clear transition between both groups, and samples showing features of A and C-type powders belong to the so-called Group A/C.

References

Abrahamsen, A.R. & Geldart, D. (1980) Behaviour of gas fluidized beds of fine powders, part 1: Homogeneous expansion. *Powder Technol.*, **26**, 35–46.

Baeyens, J. & Geldart, D. (1973) Predictive calculations of flow parameters in gas fluidized beds, and fluidization behaviour of various powders. In: *La fluidisation et ses Applications*, Toulouse, France. Cepadues Editions, Toulouse.

Botterill, J.S.M. & Teoman, Y. (1980) Fluid bed behaviour at elevated temperatures. In: *International Fluidization Conference, III*, Henniker, NH, pp. 93–100. Plenum Press, New York.

Brown, R.L. (1961) Flow properties. In: *Powders in Industry*, Society of Chemical Industry, Monograph No. 14.

Carman, P.C. (1937) Fluid flow through granular beds. *Trans. Inst. Chem. Eng. Lond.*, **15,** 150.

Carr, R.L. (1965) Evaluating the flow properties of solids. *Chem. Eng.*, **72**, 163.

Coulson, J.R., Richardson, J.F.H., & Backhurst, J.R. (1991) *Coulson & Richardson's Chemical Engineering*, Vol. 2, 4th edn, pp. 194–200. Butterworth-Heinemann, Oxford.

Donsi, G.M., Moser, S. & Massimilla , L. (1975) Solid–solid interaction between particles of a fluid bed catalyst. *Chem. Eng. Sci.*, **30**, 1533–35.

Ergun, S. (1952) Fluid flow through packed columns. *Chem. Eng. Prog.*, **48**, 89.

Geldart, D. (1972) The effect of particle size and size distribution on the behaviour of gas-fluidised beds. *Powder Technol.,* **6**, 201–215

Geldart, D. (1973) Types of gas fluidization. *Powder Technol.*, **7**, 285–92.

Geldart, D. (1986) *Gas Fluidization Technology*. Wiley, Chichester, UK.

Geldart, D. (1990) Estimation of basic particle properties for use in fluid-particle process calculations. *Powder Technol.*, **60**, 1–13.

Geldart, D. & Abrahamsen, A.R.(1981) Fluidization of fine porous powders. In: *Recent Advances in Fluidization, Chemical Engineering Progress Symposium Series*, 77(205), p. 160.

Geldart, D., Mallet, M.F. & Rolfe, N (1990) Assessing the flowability of powders using angle of repose. *Powder Handling Process.*, **2**, 341–45.

Geldart, D. & Williams, J.C. (1985) Flooding from hoppers: identifying powders likely to give problems. *Powder Technol.*, **43**, 181–83.

Geldart, D. & Wong, A.C.Y. (1985) Fluidization of powders showing degrees of cohesiveness, Part II – Rates of deaeration. *Chem. Eng. Sci.*, **40**, 653–61.

Godard, K. & Richardson, J.F. (1968) The behaviour of bubble-free fluidised beds in fluidisation. In: *Tripartite Chemical Engineering Conference*, Montreal, Canada, Institute of Chemical Engineers, London.

Grey, R.O. & Beddow, J.K. (1968/1969) On the Hausner Ratio and its relationship to some properties of metal powders. *Powder Technol.*, **2**, 63–70.

Knowlton, T.M. (1974) Minimum velocity for complete fluidization. In: *67th Annual Meeting of American Institute of Chemical Engineers*. American Institute of Chemical Engineers, Washington, DC.

Kozeny, J. (1927) *Sher. Akad. Wiss. Wien* (Abt. IIa), **136**, 271–306.

Krupp, H. (1967) Particle adhesion, theory and experiment. *Adv. Colloid Interf. Sci.,* **1**, 111.

Mathur, K. & Epstein, K.M.N. (1974) *Spouted Beds*. Academic Press, New York.

Rietema, K. (1967) Application of mechanical stress theory to fluidisation. In: *Proceedings of the International Symposium on Fluidisation*, 6–9 June, Eindhoven, Netherlands (ed. A.A.A.H. Drinkenburg), p. 154. Netherlands University Press, Amsterdam.

Rietema, K. (1991) *The Dynamics of Fine Powders*. Elsevier Science, Amsterdam.

Rietema, K.A. & Piepers, H.W. (1990) Effect of interparticle forces on stability of gas fluidized beds. *Chem. Eng. Sci.*, **45**, 1627–39.

Rumpf, H. (1990) *Particle Technology*. Chapman & Hall, London.

Train, D. (1958) Some aspects of the property of angle of repose of powders. *J. Pharm. Pharmac.* **10** Suppl., 127T–35T.

Visser, J. (1989) Van der Waals and other cohesive forces affecting powder fluidization. *Powder Technol.*, **58**, 1–10.

Wen, C.Y.Y. & Yu, H. (1966) Minimum fluidization velocities. *A.I.Ch.E. J.*, **26**, 161.

Wouters, I.M.F. & Geldart, D., (1996) Characterising semi-cohesive powders using angle of repose. *Part. Part. Syst. Charact.*, **13**, 254–59.

Yang, Z., Tung, Y. & Kwauk, M (1985) Characterising fluidization by the bed collapse method. *Chem. Eng. Commun.*, **39**, 217.

Yates, J.G. (1996) Effects of temperature and pressure on gas–solid fluidization. *Chem. Eng. Sci.*, **51**, 167–205.

Ye, M., van der Hoef, M.A. & Kuipers, J.A.M. (2004) A numerical study of fluidization behaviour of Geldart A particles using a discrete particle model. *Powder Technol.*, **139**(2), 129–39.

5 Characterisation for pneumatic conveyor design

MARK JONES

5.1 Introduction

The transport of particulate materials in pneumatic conveying pipelines is a well-established technique which is employed in a very wide range of industries. A particular feature of such systems is that the overall operation and performance of the conveying system depend very significantly on the particle and bulk properties of the material to be conveyed. In fact, the influences of the material properties of the conveyed material outweigh other influences by a very large margin, particularly for dense phase (or non-suspension flow) systems. One of the greatest difficulties in accurately predicting pneumatic conveying system performance lies in the current lack of fundamental understanding of precisely how the material properties influence the conveying flow regime within the pipeline.

This chapter attempts to identify the key material properties that influence pneumatic conveying performance and uses these properties to classify bulk materials according to their pneumatic conveying capability. The chapter goes on to look at issues regarding the characterisation of pneumatic conveying performance and provides a review of current techniques used to estimate the major design parameter, conveying line pressure drop.

5.2 Modes of flow in pneumatic conveying

There are two fundamental modes of pneumatic conveying: *dilute (or lean) phase* and *dense phase*. In dilute phase, the majority of the material is conveyed in suspension in the conveying gas (normally air); dense phase conveying refers to modes of flow where the majority of the material is not in suspension, and the material is conveyed along the pipeline as a series of fluidised dunes or as discrete plugs of material. The mode of flow is primarily dependent on the conveying gas velocity and the properties of the material being conveyed.

In dilute phase, the mechanism of flow is essentially aerodynamic drag with losses due primarily to particle–particle and particle–wall collisions. In most models used to predict pressure loss in dilute phase, the primary particle properties used include particle size and particle density. However, since the drag exerted on the particles by the gas will depend on the drag coefficient, clearly particle shape and surface texture will have an influence. In addition, the coefficient of restitution of the particles will depend on the substance from which the particle is constituted, the surface texture and the structure of the particle. For example, two particles of the same substance but having a different internal porosity may behave differently depending on how the particle has been formed. The difficulty from a characterisation viewpoint is that while there may be an intuitive acknowledgement that many of these properties will influence conveying performance, current models have no way of adequately taking these influences into account. Empirical models and correlations, however, provide indirect ways of capturing these effects.

In dense phase, the issues are far more complex, and at present, only a very limited range of particle properties are embedded in the modelling process. Hence, the approach adopted here has been to look at the material and bulk properties routinely measured and to examine how these are used in the design of pneumatic conveying systems.

The choice of whether to design for dilute phase conditions or for dense phase can be a difficult choice for the designer. In general, dilute phase conveying has a rather greater tolerance and can be safer with regard to reliability and the sensitivity of the system to changes in material properties. For this reason, dilute phase systems represent the majority of systems in industry. However, for materials that are erosive or abrasive and for materials that are fragile, dilute phase systems are generally not suitable. Lowering conveying velocities can have a very significant effect in reducing the unwanted side effects of product degradation (or attrition) and erosive wear of the system. In these situations, there is a strong case for using dense phase conveying. However, in some cases, dense phase conveying carries with it much greater risks largely due to the variations that can occur in material properties.

Not all materials are capable of dense phase conveying, and hence in the first instance, there is a need to identify those materials that are capable of dense phase conveying in a conventional system. Many workers have tried to establish methods to differentiate between those materials that exhibit dense phase capability and those that do not.

Significant confusion has existed over the definition of dense phase conveying, since there are many modes of non-suspension, low-velocity flow. However, it has long been recognised that there are two broad categories of dense phase conveying; a regime which is based largely on the fluidised nature of fine powders which has variously been termed 'moving bed' type flow or 'fluidised' dense phase; and a form of dense phase conveying based on granular materials which exhibit a high degree of permeability and which are conveyed in a plug or slug flow regime.

5.3 System design requirements

The basic design requirements in pneumatic conveying are as follows:

(1) The system is required to transport particulate material at the specified transfer rate in a reliable manner without pipeline blockage occurring.
(2) If the material to be conveyed is fragile in nature, there is a requirement to minimise attrition or particle breakage.
(3) If the material is erosive in nature, there is a requirement to minimise erosive wear of pipeline components.

A common factor in each of these requirements is that an appropriate conveying velocity is essential if these requirements are to be met. The basic mechanism of conveying is aerodynamic drag in which the gas velocity and density play a vital part. Velocity is of particular relevance, since the drag force is proportional to the square of the relative velocity. Since the conveying gas is compressible, the velocity range is an important factor in conveying. A minimum velocity will be required at the point where material is fed into the pipeline; and the maximum velocity at the end of a constant bore pipeline will depend on the conveying line pressure drop. If the velocity at the end of the pipeline is too high, this is likely to increase the potential for particle breakage and erosive wear of pipelines. Hence, the specification of the air requirements and the matching of these to the pipeline bore are the most fundamental design parameters in any system.

A fundamental decision for any proposed system is the mode of flow to be utilised. This will depend on the characteristics of the material to be handled and specific operational requirements. If erosive wear or product attrition are significant issues then minimising the conveying velocity will be beneficial in most cases. These are the two most significant reasons for considering a dense phase system. However, not all materials are capable of dense phase conveying, and hence it is important to be able to classify materials according to their potential conveying capability. From the design point of view, two questions must be answered:

(1) Does the material to be conveyed demonstrate any natural dense phase capability, or is the material limited to dilute phase in a conventional system? (In this context, a conventional system refers to a pipeline that has no flow conditioning along its length, e.g. no bypass line, boosters, etc.)
(2) If the material does exhibit dense phase capability, which of the two major modes of flow will be most suitable?

These questions are best answered by carrying out conveying trials. However, it would be useful if these questions could be answered without the need for full conveying trials by undertaking 'bench scale' tests using small quantities of representative material. It is in this context that a range of tests are used to estimate the conveying potential of a material. However, it is important to note that these classification techniques provide a guide only but are useful in terms of providing an initial assessment prior to full conveying trials.

5.4 Classification of bulk materials for pneumatic conveying

Pneumatic conveying relies on the interaction of the conveying gas and the particles. In dilute phase, the inter-particle spacing is large, and often the interaction of single particles in a gas flow can be the basis of modelling this mode of flow. However, in dense phase conveying, the close proximity of a particle with its neighbours becomes important. That is, the bulk properties in terms of gas–solid interactions become significant. Hence, it has long been recognised that air/particle characteristics such as fluidisation, permeability and de-aeration characteristics significantly influence the conveying capability of the material. Permeability represents the material's ability to allow air to flow through the interstitial voids and is basically the pathway to fluidisation. De-aeration represents the time it takes for a bulk material to reach a steady-state bed height or bulk density once the airflow is turned off. These parameters have been shown to be critical to the ability of a material to convey in a dense phase flow regime (Doig 1975).

A number of classification diagrams have been developed for determining the likely behaviour of a bulk material in a pneumatic conveying pipeline. Much of this work has its roots in the work of Geldart (1973), who classified materials according to their fluidisation behaviour. It was soon recognised that this classification might also be applied to pneumatic conveying with regard to the classification of materials in terms of their conveyability (Doig 1975). In addition, classification of materials, based on de-aeration and permeability characteristics, has also been used to identify the possible flow modes that a particulate

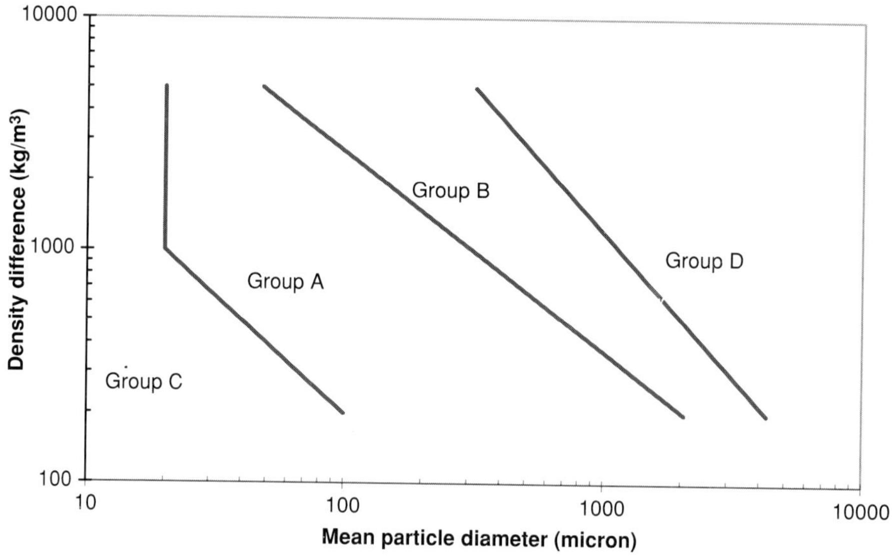

Figure 5.1 Geldart classification for fluidisation (Geldart 1973).

material may exhibit in a pneumatic conveying system (Mainwaring & Reed 1987; Jones & Mills 1990; Fargette 1998).

5.4.1 *The Geldart classification for fluidisation behaviour*

The Geldart classification (Geldart 1973) is shown in Figure 5.1 and identifies four distinct groups of bulk material, A, B, C and D, based on the density difference (between the bulk solid and the fluidising gas) and the mean particle size. Group A materials are powders which can be aerated with relative ease and exhibit good air-retention characteristics when the gas supply is shut off. Group B materials are granular in nature and generally have classical fluidisation behaviour. They fluidise readily, and the fluidised bed collapses almost instantaneously when the air supply is shut off. Group C materials are cohesive powders, which are difficult to fluidise. These powders generally lift as a plug initially until the fluidising air finds a series of weak points in the powder leading to channelling and breakup of the plug. Mechanical agitation or stirring can often lead to these powders becoming fluidised, in which case they behave in a similar manner to Group A powders. Group D materials are very coarse in nature with a high degree of permeability, requiring very large quantities of air for fluidisation to occur. They are often described as 'spoutable' materials due to their ability to maintain stable spouts when air is entered through a single point rather than distributed via a fluidisation membrane.

The basis of the boundaries of Geldart's diagram has been discussed and compared by numerous researchers, but the net effect has been that the original boundaries have changed little. Other workers have looked at including other parameters such as adhesion forces and the inclusion of cohesion (Molerus 1982) in determining some of the boundaries, while Geldart used empirical data and bubbling and fluidisation velocities to describe them. Grace

(1986) derived dimensionless parameters based on superficial gas velocity and particle diameter which described the boundaries of Geldart's bulk material classification.

For pneumatic conveying purposes, the Geldart Classification provides a very useful method of identifying which materials are likely to be good candidates for dense phase conveying and which mode of flow is likely to be best suited to the material in question. For pneumatic conveying purposes, the four Geldart groupings could be described as follows:

5.4.1.1 *Group A* The ease with which Group A materials fluidise and, perhaps more importantly, the high level of air retention capability they exhibit makes this group of materials very good candidates for the *moving bed* or *fluidised* mode of dense phase conveying. Good air-retention characteristics have been shown to be vital for reliable fluidised dense phase conveying.

5.4.1.2 *Group B* These materials show classical fluidisation behaviour but exhibit almost no air-retention capability. The size range of this group of materials often leads to poor permeability characteristics which have been shown to be important for *plug* or *slug* flow dense phase conveying. Hence, Group B powders are often not capable of being conveyed in either of the two major modes of dense phase conveying in a conventional pipeline. In some cases, the use of flow conditioning along the pipeline or conditioning the material at the feed point can offer the possibility of conveying this group of materials in dense phase.

5.4.1.3 *Group C* Materials in this group are extremely fine and often very cohesive. This can often lead to significant problems in pneumatic conveying particularly with regard to feeding and to problems of pipeline internal coating. However, if these materials can be effectively fluidised, they behave in a very similar manner to Group A materials and are very good candidates for dense phase flow.

5.4.1.4 *Group D* The large particle size of Group D materials often leads to high levels of air permeability. In other words, the conveying air can pass relatively easily through the plugs of material. This is a prerequisite for the plug and slug flow modes of dense phase; hence, materials in this group are often excellent candidates for dense phase conveying in plug or slug flow regimes provided the size range is not too wide.

5.4.2 *The use of loose-poured bulk density*

In dense phase pneumatic conveying, the behaviour of the bulk is in many ways more important than that of individual particles. For this reason, the loose-poured bulk density may provide a more useful parameter than the individual particle density when trying to assess suitability of materials for dense phase conveying. Only a limited amount of research has been conducted using 'as poured' or loose-poured bulk density (ρ_{blp}) as a parameter to characterise bulk materials for either fluidisation or pneumatic conveying. Pan (1999) investigated the use of this parameter in defining his classification based on mean particle size. The bulk density of a material tends to increase as the size distribution is increased (Heiskanen, 1993) or when the amount of fines in a mixture is increased. This is due to the smaller particles filling the interstitial voids of the larger particle sizes. This increased particle contact generally increases the adhesion and internal friction of the bulk material

Table 5.1 Properties of permeability tested bulk materials.

Particle type	Mean diameter $\times 10^{-6}$ (m)	Loose-poured bulk density (kg/m^3)	Particle density (kg/m^3)	Permeability $\times 10^{-6}$ (m/s/Pa)	Pneumatic conveying mode[a]	Investigator
Barytes	11	1590	4250	0.48	Moving bed	Jones
Cement	14	1070	3060	0.71	Moving bed	Jones
Pulverised fuel ash	25	979	2446	0.60	Moving bed	Jones
Copper ore	55	1660	3950	0.33	Moving bed	Jones
Iron powder	64	2380	5710	0.34	Moving bed	Jones
Alumina	79	1040	3600	0.42	Moving bed	Jones
Coal (pulverised)	84	393	1550	0.53	Moving bed	Jones
Flour (RHM Dem.)	90	510	1470	1.30	Moving bed	Jones
PVC powder	90	615	990	1.20	Moving bed	Jones
Zircon sand	120	2600	4600	1.30	Moving bed	Jones
Potassium sulphate	131	1260	2625	0.99	Dilute only	Jones
Coal (degraded)	146	701	1550	1.00	Dilute only	Jones
Granulated sugar (degraded)	157	656	1580	1.40	Dilute only	Jones
Pearlite	158	100	800	5.70	Dilute only	Jones
Silica sand	174	1450	2630	3.90	Dilute only	Jones
Magnesium sulphate	224	1010	2353	6.30	Dilute only	Jones
Agricultural catalyst (degraded)	270	760	4660	1.70	Dilute only	Jones
Potassium chloride	384	1010	1987	11.00	Dilute only	Jones
Granulated sugar	458	890	1580	20.00	Dilute only	Jones
Coal (as supplied)	778	870	1550	0.42	Dilute only	Jones
Agricultural catalyst (ICI)	782	767	4655	23.00	Dilute only	Jones
Polyethelene pellets	4000	540	912	420.00	Plug type	Jones
Cement	22	1030	3160	0.50	Moving bed	Mainwaring
Pulverised coal	44	610	1500	0.60	Moving bed	Mainwaring
Pearlite	200	100	800	4.00	Moving bed	Mainwaring
Flour	78	514	1470	0.50	Moving bed	Mainwaring
Pulverised fuel ash	20	980	2450	0.30	Moving bed	Mainwaring
Slate dust	500	1280	2860	0.60	Dilute only	Mainwaring
Zircon sand	115	2600	4610	1.30	Dilute only	Mainwaring
Pulverised fuel ash (grits)	700	400	2380	11.00	Dilute only	Mainwaring
Mustard seed	1650	680	1180	129.00	Plug type	Mainwaring
Polyethylene powder	825	480	990	20.00	Plug type	Mainwaring
Sand	1020	1540	2620	62.00	Plug type	Mainwaring
Polyethylene pellets	3850	558	914	80.00	Plug type	Mainwaring
Granulated sugar	720	820	1590	21.00	Plug type	Mainwaring

[a] Based on experimental observations.

creating more cohesion, especially in fine powders (Molerus, 1982). This suggests that replacing particle density with bulk density in the Geldart type diagrams may reduce the errors in predicting pneumatic conveying performance associated with bulk materials that have a large size distribution and/or significant fines content.

Williams *et al.* (2002) replaced the particle density with the loose poured bulk density in the Geldart classification. Since the bulk density of the material is so large in comparison with the air density, the density of the air can be neglected in both cases. To determine if loose-poured bulk density provides an improved indication of pneumatic conveying behaviour of a bulk material, a comparison of Geldart and Williams boundaries based on particle and loose-poured bulk density was made. The materials detailed in Table 5.1 were plotted on the Geldart and Williams diagrams, as can be seen in Figures 5.2 and 5.3, respectively. The Geldart diagram shows 77% accuracy in predicting the pneumatic conveying capability of the material. When particle density is replaced by loose-poured bulk density in the Geldart diagram, accuracy increases to approximately 83%. However, owing to the relatively small sample, it is questionable whether this result is significant, although those materials for which the Geldart chart did not predict conveying capability correctly were closer to the appropriate boundary in the loose-poured bulk density diagram. The use of loose-poured bulk density may have certain advantages over particle density for the prediction of pneumatic conveying capability, but further work is required to verify this.

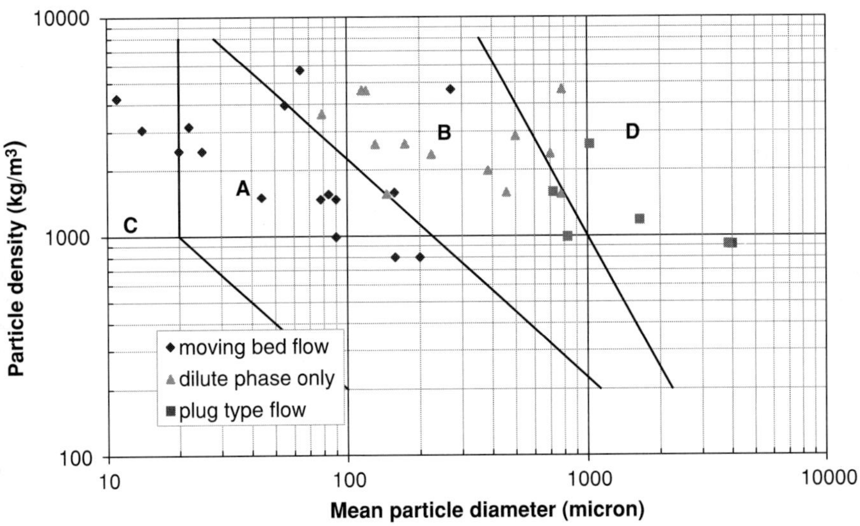

Figure 5.2 Geldart fluidisation diagram showing flow modes in pneumatic conveying.

5.4.3 *The use of permeability and de-aeration*

Numerous pneumatic conveying classifications have been developed to determine the likely mode of flow for a bulk material (Hong *et al.* 1995). Dixon (1979) developed the slugging diagram, which is loosely based on, and has the same axes as, Geldart's diagram and classifies bulk materials according to their ability to slug in vertical columns. Mainwaring

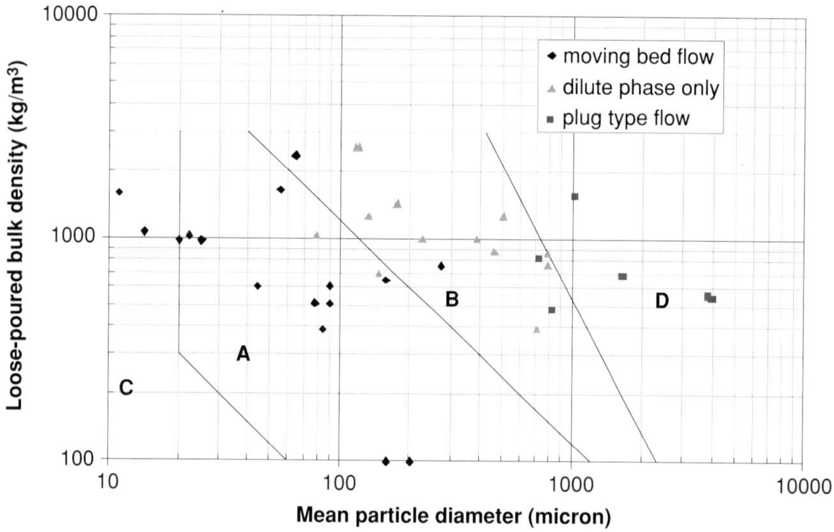

Figure 5.3 Modified Geldart fluidisation diagram showing flow modes in pneumatic conveying (Williams *et al.* 2002).

and Reed (1987) noticed that dense phase modes of flow were observed in all the powder classifications of Geldart and Dixon, which led them to look for alternative parameters which more closely match conveying conditions in the pipeline. They proposed the use of permeability and de-aeration as parameters which more closely represented material behaviour in the conveying pipeline.

To obtain these parameters, a fluidisation column was used, as shown in Figure 5.4. The Perspex tube used by Mainwaring and Reed (1987) was 138 mm in diameter, and the distance between the pressure tappings was 150 mm. Air is supplied in a controlled and measured manner through the plenum chamber and the porous membrane gas distributor. Figure 5.5 shows the variation of pressure gradient with superficial gas velocity for a given material. It can be seen that during the initial stages of the test, the pressure gradient increases as the flow of gas increases through the interstitial spacing between the particles in the fixed bed of material. As the pressure gradient increases, a point will be reached where the pressure gradient balances the weight of the bed. At this point, the material will begin to dilate leading to increased inter-particle spacing. This is the point at which the minimum fluidising velocity is reached, and under ideal conditions, the pressure gradient will remain constant as the superficial air velocity is further increased. The permeability factor, P_f, is defined as the inverse of the gradient of the straight-line relationship between the pressure gradient and the superficial gas velocity prior to the minimum fluidising conditions. Hence:

$$P_\mathrm{f} = \frac{U_\mathrm{g}}{\Delta P/L} = \frac{U_\mathrm{g} L}{\Delta P}. \tag{5.1}$$

The de-aeration of a fully fluidised bed will depend on the material characteristics. In pneumatic conveying, the ability for a material to retain air and hence remain fluidised is recognised as important to conveyability. A measure of the rate at which the material de-aerates

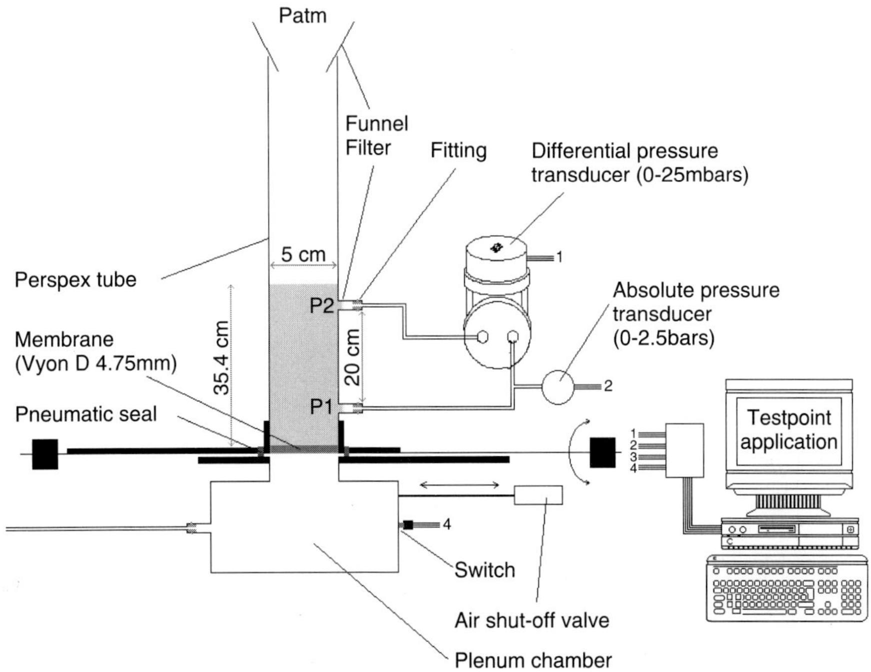

Figure 5.4 Fluidisation column.

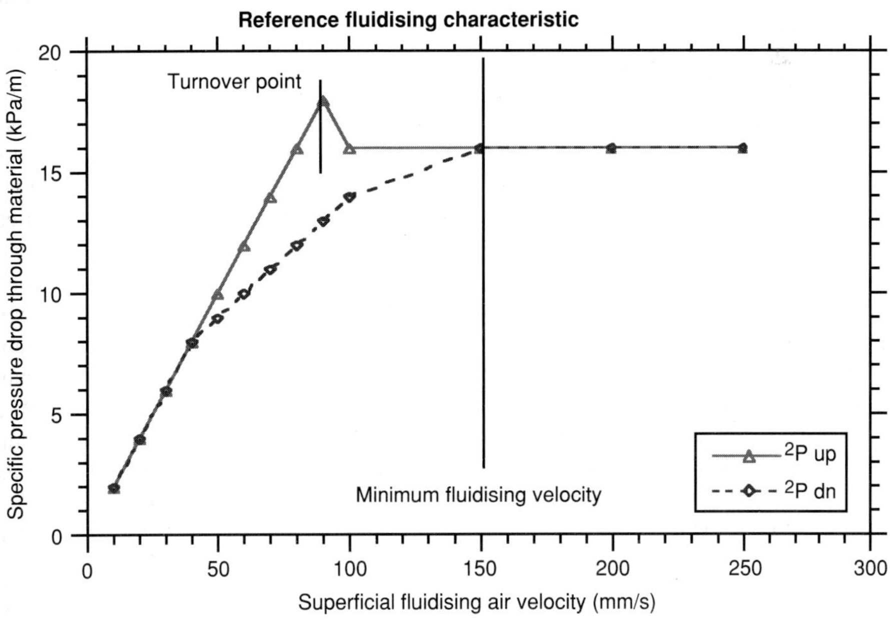

Figure 5.5 Typical fluidisation curve.

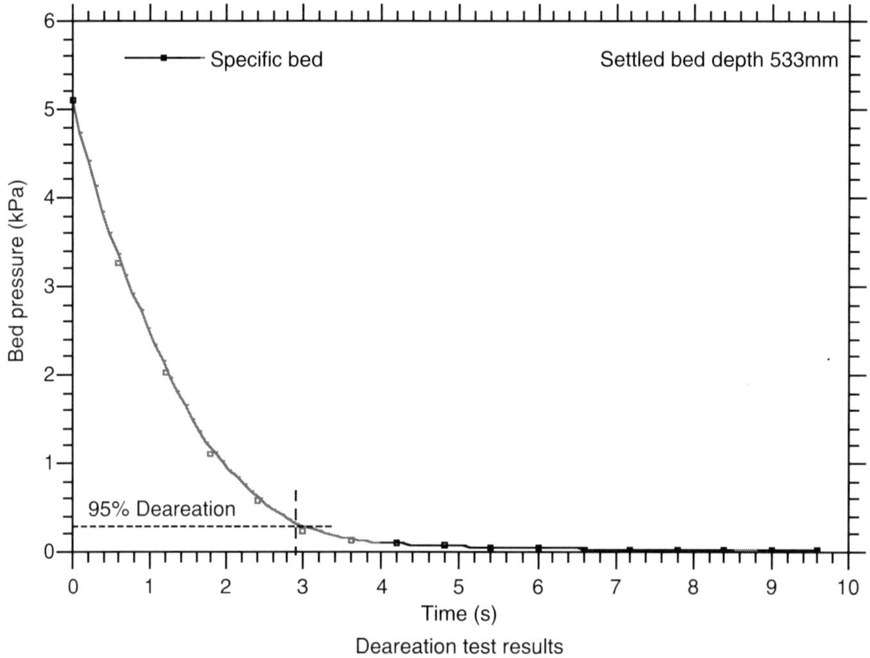

Figure 5.6 De-aeration curve for burnt lime.

is therefore useful. Mainwaring and Reed (1987) measured de-aeration by monitoring the pressure gradient decay in the bed with time and assumed an inverse proportionality. Hence, the de-aeration factor is defined as:

$$A_{\mathrm{f}} = \frac{\Delta P}{L} t_{\mathrm{d}}. \tag{5.2}$$

An example of the de-aeration curve is shown in Figure 5.6.

Mainwaring and Reed (1987) concentrated on the plug and slug flow conveying regimes and used two graphs to classify materials according to their observed conveying mode in a pneumatic conveying pipeline. The first graph (Figure 5.7) relates permeability factor to the pressure gradient in a bed of material under the minimum fluidising condition. An empirical boundary was established which separates materials into two categories: materials above the boundary which have high values of permeability factor are considered good candidates for the slug or plug flow mode of dense phase, while materials below this boundary are limited to the dilute phase only in a conventional pipeline or will convey in the fluidised moving bed type flow regime of dense phase. The second graph (Figure 5.8) relates de-aeration factor normalised by particle density to the pressure gradient at minimum fluidising conditions. In this case, materials above the empirical boundary exhibit high levels of air retention capability and are good candidates for fluidised dense phase conveying, while materials falling below the boundary are candidates for plug or slug flow or dilute phase only.

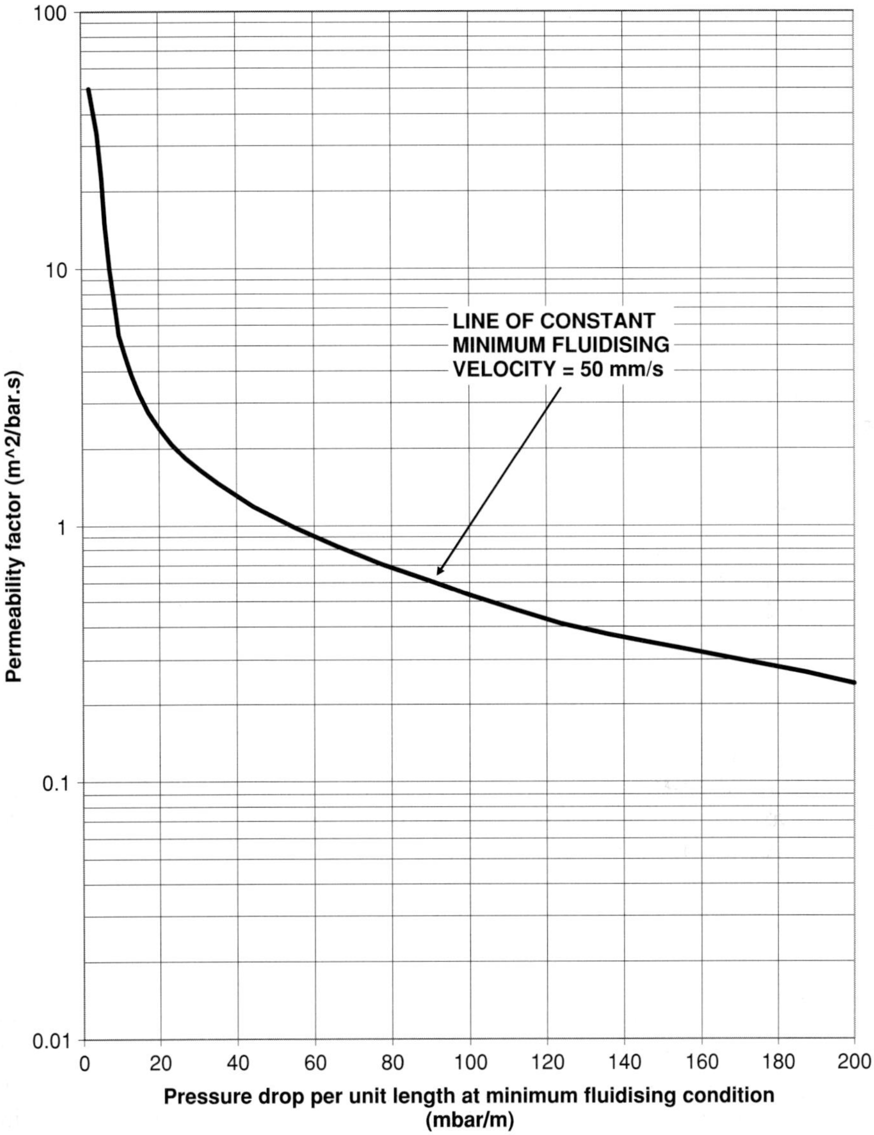

Figure 5.7 Permeability factor vs. pressure drop per unit length (Mainwaring & Reed 1987).

Jones (1988) also used permeability and de-aeration analysis to develop a model for classifying materials for pneumatic conveying, as shown in Figure 5.9. The main difference between the models of Jones and Mainwaring is that the classification of each material is determined from only one graph. Also, the de-aeration factor does not consider particle density and is determined from the time it takes for the material to settle from the 'as poured' state to the 'consolidated state', using vertical vibration as the forcing mechanism. Jones also carried out an extensive range of conveying trials to establish actual conveying

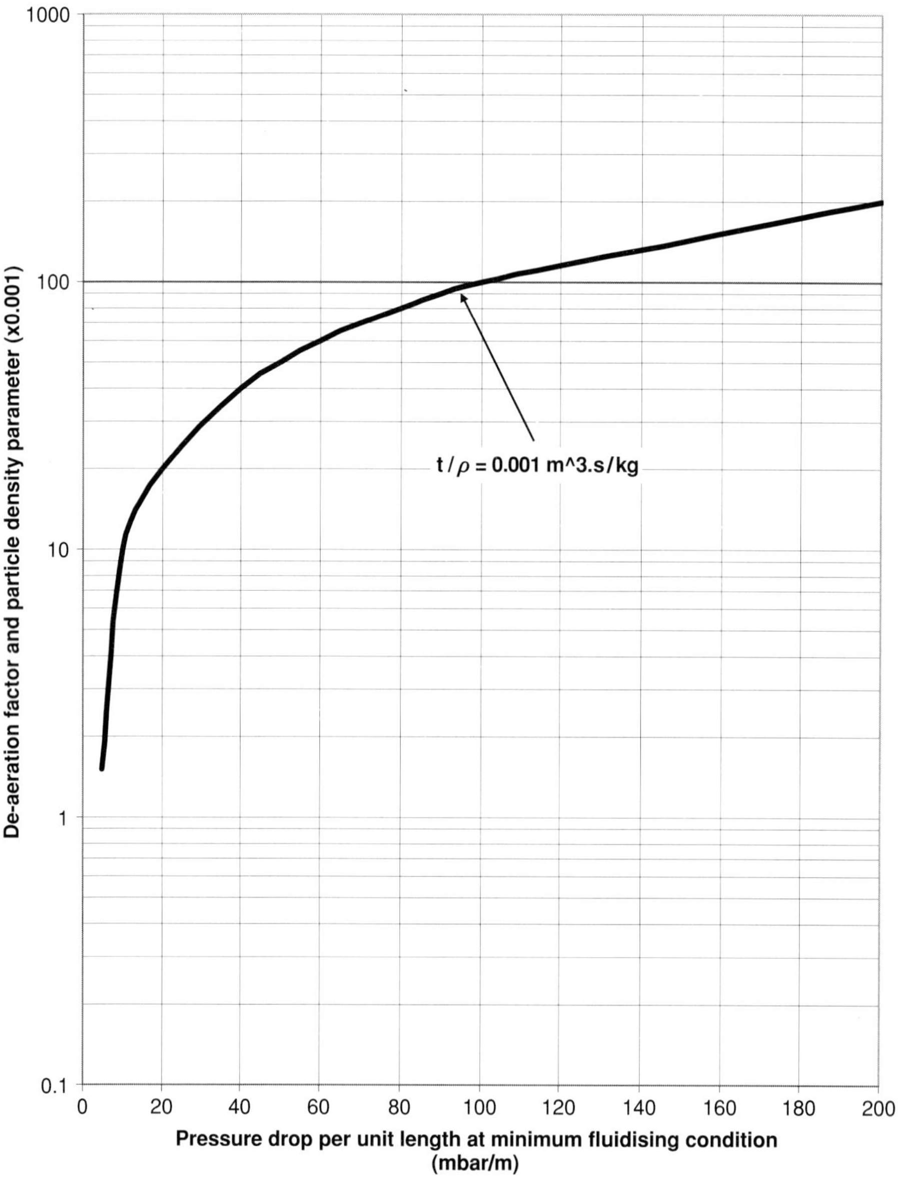

Figure 5.8 De-aeration/particle density factor vs. pressure drop per unit length (Mainwaring & Reed 1987).

performance and whether or not each material exhibited dense phase capability. This model also classifies bulk materials into the same three categories as the model of Mainwaring.

Although the permeability factor of a bulk material is generally well defined, the de-aeration value is difficult to measure. Different methods in determining de-aeration values also make it difficult to compare the pneumatic conveying charts of Mainwaring and Jones. Again, loose-poured bulk density was investigated to determine if this bulk

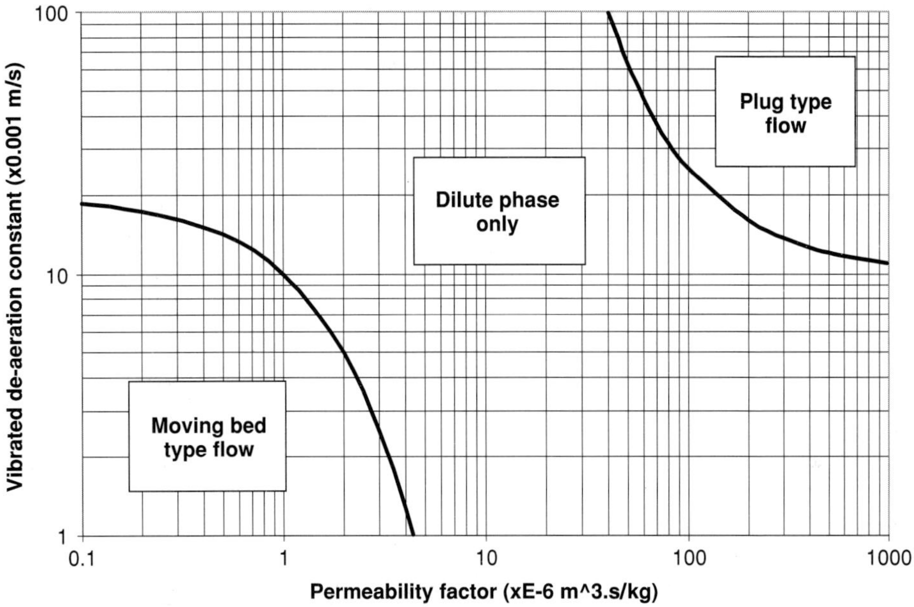

Figure 5.9 Jones classification for pneumatic conveying (Jones 1988).

material characteristic could help determine its possible dense phase pneumatic conveying flow mode.

Using the material data in Table 5.1, Williams *et al.* (2002) plotted the loose-poured bulk density vs. permeability, as can be seen in Figure 5.10. This shows a well-defined boundary (line X–X) between moving bed type flow and dilute phase only. The boundary between

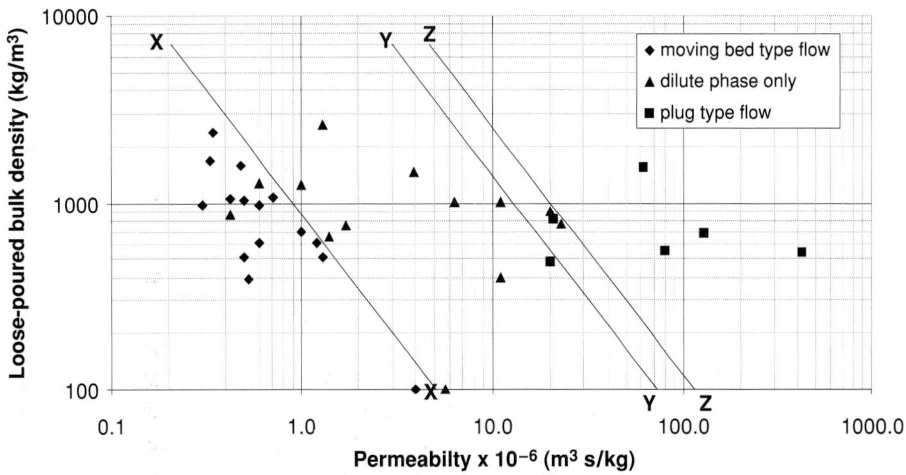

Figure 5.10 Flow-mode diagram for bulk solid materials in pneumatic conveying (Williams *et al.* 2002).

Table 5.2 Equations of lines detailed in Figure 5.10.

Line	Equation
X–X	$c\rho_{\text{blp}}^{3/4} = 160$
Y–Y	$c\rho_{\text{blp}}^{3/4} = 2300$
Z–Z	$c\rho_{\text{blp}}^{3/4} = 3600$

dilute phase only and plug type flow can also be defined somewhere between the lines Y–Y and Z–Z. The equations of these three lines are detailed in Table 5.2. The accuracy of this diagram in predicting pneumatic conveying performance with these materials is approximately 88.5%.

Clearly, there is no universally accepted method of predicting the conveying potential of a bulk solid. However, a range of techniques have been presented here, all of which are useful in obtaining a first indication of likely conveying capability. However, there is no alternative to pilot testing of the material to establish the conveying capability of a bulk solid for design purposes.

5.5 Experimental determination of conveying characteristics

The primary specification for any pneumatic conveying system is the conveying capacity or transfer rate. However, a feature of pneumatic conveying systems, which causes considerable confusion in the mind of users, is that there is no single set of operating conditions that will achieve the desired rate. In other words, there is always a choice to be made in terms of operating conditions. The starting-point for any system is the air-only pressure drop which represents the pressure drop required to drive a given air mass flow rate through the pipeline when no solids are conveyed. Any additional pressure drop available can then be used to convey material. A very powerful representation of the relationship between the air mass flow rate, the solids mass flow rate and the conveying line pressure drop for a given pipeline can be given in graphical form, which is termed the 'conveying characteristics'. This is essentially a 'performance map' for the material and will differ very significantly for different materials.

Conveying characteristics are plotted in two ways, as shown in Figures 5.11 and 5.12. The axes used in Figure 5.11 show clearly the air-only pressure drop as the curve representing a solids mass flow rate of zero. The family of curves represent a series of constant solids mass flow rates; increasing the solids flow rate obviously results in an increasing conveying line pressure drop requirement. An alternative presentation of the same information is provided in Figure 5.12. In this case, the vertical axis is the solids mass flow rate with the family of curves representing the conveying line pressure drop. An advantage of presenting the conveying characteristics in this second manner is that it is very easy to apply a second set of curves to the characteristics representing the solids loading ratio (m^*). The solids loading ratio is simply the ratio of the mass flow rate of solids to the mass flow rate of conveying gas:

$$m^* = \frac{m_s}{m_f}. \tag{5.3}$$

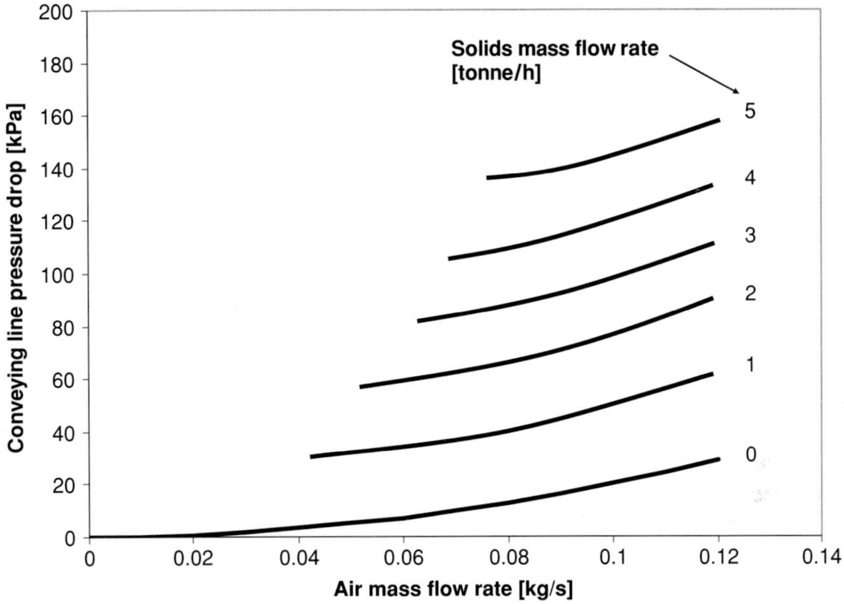

Figure 5.11 Pressure-drop analysis for potassium chloride.

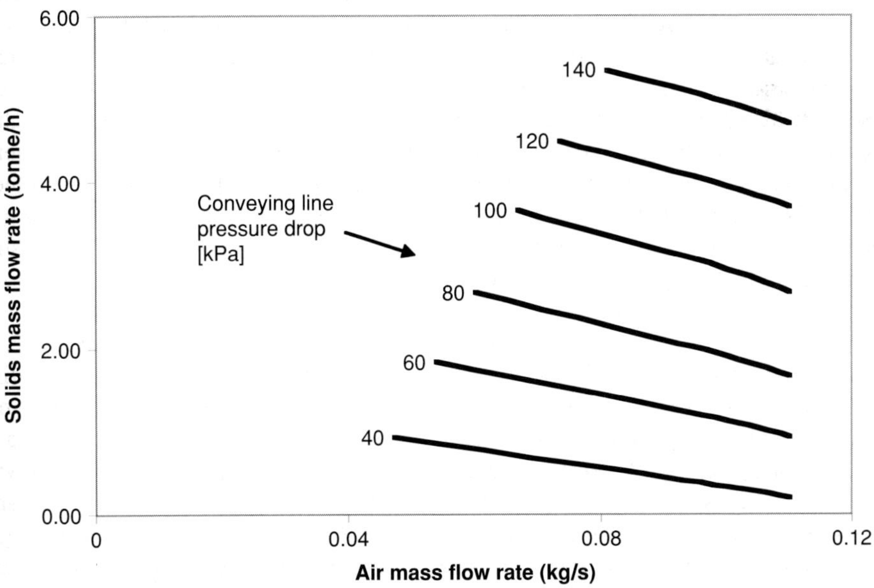

Figure 5.12 Conveying characteristics for potassium chloride.

This ratio is a useful parameter which gives some idea of the likely mode of conveying. In general, dilute phase conveying occurs at low solids loading ratios typically up to a maximum of 10–15. Solids loading ratios for fluidised dense phase can often exceed 100, while in slug/plug flow, solids loading ratios are typically in the range of 30–80.

To produce the conveying characteristics for a material, a series of pilot conveying trials are undertaken. Clearly, the resulting conveying characteristics will be particular to the bulk solid conveyed and the pipeline layout in which the tests were carried out. Provided the geometry of the pipeline is known, scaling is then possible for alternative pipeline geometries. However, scaling is not possible for alternative bulk solids. Even a change in the size distribution for the same bulk solid may significantly alter the conveying characteristics of the material and lead to a change in pneumatic conveying system performance.

It is important, when testing, to ensure that the widest range of conveying conditions are tested, and hence it is usual to undertake such tests using a blow tank system, as shown in Figure 5.13. Blow tank systems are batch systems, and hence a series of test runs are undertaken. It is the steady-state pipeline conditions which are of interest for the construction of the pipeline conveying characteristics.

Typical conveying test data are shown in Figure 5.14; a trace such as this will provide one data point (solids flow rate, air mass flow rate and conveying line pressure drop) on the conveying characteristic. The air mass flow rate is normally set for a test using critical flow nozzles; the conveying line pressure drop and mass collected in the receiving vessel are recorded to provide a single data point on the conveying characteristic. The number of data points used to develop the conveying characteristic depends on the conveying capability of the bulk solid. For a bulk solid that has only dilute phase capability, 15–20 data points are usually sufficient. For materials capable of dense phase conveying, 20–50 data points are usually obtained, depending on the mode of flow. The data points are plotted on a graph of solids mass flow rate against air mass flow rate with each plotted point annotated with the conveying line pressure drop. The family of curves representing a series of conveying line pressure drop contours are interpolated from the data.

The conveying characteristics presented in Figure 5.12 were obtained with a coarse grade of potassium chloride with a mean particle size of 384 μm. The pipeline layout consisted of 50 m of horizontal pipeline with nine long radius bends, each with a diameter of bend to pipe bore ratio (D/d) of 24 and a pipeline internal bore of 53 mm. This material has a wide size distribution, with 95% of the material by mass falling in the range of 100–1100 μm. The size range of this material leads to a relatively poor air permeability and air-retention capabilities and experimental testing of the material demonstrated that the material was not capable of dense phase flow in a conventional pipeline.

The limits of the conveying characteristic are imposed partly by the conveying hardware and partly by the bulk solid being conveyed. The limit to the right-hand side of the characteristic is imposed by the volumetric capacity of the air mover. Clearly, increasing the air flow available would extend the characteristic to the right. However, there is little advantage in this, since it is clear that the pressure contours have a negative slope. Increasing the air flow rate would reduce the solids mass flow rate capacity for a given conveying line pressure drop.

The limit of the characteristic in the vertical direction is based on a combination of the maximum pressure available from the air mover and the capacity of the feeder to provide the solids mass flow rate. The limit to the left of the characteristic is based on the characteristics

Figure 5.13 Pilot test plant using a blow tank as a feeder.

of the material to be conveyed. This limit is referred to as the minimum transport velocity which is clearly a critical parameter with regard to system design. Attempting to convey material in the area to the left of this conveying limit will result in pipeline blockage.

Figure 5.15 shows the conveying characteristics for the potassium chloride but with lines of constant solids loading ratio superimposed. It can be seen that over the range of the conveying envelope, the solids loading ratio reached maximum values of only 15. In addition, the limit of conveying shown on the conveying characteristic equates to a minimum velocity of approximately 15 m/s. These are typical limits for dilute phase conveying.

Not all materials are limited to such low values of solids loading ratio and such high values of minimum transport velocity. The conveying characteristics for pulverised fuel ash (PFA) conveyed in the same pipeline as the potassium chloride are shown in Figure 5.16.

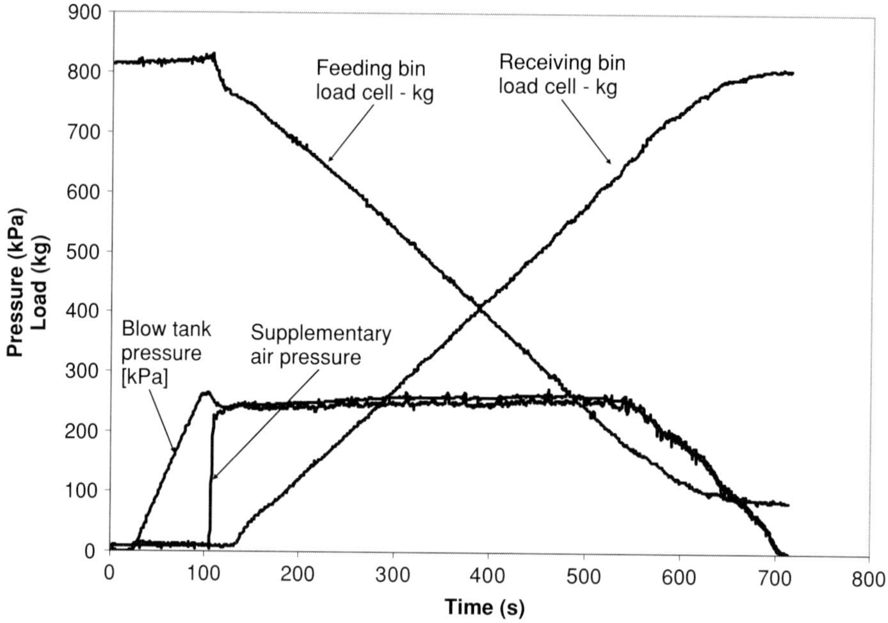

Figure 5.14 Typical conveying cycle test data.

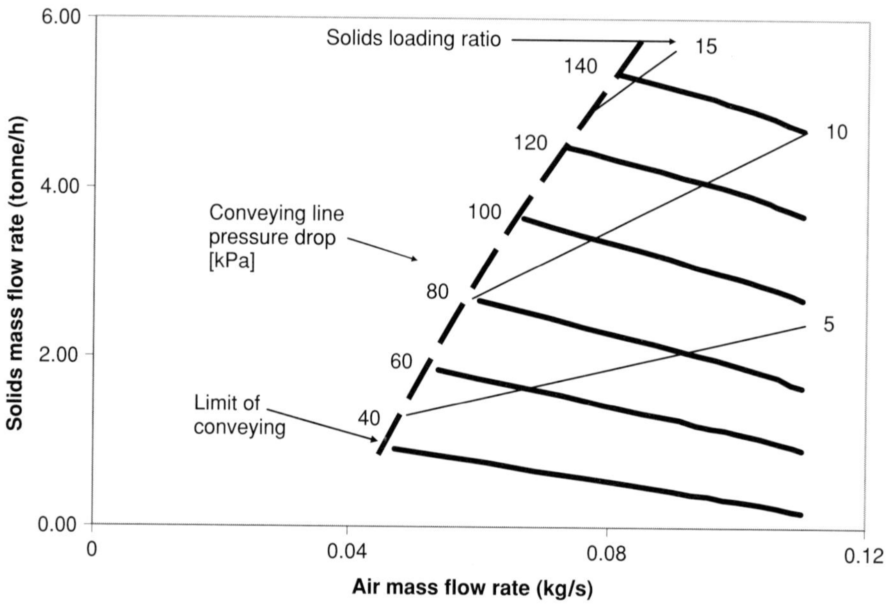

Figure 5.15 Conveying characteristics for potassium chloride with conveying limit superimposed.

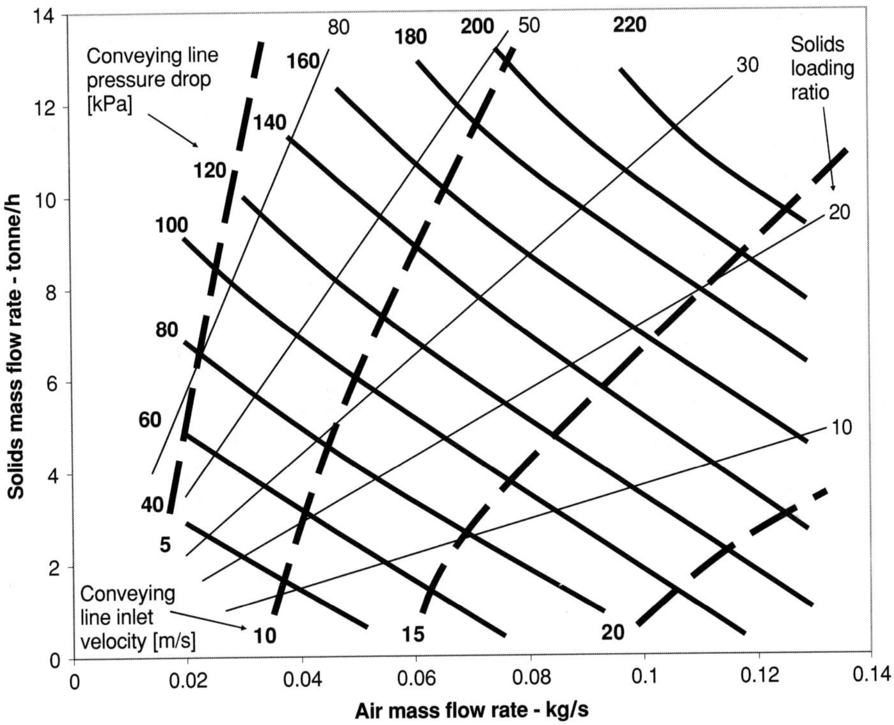

Figure 5.16 Conveying characteristics for pulverised fuel ash.

In this case, solids loading ratios in excess of 80 have been achieved, and the minimum transport velocity is below 5 m/s. In the case of the PFA, the mean particle size was 25 μm with 95% of the material by mass between 5 and 120 μm. This ash exhibits excellent air retention properties and was found to be conveyed very successfully in a fluidised dense phase flow regime.

The conveying characteristics produced from pilot scale trials are particularly useful in providing a visual representation of the conveying envelope and are extremely powerful in terms of the information which can be extracted from them. For example, if the conveying requirement is for 6 tonne/h of PFA, this could be achieved with a variety of air flow rate and pressure drop combinations. From Figure 5.16, it can be seen that this could be achieved using an air mass flow rate of 0.025 kg/s and a pressure drop of about 80 kPa or with an air mass flow rate of 0.1 kg/s and a pressure drop of 150 kPa or with a range of combinations between these two extremes. From this information, we can use the ideal gas law to establish the superficial air velocities at the feed point and at the exit of the pipeline.

$$pV = m_f RT. \tag{5.4}$$

The volumetric flow rate, V, can be written as:

$$V = U_f A = \frac{U_f \pi D^2}{4}, \tag{5.5}$$

where U_f is the velocity of the fluid, and A is the cross-sectional area of the conveying duct. Hence, the superficial velocity is obtained by substituting Equation (5.5) into Equation (5.4) thus:

$$U_f = \frac{4\, m_f RT}{p\pi D^2}. \tag{5.6}$$

Equation (5.6) can be used at any point in the pipeline provided that the velocity, U_f, is based on the static pressure, p, at that point. For pneumatic conveying, either the inlet or exit conditions would typically be used.

However, a set of characteristics are particular to both the material conveyed in the pilot trials and the pipeline layout in which the tests were carried out. Clearly, for the conveying data to be useful, scaling techniques are essential. While it is possible to scale with a reasonable degree of confidence from one pipeline layout to another, scaling from one bulk solid to another is not possible. Even a different grade of the same bulk solid can often lead to significant variations in system performance. To illustrate this point, Figure 5.17 shows the 150 kPa conveying line pressure drop curves for a series of materials all conveyed in the same pipeline. It is clear that the solids flow rate for a given air flow rate and pressure drop can vary very significantly particularly for dense phase conveying. This illustration highlights the folly of basing system design on experience of handling a different bulk solid.

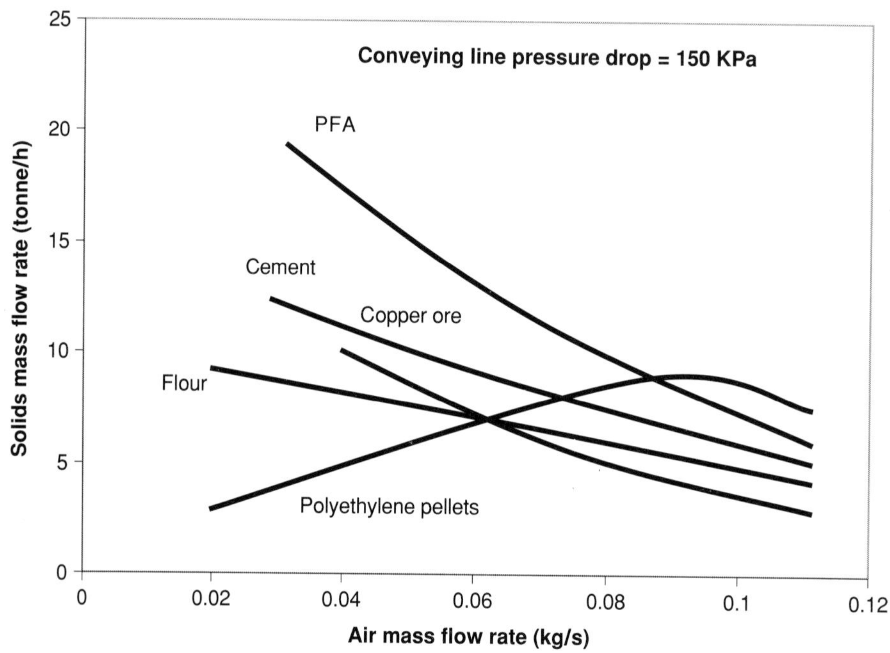

Figure 5.17 Comparison of conveying potential for a range of bulk solids.

5.5.1 *Scaling parameters*

Scaling rules were developed by Mills (Mills *et al.* 1982; Mills 1990) for use in scaling pneumatic conveying characteristics. Two rules were proposed: one for conveying distance and one for pipeline cross-sectional area. The solids mass flow rate with respect to conveying distance can be scaled up with a reasonable degree of accuracy using a reciprocal law, provided extrapolation is not too great, as follows:

$$m_s \propto \frac{1}{L}, \tag{5.7}$$

where m_s is the solids mass flow rate, and L is the conveying distance. An important condition is that the air mass flow rate and the conveying line pressure drop remain constant, hence:

$$m_{s1}L_1 = m_{s2}L_2, \tag{5.8}$$

where subscripts 1 and 2 refer to the appropriate conveying distance. Ideally, the conveying distance should be thought of as the equivalent length. However, it should be noted that the equivalent length per bend in pneumatic conveying is significantly greater than that experienced in single phase flow. For dilute phase conveying, in particular, the equivalent length can be up to 15 m per bend.

The scale-up of solids mass flow rate with respect to pipe bore suggested by Mills is as follows:

$$m_p \propto A \propto D^2 \tag{5.9}$$

and

$$m_a \propto A \propto D^2, \tag{5.10}$$

where D is the pipeline bore, and m_a is the air mass flow rate. It is important that these scaling rules are based on the conveying line pressure drop remaining constant. By scaling the air mass flow rate as well as the solids flow rate, the velocity conditions in the pipeline are maintained in both the test and scaled pipelines.

Experience has shown that these scaling parameters work well provided that extrapolation is limited to scale factors of about 20. Later work by Wypych and Arnold (1987) suggests that, when scaling to larger bore pipelines, these scaling rules are too conservative for practical design. They suggested that the exponent could be raised to 2.2, and in the case of fine, air retentive powders, this exponent could be raised as high as 2.8.

5.6 Approximate models for dilute phase conveying

One of the most important design parameters in any pneumatic conveying system is the conveying line pressure drop. This is clearly a major system design parameter which influences the overall performance of the conveying system. For a given pipe size and air mass flow rate, the solids transfer rate will be directly dependent on the pressure drop available. Unfortunately, the exact relationship between solids transfer rate and pressure drop is largely dependent on the properties of the material being conveyed. However, for dilute phase conveying, we are able to estimate the pressure loss/transfer rate based on some basic empirical relationships. While this is not suitable as the only basis for design of a system, these 'rules of thumb' allow a reasonable estimate of pressure drop to be obtained. This can be used either to obtain a rough estimate of the likely capability of a proposed system or to troubleshoot an existing system.

First of all, it is important to establish our definition of a dilute phase system. The most widely accepted definition of dilute phase is that the majority of the material being conveyed is in suspension in the conveying gas. In practice, most dilute phase systems operate with a solids loading ratio of up to 10. The solids loading ratio is simply the ratio of the mass flow rate of solids to the mass flow rate of conveying gas.

The basis of the approach is to assume that the overall pressure drop is based on two components: the pressure drop due to the gas alone and the additional pressure drop due to the presence of the solids, hence:

$$\Delta p_{\text{conveying line}} = \Delta p_{\text{gas}} + \Delta p_{\text{solids}}. \tag{5.11}$$

A simple approach is to assume that the extra pressure drop due to the solids is a function of the gas pressure drop; hence, Equation (5.11) can be written thus:

$$\Delta p_{\text{conveying line}} = \Delta p_{\text{gas}}(1 + \alpha). \tag{5.12}$$

There are many different empirical correlations for α based on a wide variety of material properties. However, the simplest form of α which has been shown to provide reasonable results with a wide range of materials is where α is equated to the solids loading ratio, m^*, hence:

$$\Delta p_{\text{conveying line}} = \Delta p_{\text{gas}}(1 + m^*), \text{ where } m^* = m_s/m_f \tag{5.13}$$

and

$$\Delta p_g = \frac{\lambda L_{eq}\, \rho_{av}\, U_{av}^2}{2D}. \tag{5.14}$$

Equation (5.14) is the Darcy equation, which is for non-compressible isothermal flow conditions. However, provided average velocity and density conditions are used, and the Mach number is kept below 0.2, the equation is found to give a good agreement with compressible flow theory, which is rather more complicated to use. This very simple method of estimating the conveying line pressure drop works well for the dilute phase, but it should be pointed out that no material properties are accounted for in the method, and so this can only be used as a first approximation. An example best illustrates the use of this method:
Pipeline data:

(1) pipeline diameter, $D = 100$ mm
(2) horizontal pipeline length, $L_h = 80$ m
(3) vertically up pipeline length, $L_v = 20$ m
(4) number of bends, $N = 5$
(5) bend factor, $K = 0.75$ (standard radius bend)

Assumed inputs:

(1) solids loading ratio, $m^* = 5$
(2) average velocity, $U_{av} = 24$ (typical of dilute phase systems)

(3) average gas density, $\rho_{av} = 1.7\ kg/m^3$
(4) assumed gas friction factor, $\lambda = 0.02$ (from Moody chart – typical value for pneumatic conveying)
(5) inlet velocity (or pick-up velocity), $U_i = 18$ m/s

Step 1: Determine the equivalent length of the pipeline, L_e

$$L_e = L_h + 2L_v + \frac{KDN}{\lambda}$$

$$L_e = 80 + 2(20) + \frac{0.75(0.1)(5)}{0.02}$$

$$L_e = 138.75\,\text{m}.$$

Step 2: Determine the air only pressure drop, Δp_g

$$\Delta p_g = \frac{\lambda L_{eq} \rho_{av} U_{av}^2}{2D}$$

$$\Delta p_g = \frac{0.02(138.75)(1.7)(24)^2}{2(0.1)}$$

$$\Delta p_g = 13.6\,\text{kPa}.$$

Step 3: Determine the conveying line pressure drop

$$\begin{aligned}\Delta p_{\text{conveying line}} &= \Delta p_{\text{gas}}(1 + m^*)\\ &= 13.6(1+5)\\ &= 82\,\text{kPa}.\end{aligned}$$

Average values of the gas density and velocity are used in this procedure which themselves depend on the overall pressure drop in the system. Hence, the calculation is iterative, which requires that the answers be checked to ensure that the initial inputs are reasonable.

Check on average density:

$$\rho_{av} = \frac{\rho_i + \rho_o}{2},$$

where $\rho_o = 1.2\ \text{kg/m}^3$ at standard pressure and 20°C and

$$\rho_i = \frac{p_i}{p_o}\rho_o = \frac{(101 + 82)}{101} 1.2 = 2.17\ \text{kg/m}^3,$$

hence

$$\rho_{av} = \frac{1.2 + 2.17}{2} = 1.69\ \text{kg/m}^3.$$

This compares with the input value of 1.7 kg/m^3. Note that the subscripts i and o refer to the inlet and outlet of the pipeline.

Check on average velocity:

$$U_{av} = \frac{\rho_i}{\rho_{cv}} U_i \text{ from continuity}$$

$$U_{av} = \frac{2.17}{1.7} 18 = 22.9\ \text{m/s}.$$

This compares with a velocity input of 24 m/s.

This algorithm could simply be put into a spreadsheet such that a series of iterations could be performed until the input and output converge. However, this simple algorithm can be calculated manually to obtain a rough approximation. It provides a very quick method of obtaining a first approximation to conveying line pressure drop in dilute phase conveying. However, it should be noted that this technique does not adequately take into account the material properties and hence can only be used to provide approximate figures for initial design or troubleshooting of existing systems.

A development of this approach is to add an additional solid friction factor term to the Darcy equation (Weber 1982; Chambers & Marcus 1986) such that the friction loss term in Equation (5.14) becomes:

$$\Delta p = \frac{(\lambda_f + m^* \lambda_s) L \rho_{av} U_{av}^2}{2D} \tag{5.15}$$

where λ_f is the fluid friction factor; λ_s is the solids friction factor; and m^* is the solids loading ratio.

Again, the calculation procedure is based on the average density and velocity conditions. Additional pressure drop terms are also used with this method to account for initial acceleration of the material, the bend effects and a potential energy term to account for changes in elevation. These terms are as follows:

Acceleration pressure drop:

$$\Delta p_{accel} = \left(1 + 2m^* \frac{U_s}{U_f}\right) \frac{\rho_i U_i^2}{2}. \tag{5.16}$$

Bend pressure loss:

$$\Delta p_b = NK(1 + m^*) \frac{\rho_{av} U_{av}^2}{2}. \tag{5.17}$$

Potential energy loss:

$$\Delta p_z = m^* \rho_{av} g L_v \frac{U_f}{U_s}. \tag{5.18}$$

However, the key to this method is the solids friction factor, which is used to evaluate the overall conveying line pressure drop. Many correlations exist; however, one of the more recent correlations that have been shown to give good agreement in practice is that of Weber (1982). Two correlations are proposed depending on the mean particle size. These are as follows:

$$\lambda_s = \frac{2.1 Fr_s^{0.25} (D/d)^{0.1}}{(m^*)^{0.3} Fr} \quad \text{for } d < 0.5 \text{ mm} \tag{5.19}$$

$$\lambda_s = \frac{0.082 Fr_s^{0.25} (D/d)^{0.1}}{(m^*)^{0.3} Fr^{0.86}} \quad \text{for } d > 0.5 \text{ mm}, \tag{5.20}$$

where

$$Fr_s = \frac{w_{fo}^2}{gd} \tag{5.21}$$

and

$$Fr = \frac{U_{av}^2}{gD} \tag{5.22}$$

$$w_{fo} = \sqrt{\frac{4gd(\rho_p - \rho_g)}{3C_d\rho_E}} \tag{5.23}$$

$$\frac{U_s}{U_f} = 1 - 0.008\, d_i^{0.3} \rho_b^{0.5} \tag{5.24}$$

The coefficient of drag, C_d, used in Equation (5.23) is based on the particle Reynolds number, $\mathrm{Re_d}$, and is given in Equations (5.25) to (5.27) for various ranges of particle Reynold number.

$$C_d = (24/\mathrm{Re_d})(1 + 3\,\mathrm{Re_d}/16) : \mathrm{Re_d} < 1.0 \tag{5.25}$$

$$C_d = 0.4 + 26/\mathrm{Re_d^{0.8}} : 1 < \mathrm{Re_d} < 10^4 \tag{5.26}$$

$$C_d = 0.4 : \ 10^4 < \mathrm{Re_d} < 10^5 \tag{5.27}$$

$$\mathrm{Re_d} = \frac{\rho_g w_{fo} d}{\mu}. \tag{5.28}$$

5.7 Pressure-drop prediction for fluidised dense phase conveying

The Weber (1981) correlation for the solids friction factor is only valid for dilute phase for solids loading ratios up to about 10. However, the analysis can also be used with experimental data to 'back calculate' the solids friction factor and to establish the relationship between the solids loading ratio, m^*, solids friction factor, λ_s, and the Froude number, Fr (Chambers & Marcus 1986). The relationship developed from the back calculation can then be used to estimate the pressure loss in another pipeline layout. The form of the relationship is normally as follows:

$$\lambda_s(m^*)^a = bFr. \tag{5.29}$$

Establishing the constants a and b is essentially a 'trial and error' process. Data for a particular grade of pulverised fuel ash are plotted in Figure 5.18 for three values of the m^* exponent; in this case, an exponent of 0.5 provides the best correlation.

This approach has been found to provide good pressure drop predictions for a wide range of m^* values. Chambers and Marcus (1986) report that this technique has been used for materials of mean particle size between 16 μm and 3 mm; mass flow ratios (m^*) from 2 to 530 and conveying distances between 25 and 1600 m. However, the authors conjectured that within the model, there are a number of compensating factors which lead to its success. It is also recognised that the use of this technique to compare individual loss terms would not be as successful as for the total pressure loss.

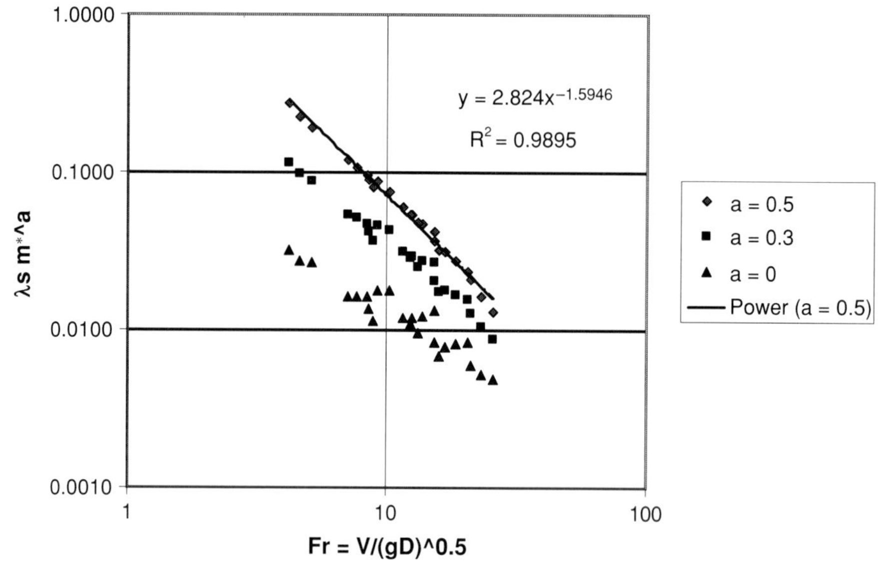

Figure 5.18 Correlation for solids friction factor and solids loading ratio.

5.8 Pressure-drop prediction for low-velocity slug flow

The conveying of materials in a low-velocity slug flow mode has particular attractions for granular materials with a narrow size distribution. Typical of this class of materials are products such as polyethylene pellets (normally about 4 mm in size) or various forms of grain. These materials are free-flowing, generally have a high degree of permeability to the conveying gas flow and are often susceptible to particle breakage. However, this form of dense phase conveying is used rather less than fluidised dense phase in industrial applications. This may be due to the relatively limited solids flow rates that can be achieved.

Pan and Wypych (1997) developed a useful model to predict the pressure loss in a slug flow system. This involves the use of established particle mechanics combined with the work of Ergun (1952) related to the flow of gas through fixed beds of particles. The method involves an analysis of the forces acting on a moving slug of material in a pipeline due to wall friction and air pressure. From this process, an equation is derived to calculate the pressure gradient across the moving slug of material as follows:

$$\frac{\Delta p}{L} = 2.168\rho_b\mu_w k_w \left(\frac{g}{D}\right)^{0.5} U_p + \rho_b g \mu_w(2 - 1.175k_w). \tag{5.30}$$

This expression is based on a number of assumptions. It assumes that the radial and axial stresses are functions of the horizontal direction only and that the materials exhibit no cohesion. Hence, the pressure gradient is assumed to be constant throughout the slug. This equation contains two unknowns, the pressure gradient and the plug velocity, and therefore a second equation is necessary. Ergun (1952) developed an equation for pressure gradient through a packed bed of bulk solids of the form:

$$\frac{\Delta p}{L} = aU_{\text{rel}} + b\rho_{\text{f}}U_{\text{rel}}^2, \tag{5.31}$$

where U_{rel} is the relative velocity between the gas and the plug

$$U_{\text{rel}} = U_{\text{f}} - U_{\text{p}} \tag{5.32}$$

and

$$a = \frac{150\mu(1-\varepsilon)^2}{d_{\text{p}}^2\varepsilon^3} \quad \text{and} \quad b = \frac{1.75(1-\varepsilon)}{d_{\text{p}}\varepsilon^3}$$

By substituting Equation (5.32) into Equation (5.31), the following expression is derived:

$$\frac{\Delta p}{L} = b\rho_{\text{f}}U_{\text{p}}^2 - (a + 2b\rho_{\text{f}}U_{\text{f}})U_{\text{p}} + aU_{\text{f}} + b\rho_{\text{f}}U_{\text{f}}^2. \tag{5.33}$$

Hence, using Equation (5.30) and Equation (5.33), the pressure gradient, $\Delta p/L$, and the plug velocity, U_{p}, for cohesionless materials can be determined from the following quadratic expression:

$$A_1U_{\text{p}}^2 + A_2U_{\text{p}} + A_3 = 0, \tag{5.34}$$

where

$$A_1 = b\rho_{\text{f}}$$

$$A_2 = -\left[a + 2b\rho_{\text{f}}U_{\text{f}} + 2.168\rho_{\text{b}}\mu_{\text{w}}k_{\text{w}}\left(\frac{g}{D}\right)^{0.5}\right]$$

$$A_3 = aU_{\text{f}} + b\rho_{\text{f}}U_{\text{f}}^2 - \rho_{\text{b}}g(2 - 1.175\mu_{\text{w}}k_{\text{w}}).$$

Since U_{f}–U_{p} is greater than zero:

$$U_{\text{p}} = \frac{-A_2 - \left(A_2^2 - 4A_1A_3\right)^{0.5}}{2A_1}. \tag{5.35}$$

However, in order to use these equations, there is a need to determine the constants a and b. From practical tests, it has been found that there is a good linear relationship between the ratio of the pressure gradient to the mean superficial air velocity and the air mass flow rate. Hence, an alternative form of the Ergun equation can be stated thus:

$$\frac{\Delta p}{LU_{\text{f}}} = a + b\frac{m_{\text{f}}}{A}. \tag{5.36}$$

The constants a and b are derived from a simple permeability test undertaken using a fluidising column. However, to obtain values of a and b in the range of superficial gas velocities anticipated in an aerated plug, the material sample to be tested is restrained in the test apparatus, since the superficial air velocity will be above the minimum fluidising velocity.

Mi *et al.* (1994) found that the sum of the pressure drops across each of the individual plugs provides a reasonable estimate for the overall pressure drop provided that average conditions (based on the average gas density) are used. Therefore, the total pressure drop is given by:

$$\Delta p = \frac{\Delta p}{L} L_s, \tag{5.37}$$

where L_s is the sum of the individual slug lengths.

Clearly, L_s is a transient parameter which will vary with conveying conditions in the pipeline. Konrad *et al.* (1980) estimated the total length of slugs using the following correlation:

$$L_s = \frac{m_s L}{A(1-\alpha)\rho_b U_p}, \tag{5.38}$$

where

$$\alpha = \frac{A_{st}}{A} = \frac{0.542(gD)^{0.5}}{U_p}.$$

This method has been used on a variety of materials and has shown some encouraging results. However, it is important to acknowledge that the results of such analyses can only be used as a guide to conveying performance.

References

Chambers, A.J. & Marcus, R.D. (1986) Pneumatic conveying calculations. In: *2nd International Conference on Bulk Materials Storage, Handling and Transportation*, Wollongong, 7–9 July.

Dixon, G. (1979) The impact of powder properties on dense phase flow. In: *Proceedings of the International Conference on Pneumatic Conveying*, London.

Doig, I.D. (1975) Design and performance aspects of dense phase pneumatic conveying systems. *South Afr. Mech. Eng.*, **25**, 394–403.

Ergun, S. (1952) Fluid flow through packed columns. *Chem. Eng. Prog.*, **48**, 89–94.

Fargette, C (1998) *Bench scale tests for the assessment of pneumatic conveying behaviour of powders*. M.Phil. thesis, Glasgow Caledonian University, UK.

Geldart, D. (1973) Types of gas fluidisation. *Powder Technol.*, 7, 285–92.

Grace, J.R. (1986) Contacting modes of behaviour classification of gas–solid and other two-phase suspensions. *Can. J. Chem. Eng.*, **64**, 353–63.

Heiskanen, K. (1993) *Particle Classification*. Chapman & Hall, London.

Hong, J., Shen, Y. & Tomita, Y. (1995) Phase diagrams in dense phase pneumatic transport. *Powder Technol.*, **84**, 213–19.

Jones, M.G. (1988) The Influence of Bulk Material Properties on Pneumatic Conveying Performance. Ph.D. thesis, Thames Polytechnic, London.

Jones, M.G. & Mills, D. (1990) Product classification for pneumatic conveying. *Powder Handling Process.*, **2**, 117–22.

Konrad, K., Harrison, D., Nedderman, R.M. & Davidson, J.F. (1980) Prediction of the pressure drop for horizontal dense phase pneumatic conveying of particles. In: *Proceedings of the 5th International Conference on Pneumatic Transport of Solids in Pipes (Pneumotransport 5)*, London, UK, pp. 225–44. BHRA Fluid Engineering, Cranfield, UK.

Mainwaring, N.J. & Reed, A.R. (1987) Permeability and air retention characteristics of bulk solid materials in relation to modes of dense phase pneumatic conveying. *Bulk Solids Handling*, **7**, 415–25.

Mi, B., Wypych, P.W. & Pan, R. (1995) Low velocity pneumatic conveying of powders. Mech. Eng. Trans. IEAust, ME20, **3**, 187–91.

Mills, D. (1990) *Pneumatic Conveying Design Guide*. Butterworths, London.

Mills, D., Mason, J.S. & Stacey, R.B. (1982) *Solidex 82 Conference*, Harrogate, UK, pp. C1–C75.

Molerus, O. (1982) Interpretation of Geldart's Type A, B, C and D Powders taking into account interparticle cohesion force. *Powder Technol.*, **33**, 81–87.

Pan, R. (1999) Material properties and flow modes in pneumatic conveying. *Powder Technol.*, **104**, 157–63.
Pan, R. & Wypych, P.W. (1997) Pressure drop and slug velocity in low-velocity pneumatic conveying of bulk solids. *Powder Technol.*, **94**, 123–32.
Weber, M. (1981) Principles of hydraulic and pneumatic conveying in pipes. *Bulk Solids Handling*, **1**, 57–63.
Weber, M. (1982) Correlation analysis in the design of pneumatic transport plant. *Bulk Solids Handling*, **2**, 231–33.
Williams, K.C., Jones, M.G. & Pan, R. (2002) Bulk material classifications for the design of pneumatic conveying systems. In: *Proceedings of Hydrotransport 15*, pp. 417–30. BHR Group, UK.
Wypych, P.W. & Arnold, P.C. (1987) On improving scale-up procedures for pnuematic conveying. *Powder Technol.*, **50**, 281–94.

Nomenclature

A	Cross-sectional area (m^2)
A_f	De-aeration factor (Pa.s/m)
C	Permeability factor (m/Pa.s $\times 10^{-6}$)
C_d	Drag coefficient (–)
d	Particle diameter (or size) (m)
D	Pipeline internal diameter (m)
d_i	Mean particle diameter (or size) (mm)
d_p	Median particle diameter (m $\times 10^{-6}$)
Fr	Froude number
g	Acceleration due to gravity (9.81 m/s^2)
K	Bend head loss factor (–)
k_w	Ratio of radial stress to lateral stress (–)
L	Length (m)
L_{eq}	Equivalent length (m)
L_s	Sum of individual slug lengths (m)
m^*	Solids loading ratio or mass concentration (m_s/m_f)
m_f	Mass flow rate of gas (kg/s)
m_s	Mass flow rate of solids (kg/s)
N	Number of bends (–)
p	Static pressure (Pa)
P_f	Permeability factor (m^2/Pa.s)
R	Characteristic gas constant – 287 J/kg.K for air
Re	Reynolds number in pipeline ($\rho D U_f/\mu$)
Re_d	Particle Reynolds number ($\rho d w_{fo}/\mu$)
T	Absolute temperature (K)
t_d	De-aeration time
U_{av}	Average superficial gas velocity (m/s)
U_f	Fluid (or gas) velocity (m/s)
U_f	Fluid (or gas) velocity (m/s)
U_g	Gas (or fluid) velocity (m/s)
U_p	Plug velocity (m/s)
U_{rel}	Relative velocity between plug and gas (m/s)
U_s	Solids velocity (m/s)
V	Volumetric flow rate (m^3/s)
v_{mb}	Minimum bubbling velocity (m/s)

v_{mf} Minimum fluidisation velocity (m/s)
w_{fo} Terminal velocity of particle (m/s)

Greek symbols
ε Bulk voidage, $\varepsilon = 1 - (\rho_b/\rho_s)$
ρ_{av} Average gas (or fluid) density (kg/m^3)
ρ_b Bulk density (kg/m^3)
ρ_{blp} Loose-poured bulk density (kg/m^3)
ΔP Pressure drop (Pa)
λ Fluid friction factor (–)
λ_f Fluid friction factor (–)
λ_s Solids friction factor (–)
μ Gas viscosity (Pa.s)
μ_w Wall friction coefficient (–)
ρ_g Gas (or fluid) density (kg/m^3)
ρ_p Particle density (kg/m^3)

6 Characterisation of explosibility

GEOFF LUNN

6.1 Introduction

A flammable solid such as a lump of coal burns in air relatively slowly because the area exposed to the oxygen in the air is not great compared with the volume of the solid. When the same solid is ground into a powder, however, this area is increased substantially, and the rate of combustion can consequently become very fast. If this powder is dispersed in air, or any other atmosphere that will support a reaction with it, the cloud formed is potentially explosive in the same way that a mixture of air and propane is. If the cloud is confined, an explosion can generate destructive pressures.

In the process industries, powder handling is used extensively in a range of activities, including food and animal feed production, woodworking, chemicals and pharmaceuticals processing, coal and metal powder treatment. Eighty per cent of the powders used in industry are flammable, and fire and explosion hazards arise whenever and wherever these powders give rise to flammable dusts.

Dust explosions can be comparable in their effects to blasts of explosives (Uth 1999), and can cause injury and loss of life, destruction of infrastructure and damage to the environment. People close to an explosion are under obvious danger from blast waves, debris, flame and thermal radiation. Escalation of the event through secondary explosions may instigate fires or the release of dangerous substances that may pose a threat over a wide area. Such events have been classified as Major Accidents, meaning a serious hazard to people and the environment, and some as Major Accidents with serious effects, meaning releases of 1200 kg or more (Uth 1999).

Some dust explosions cause little damage. Figure 6.1 shows an explosion in a dust collector that has been safely vented into the open air. The overpressure generated in this example is no more than a fraction of a bar, and the housing of the collector has suffered no damage. Some explosions, however, can travel for long distances, generating overpressures of several bar and reaching speeds of several hundred metres per second. As it moves along, the explosion causes movement in the air ahead of it that is sufficient to lift dust into the air. The explosion then propagates into a continually created dust cloud, accelerating and producing ever-higher pressures as it goes. This demonstrates an important difference between flammable dusts and flammable vapours: vapours can diffuse away, but dusts settle out. In many industrial incidents, a relatively mild explosion, the primary explosion, has developed into a very destructive secondary explosion by involving the dust that has accumulated in the plant.

One example of a large incident was the explosion at a grain-storage complex in Blaye, France, in 1997, in which 11 people were killed. This explosion probably started in the dust collection system associated with 40 linked grain storage silos. It spread from the collector into the building and travelled along the gallery above the silos, into many of the cells and

Figure 6.1 Vented explosion in a dust collector. Reproduced with the permission of DCE Donaldson Ltd.

down the towers at either end of the building. Over 25 of the silos were destroyed, together with much of the rest of the complex.

It is often not realised how destructive a dust explosion can be, and every effort must be made either to prevent such explosions or to provide protection against their effects. The risk of an explosion arises when a flammable dust is dispersed, at a concentration that will support self-propagating combustion, into an atmosphere (usually air) that will support this combustion and when an effective ignition source is present.

In Europe, directives concerned with explosion risks in industry (The Machinery Directive 1998; The Potentially Explosive Atmospheres Directive 1994; Directive 1999/92/EC of the European Parliament and of the Council of 16 December 1999), and put into UK law by the Dangerous Substances and Explosive Atmospheres Regulations (DSEAR) (2002), have introduced a set of Essential Health and Safety Requirements (EHSRs) that in general terms give the precautions to be taken against the risk of an explosion and the order in which they should be applied:

(1) prevent the formation of explosive atmospheres either by avoiding a flammable cloud or by rendering the atmosphere in which the cloud is dispersed incapable of supporting combustion,
(2) prevent the ignition of explosive atmospheres by control of potential ignition sources,
(3) limit the range of flames and pressures to a sufficient level of safety or halt the explosion immediately.

The first two prevent an explosion, and the third protects against the effects of an explosion it is assumed will occur. There are a number of prevention and protection techniques, each of which requires knowledge of the relevant ignitability and explosibility characteristics of the dust. These characteristics are measured in tests that are often laid down in national or international standards.

Compendia of dust ignitability and explosibility characteristics are available (Beck *et al.* 1997). These can be used to obtain a value of a particular characteristic for a dust, but care is required to ensure that the value chosen is a worst-case example and not likely to be exceeded by the dust in the plant. These characteristics vary, sometimes substantially, with the dust, the dust composition, the particle size, the particle-size distribution, the moisture content and the presence of additives. It is not sufficient to obtain from published lists a value of a characteristic for sugar, say, and then assume it applies to every sugar dust.

It is best practice to take measurements. It is important, however, when conducting these tests, that the dust sample is relevant to the process conditions; it should be consistent with the finest dust in the process, but if this is not known, a sample passing a 63 μm sieve can be used. The sample must also be consistent with the driest material in the process, but if this is not known, the sample can be dried to constant weight at 50°C under vacuum or 75°C at atmospheric pressure.

6.2 Identification of flammable dusts

Not all dusts encountered in powder handling industries are capable of forming an explosive atmosphere if dispersed in air, but it is important to know which are. Although there is no British Standard that describes it, a sequence of tests is accepted in the UK as effectively identifying dusts that can form an explosive atmosphere in air under process conditions. A sequence of similar tests is used elsewhere.

If a dust is characterised as flammable in these tests, when they are operated in a manner that simulates the relevant plant/process conditions, then potentially explosive atmospheres are a hazard in the powder handling plant where this dust occurs. Such plant must satisfy European and UK requirements for ensuring that explosions are either prevented or the consequences minimised. These requirements can be found in the ATEX Directives (The Potentially Explosive Atmospheres Directive 1994; The Machinery Directive 1998; Directive 1999/92/EC of the European Parliament and of the Council of 16 December 1999) and the UK Dangerous Substances and Explosive Atmospheres Regulations (DSEAR) (Dangerous Substances and Explosive Atmosphere Regulations 2002).

A 1.2 litre vertical tube apparatus, similar to the Hartmann apparatus is used to identify flammable dusts at atmospheric pressure and room temperature. The apparatus used for this test in the UK is an open tube made of Pyrex, shown in Figure 6.2.

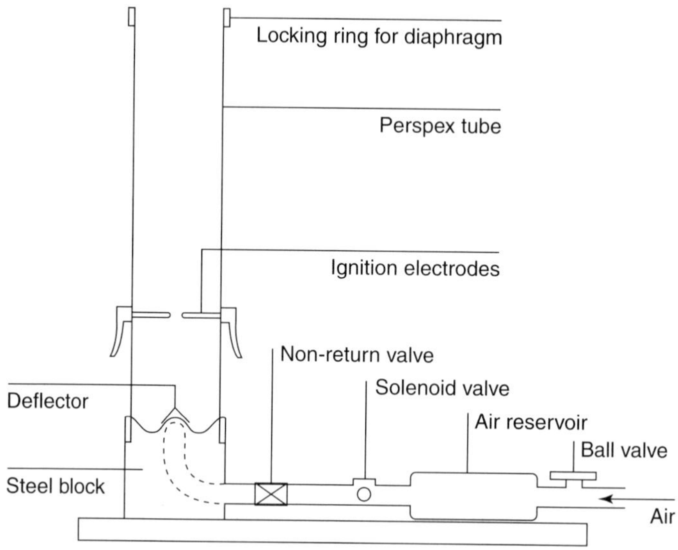

Figure 6.2 Vertical tube (1.2 l) dust flammability apparatus. Reproduced with permission.

A known quantity of dust is placed at the bottom of the tube in a specially designed receptacle and dispersed by a blast of air around either a spark or a hot coil. The amount of dust is varied so that a range of dust concentrations, across a credible Explosion Range, is covered in the tests. If any of the dust dispersions ignites, the dust is classified as Group A: flammable dust; otherwise the dust is classified as Group B: non-flammable. This test is sufficient to classify the dust under atmospheric conditions for small and medium ignition sources, but if the process involves temperatures above normal room temperature, pressures above atmospheric and large heat sources, eg. flames, the test is not a sufficient guide to a dust's potential flammability. Dusts classified as Group B: non-flammable, at atmospheric conditions, may be flammable at process conditions.

The Godbert–Greenwald furnace (BSEN 50281-2-1:1999) can be used to test whether a Group B dust is flammable when the surroundings are at an elevated temperature. This apparatus is designed for measurement of the Minimum Ignition Temperature of a dust cloud and its description is given in Section 6.3.3.1. The normal test procedure is a relatively onerous one when testing for flammability only, in that the dust is injected into a heated atmosphere surrounded by a large ignition source – the furnace wall.

More realistically, the furnace can be modified so that a localised ignition source – a continuous spark, for instance – is positioned inside the furnace tube. The air in the tube can be heated to the required temperature before the dust is dispersed. The likelihood of an ignition can then be tested at the process temperature. Accurate measurements of the temperature are necessary if the result is to be applied to known process conditions.

This sequence of tests and the Group A/Group B classification are rarely, if ever, used outside the UK. An alternative sequence used in other countries in Europe involves a Modified Hartmann Apparatus and either the 1 m^3 vessel or the 20 litre sphere. The Modified Hartmann Apparatus consists of the 1.2 litre vertical tube with a hinged cover added (Dust Explosion Test). The ignition source is either a continuous spark or a hot coil. The main series of tests is done with the continuous spark but if no ignition occurs, tests at the higher dust concentrations are repeated using the hot coil. The dust is identified as flammable at room conditions if either a visible flame is seen or the instrumentation indicates the hinged cover has opened. If tests in the Modified Hartmann Apparatus show no indication that the dust is flammable, further tests are carried out in either the 1 m^3 vessel or the 20 litre sphere test vessels. Alternatively, these vessels can be used to test for dust flammability without prior screening in the Modified Hartmann Apparatus.

The 1 m^3 vessel and 20 litre sphere are used primarily for measuring explosibility characteristics such as the Maximum Explosion Pressure and the Maximum Rate of Pressure Rise; they are described in Section 6.4.1.1. When the 1 m^3 vessel is used to test for dust flammability, the ignition source comprises two 5 kJ pyrotechnic igniters (Cesana & Siwek 2001b). If the explosion pressure, including the effect of the igniters, is less than 0.5 bar over concentrations across a credible Explosion Range, then the dust is considered to be non-flammable.

When the 20 litre sphere is used to test for dust flammability, the ignition source comprises two 1 kJ pyrotechnic igniters. If the explosion pressure, including the effect of the igniters, is less than 0.5 bar or, when corrected to take into account the effect of the igniters, is less than 0.2 bar, the dust is considered to be non-flammable. An explosion may occur, however, if a more energetic ignition source is used, and the test may need to be done with a higher energy source (10 kJ) in some circumstances (Cesana & Siwek 2001b). Some models of the 20 litre sphere can be operated at elevated temperatures and pressures and so dust flammability can be assessed at conditions other than atmospheric conditions in this test.

6.3 Explosion prevention

Explosion prevention is achieved either by avoiding an explosive atmosphere or by eliminating ignition sources.

6.3.1 *Avoidance of an explosive atmosphere*

One method of avoiding an explosive atmosphere is substitution of a flammable dust by a non-flammable dust. In most cases, however, this will be impractical. An alternative is to render the atmosphere non-explosive by limitation of either the concentration of the flammable substance or the concentration of oxygen.

6.3.1.1 *Limitation of the flammable concentration* A cloud of dust in air will propagate an explosion only if its concentration is within a certain range. If the flammable concentration is to be limited, then a measurement of the Explosion Limits and the Explosion Range is required. Definitions of the Explosion Range and the Explosion Limits are (BSEN 1127-1:1998):

(1) Explosion Range: range of the concentration of a flammable substance in air within which an explosion can occur,
(2) Lower Explosion Limit: the lower limit of the explosion range,
(3) Upper Explosion Limit: the upper limit of the explosion range.

The Lower and Upper Explosion Limits are, respectively, the concentrations of the dust below and above which the atmosphere is not explosive.

The Lower Explosion Limit is measured in the 20 litre test apparatus with two 5 kJ chemical igniters as the ignition source (Cesana & Siwek 2001b). The atmosphere is considered to be non-explosive when either the maximum explosion pressure is less than 2 bar or the corrected maximum explosion pressure is less than 0.4 bar. When the 1 m^3 vessel is used with two 5 kJ chemical igniters as the ignition source, the atmosphere is considered to be non-explosive when the maximum explosion pressure is less than 0.4 bar.

The Upper Explosion Limit of a dust is not normally measured owing to the difficulties in dispersing the large concentrations often required and because it has little practical relevance to explosion prevention – when dust settles out from a dust cloud with a high concentration, this concentration decreases and enters the Explosion Range. Some models of the 20 litre sphere can be operated at elevated temperatures and pressures, and so Explosion Limits can be measured at conditions other than room temperature and pressure.

An American Standard (American Society for Testing Materials 2000) specifies a 20 litre test vessel with a chemical igniter ignition source of either 2.5 or 5 kJ. The atmosphere is considered non-explosive if the maximum explosion pressure is less than 2.0 bar, and the K_{St} value is less than 1.5 bar m/s. The K_{St} value is a measure of the dust's explosibility and is discussed in Section 6.4.1.2. This latter criterion indicates that there has been some real propagation of the flame and that the pressure rise is not due simply to burning of dust in the igniter flame. It also compensates for possible overdriving effects in the 20 litre vessel due to the energy of the ignition source.

Explosion Limit measurements are used where the basis of safety is avoidance of explosive atmospheres by control of dust concentration. Avoidance of a potentially explosive atmosphere is theoretically possible in parts of a plant if the flammable concentration can be kept at all times outside the Explosion Range. The Explosion Limits are, however, measured in a homogeneous atmosphere, whereas, in practice, although the total amount of dust in a particular part of the plant might be insufficient to produce an explosive concentration overall, an uneven distribution can lead to local concentrations in the Explosion Range. If this prevention technique is to be applied, an accurate assessment of dust concentrations must be made and the risks of concentrations entering the Explosion Range quantified.

With few exceptions, Lower Explosion Limits for dusts are in excess of 50 g/m^3. This concentration is clearly visible and, unlike gases and vapours, a flammable dust cloud cannot go undetected. The Lower Explosion Limit of a dust is a fairly clean-cut value, as is the case with gases. The Upper Explosion Limit of a dust can be at a very high concentration and is often not clearly defined. This is because many dusts burn through the production of gaseous volatiles; the flame burns through the volatiles essentially as a gaseous phase combustion leaving partially reacted particles in its wake (Hertzberg & Cashdollar 1987). Only at high dust concentrations does the quenching effect of the extra particles have such an effect on the flame that the atmosphere becomes non-explosive, and the Upper Explosion Limit is reached, often at several kg/m^3.

Within the Explosion Range, there is an optimum dust concentration at which the dust explosibility – essentially a measure of the rate of combustion – is greatest. This concentration is often much higher than the nominal stoichiometric concentration of the dust at which the entire amount of dust is just sufficient to react with all the available oxygen. This is because of the way many dusts combust through the production of volatiles. As additional dust is introduced, the explosibility either remains the same as at the optimum concentration or declines relatively slowly, until the concentration is relatively high. Eventually, an Upper Explosion Limit is reached.

The Lower Explosion Limit decreases as temperature increases essentially linearly if the temperature is expressed on a log scale (Bartknecht 1981). The effect of initial pressure on the Lower Explosion Limit of a dust cannot be predicted. For some dusts, the Lower Explosion Limit may decrease as pressure increases (Bartknecht 1981), but for others, as measurements by Wiemann show for brown coal, the Lower Explosion Limit increases as the initial pressure increases (Wiemann 1987).

The dust concentration inside a powder handling plant can change quite drastically, and if the basis of safety is to keep concentrations below the Lower Explosion Limit, the propensity for dusts to settle out onto horizontal or near horizontal surfaces inside plant has to be taken into account. Dust layers are easy to disturb and can be dispersed by relatively gentle movements of air. Thus, parts of the plant where airborne concentrations are below the Lower Explosion Limit cannot be guaranteed to be safe if the concentration can be increased due to disturbances of deposited dust.

Very rich dust clouds with concentrations above the Upper Explosion Limit are unstable in that at start up, upon shut down and in an emergency, the dust concentration will fall through the Explosion Range as dust particles settle out. During these periods, a potentially explosive atmosphere will exist. If avoidance of explosive atmospheres is to be by control of dust concentration, care must be taken to ensure that a realistic assessment of likely dust concentrations is made.

6.3.1.2 *Limitation of the oxygen concentration* When an inert gas is added to air, the oxygen concentration decreases. Eventually, the oxygen content becomes so low that an explosion cannot be sustained. This is the basis of an explosion prevention technique known as Inerting, and its application requires knowledge of the Limiting Oxygen Concentration of the dust.

The Limiting Oxygen Concentration is defined as (BSEN 1127-1:1998) 'the maximum oxygen concentration in a mixture of a flammable substance and air and an inert gas, in which an explosion will not occur, determined under specified test conditions'. The Limiting Oxygen Concentration is measured in either the 1 m^3 vessel or the 20 litre sphere test apparatus (Cesana & Siwek 2001b). In the 1 m^3 vessel, the ignition source consists of two 5 kJ chemical igniters, and the atmosphere is considered non-explosive if the maximum explosion pressure is less than 0.3 bar. In the 20 litre vessel, the ignition source consists of two 1 kJ chemical igniters, and the atmosphere is considered non-explosive if either the maximum explosion pressure is less than 0.5 bar or the corrected maximum explosion pressure is less than 0.2 bar.

Some models of the 20 litre sphere can be operated at elevated temperatures and pressures, and so Limiting Oxygen Concentrations can be measured at conditions other than room temperature and pressure. Alternative test equipment can be used if conformity with the standard apparatus can be demonstrated according to the procedure in the standard.

Typical Limiting Oxygen Concentration values for dusts are in the range 6–15% (v/v) when nitrogen is the inert gas. In practice, a safety margin is applied so that the maximum permissible level of oxygen concentration in the plant lies below the measured Limiting Oxygen Concentration. The difference between the Limiting Oxygen Concentration and the permissible oxygen concentration will depend on the size of the plant, the explosibility of the dust, the level and accuracy of oxygen monitoring in the plant and the reliability and rapidity of shutdown procedures. Typical safety margins are in the range 2–5% (v/v). The lower the risk of the Limiting Oxygen Concentration being exceeded, the lower the safety margin need be. Guidance on the design of inert gas systems is given in Barton (2002). The value of the Limiting Oxygen Concentration is dependent on both the dust and the gas used in the inerting process.

The Limiting Oxygen Concentration decreases essentially linearly as the temperature increases (Bartknecht 1981). The effect of an increase in the initial pressure on the Limiting Oxygen Concentration depends on the dust. In some instances, there is a decrease as pressure increases (van Laar 1996). This generally occurs when an increase in the initial pressure causes a decrease in the Lower Explosion Limit. When an increase in the initial pressure causes an increase in the Lower Explosion Limit, an increase in the Limiting Oxygen Concentration can be expected. These are not hard and fast rules, however: for some dusts, the opposite can occur.

6.3.2 *Elimination of potential ignition sources*

If all possible ignition sources can be eliminated, an explosive atmosphere cannot under any circumstances be ignited. The list of potential ignition sources is long, however, and includes flames, static electricity, electrical sparking, mechanically generated sparks, heated surfaces and thermal decomposition of the dust. BSEN 1127-1:1998 discusses these and other causes of ignition.

Elimination of all potential ignition sources is often not possible, and although this method of explosion protection may appear feasible, its application requires care and a thorough knowledge of the plant, equipment and process. Although it should always be attempted, to reduce the frequency of ignitions, elimination of potential ignition sources may not be a sufficient basis for safety.

Explosive dust clouds can be ignited only if the energy available in the ignition source is sufficient to set off a self-sustaining combustion. If the energy in the ignition source is below this critical value, the explosive atmosphere will not ignite. Ensuring that all potential ignition sources do not exceed the critical value is the key to this prevention technique.

6.3.2.1 *Ignition by electrical discharges* Electrical discharges are potential ignition sources for dusts, but the discharge must be capable of releasing sufficient energy into the explosive atmosphere if an ignition is to occur. The necessary minimum energy depends on the dust and how the discharge is produced – whether it is, for example, due to static electricity or to electrical circuitry. For instance, a capacitative electrical circuit is specified for testing when the ignition energy data are to be used in the assessment of electrostatic ignition risk, because many insulated conductors are essentially capacitative. Other circuits will produce ignitions at much lower energies.

To measure this minimum energy, discharges are passed between two electrodes and the minimum energy for ignition determined. The igniting power of a discharge depends on its total energy and the distribution of the energy with respect to space and time. The ignition energy depends on the circuitry from which the discharge emanates and on physical factors such as the electrode size and shape and the spark gap dimensions.

The Minimum Ignition Energy (MIE) is measured in a standard test using a specified circuit and is defined as (BSEN 1127-1:1998) 'the lowest electrical energy stored in a capacitor that upon discharge is sufficient to effect ignition of the most ignitable atmosphere under specified test conditions'.

The most ignitable atmosphere of a flammable substance is that concentration within the Explosion Range that has the least ignition energy. Two types of test equipment for determination of the Minimum Ignition Energy have been in common use.

(1) Equipment based on British Standard BS 5958: 1991: Sparks are released from the high-voltage electrode (10–30 kV) across a spark gap of not less than 2 mm. The voltage is slowly increased until a discharge is produced. The energy is considered to be that of the capacitance circuit. A typical value for inductance in the circuit is 12 μH.

(2) Kuhner MIKE 3 Equipment (Cesana & Siwek 2001a): This differs from the BS 5958:1991 equipment in that the electrode is more pointed. The electrode gap is 6 mm. A moving electrode system is used to trigger the sparks of 10 mJ and above. For 1 mJ and 3 mJ sparks, a high-voltage relay is used to trigger the discharge at 15 kV. The inductance can be 20 or 1020 μH. The former is used to assess electrostatic hazards.

A comparative study of data from the two tests for 17 powders showed that in no case did the BS 5958:1991 test produce the lower Minimum Ignition Energy (Bailey *et al.* 2001). The results were in agreement for four powders, and the MIKE 3 test produced the lower Minimum Ignition Energy for 13 powders. The BS 5958:1991 test failed to detect sensitive materials (i.e. Minimum Ignition Energy $<$10 mJ) in six cases.

The European standard for minimum ignition energy measurements for dusts, BSEN 13821:2002, recommends the 1.2 litre vertical Hartmann tube as the test vessel, although other vessels can be used, provided the volume is greater than or equal to 1 litre. The electrical circuitry, electrode dimensions and spark gap are comparable with those in the MIKE 3 test. If other vessels or test procedures are used, conformity with the standard test has to be demonstrated as described in the standard.

The Minimum Ignition Energy decreases substantially with temperature (Bartknecht 1981), essentially linearly on a log–log scale. Experiments on several dusts indicate that the Minimum Ignition Energy value tends to a value of 0.088 mJ at 1000°C, regardless of the dust. The moisture content can have a substantial effect on the Minimum Ignition Energy (Zeeuwen & van Larr 1984).

6.3.2.2 *Ignition by static electricity* For most dusts, ignition is made easier if the duration of the spark can be increased by installation of a suitable induction in the circuit. The inductance of the discharge circuit in the standard test is at two levels, 1–2 mH and 20 μH. The inclusion of a 1–2 mH inductance in the discharge generating circuit in the test method for measuring Minimum Ignition Energy ensures that the ignition energies measured in this test are the lowest values likely to be measured but should not be used in the assessment of electrostatic hazards. The ignition energy relevant to electrostatic discharges is measured with an induction of around 20 μH in the discharge-generating circuit, that is, an essentially purely capacitative discharge.

The theoretical spark energy, neglecting external circuit losses, is given by $\frac{1}{2}CV^2$, where C is the capacitance, and V is the voltage difference. Detailed guidance on this is given in British Standard BS 5958:1991 and PD CLC/TR SO4O4:2003. Recommended safety measures in BS 5985:1991 are based on Minimum Ignition Energy data obtained with the test described in the standard. The recommendations in PD CLC/TR SO4O4:2003 are based on the BSEN 13821:2002 type test, that is the MIKE 3, or a test giving comparable values.

At some point in the future, all data can be expected to have been determined using the European standard for minimum ignition energy measurements for dusts test (BSEN 13821:2002). Until then, it is recommended (Gibson & Lunn 2003) that:

(1) new data should be obtained with the BSEN 13821:2002 test and interpreted in accordance with PD CLC/TR SO4O4:2003,
(2) interpretation of existing BS 5958:1991 test data should be based on guidance in BS 5958.

There are several different types of electrostatic discharge (International Social Security Association 1996b). Spark discharges occur between two conductive objects, and almost all the stored energy is released rapidly into a narrow channel. Spark discharges are capable of igniting dry, flammable dusts. Brush discharges occur when a conductive, grounded and curved object is subject to a high electrical field. The discharge takes on the appearance of multiple discharges distributed over a wide channel. Brush discharges are capable of igniting only dusts that are very sensitive to ignition. If the radius of curvature is very small, a continuous weak discharge at the tip, known as a corona discharge, can occur. Corona discharges are much weaker than brush discharges. Propagating brush discharges occur over the surface of a thin sheet of insulating material that is charged on both sides with opposite polarity. Because a large part of the energy can be released, propagating brush discharges are capable of igniting dry, flammable dusts. Cone discharges occur when large containers are filled with a highly insulating powder, and charges appear on the surface of the material. Cone discharges may be capable of igniting dry flammable dusts, depending on the circumstances.

6.3.2.3 *Ignitions due to electrical equipment* Sparks or arcs generated when live currents are broken are transient inductive sparks, and the theoretical energy is given by $\frac{1}{2}LI^2$, where L is the circuit inductance, and I is the current in the circuit. Discharges from components with significant inductance or resistance generally produce discharges with a longer duration than purely capacitative discharges, with the result that less total energy is necessary to ignite a specific dust. Boyle and Llewellyn (1950) showed that the minimum capacitor energy capable of igniting a given powder decreased substantially when a series resistance was included in the circuit. Because the series resistance absorbs a large fraction of the capacitor energy during discharge, the energy going into the spark gap is only a small part of the theoretical stored energy, generally about 5–10% (Eckhoff 1997). Thus, in Boyle and Llewellyn's experiments, at a series resistance of 10^4–10^5 ohms, for some dusts the ignition energy was reduced to approximately 1% of the energy necessary without the resistance. Line *et al.* (1959) measured similar changes. Smielkow and Rutkowski (1971) increased the spark duration either by adding a large inductance or by adding a large resistance to the circuit, and again the theoretical ignition energy from the capacitor decreased by a factor of ten, approximately.

A provisional test apparatus has recently been developed to measure the ignition energies of dusts using a 'make and break' sparking apparatus (Testing Methods for Electrical Apparatus Installed in a Dusty Environment with a Potential Risk of Explosion (EADE)). The test apparatus is a combination of the vertical, 1.2 litre Hartmann tube with a dust dispersion chamber based on the principles and, where practicable, the dimensions of the MIKE 3 apparatus, and the break-flash apparatus (BSEN 50 020:2002) as modified in EU project SMT4-CT98-2217: Gas Ignition test for Heavy Current, Low Voltage Circuits (EU Funded Project SMT4-CT98-2217 1999). Successive blasts of air at timed intervals disperse dust placed in the base to produce a cloud around the spark test apparatus disc and the location of the sparks.

Tests with three dusts, lycopodium, calcium stearate and sulphur, were performed, aimed at devising a workable test methodology. The ignition results for lycopodium and calcium stearate, when compared with the published current–voltage ignition curves for gases (BSEN 50 020:2002) were higher than the curves for methane, that is the dusts were more difficult to ignite. Sulphur, however, being much easier to ignite, fell close to the ignition curve for IIB gases, typically ethylene.

Harper *et al.* (1997) have used a similar apparatus to measure minimum igniting currents of ten dusts with low ignition energies, as measured with capacitative sparks. Of the dusts tested, only sulphur had a minimum igniting current lower than that of methane. With a resistive circuit, at low voltage, the minimum igniting current measurements for sulphur followed the gas Group IIB data and then passed through the gas Group IIA, to gas Group I, as the voltage increased. In an inductive circuit, the results for sulphur fell between gas Groups IIB and IIA.

These proposed tests are under discussion by standards bodies.

6.3.3 *Hot surface ignition*

Hot surfaces or bodies are potential ignition sources for both dusts and gases. Several Ignitability Characteristics are required if elimination of hot surface ignition sources is to be applied successfully:

(1) the Minimum Ignition Temperature of an explosive atmosphere,
(2) the Minimum Ignition Temperature of a dust layer.

The values obtained from standard tests are, however, essentially approximations because the shape and size of the heating space, the method and duration of heating, and the ignition source can all have an effect on the onset of self-sustained combustion.

6.3.3.1 *Ignition of explosive atmospheres by hot surfaces* The Minimum Ignition Temperature of a dust cloud is defined as (BSEN 1127-1:1998) 'the lowest temperature of a hot surface on which the most ignitable mixture of the dust with air is ignited under specified test conditions'.

The Minimum Ignition Temperature of an explosive atmosphere is often said to be the lowest temperature at which the atmosphere will ignite spontaneously without the presence of a 'localised' ignition source. And although the strict definition specifies ignition against a heated surface, and hot surfaces are used in the test equipment to produce the high temperatures required for ignition, the cloud itself is at an elevated temperature when ignition occurs. An American standard (American Society for Testing Materials 1997) describes the test method therein as one that determines the minimum temperature at which a given dust cloud will autoignite when exposed to air heated in a furnace at local atmospheric pressure.

There are two furnace tests used: the Godbert–Greenwald Furnace and the BAM oven. Minimum Ignition Temperature values obtained through the Godbert–Greenwald Furnace are not the same as for the BAM oven. The value from the latter is usually the lower. The Godbert–Greenwald Furnace is a 400-mm-long, 36-mm-wide vertical tube, the surface of which is heated (BSEN 50281-2-1:1999). Dust is blown into the tube from a dust reservoir by a blast of air, and any visible signs of an explosion are observed. The dust concentration and the temperature of the furnace are varied until the minimum temperature at which an ignition occurs is obtained. The apparatus is shown in Figure 6.3.

In the Godbert–Greenwald apparatus, room-temperature dust is injected into heated air, and the dust cloud is surrounded by a heated surface. Ignition can occur in two ways: either particles ignite as they come into contact with the heated wall, and this ignition then transmits to the rest of the cloud; or the cloud itself undergoes volumetric ignition because the dust is injected into air already at an elevated temperature. The residence time of the cloud inside the vertical furnace tube is relatively short, however, because the dust falls through. As a result, the ignition, when it occurs, follows rapidly on dust injection.

The BAM furnace is a 170-mm-long horizontal, heated tube with a heated plate inside (Bartknecht 1993). The dust is blown against the heated plate, and the lowest temperature at which the dust either ignites or decomposes to produce flame within 10 s is taken to be the Minimum Ignition Temperature. The dust concentration and the temperature of the furnace are varied until the minimum temperature at which an ignition occurs is obtained. The apparatus is shown in Figure 6.4.

In the BAM oven, the room-temperature dust is injected into heated air, directly against a heated plate. Again, the apparatus is designed to ensure that the dust cloud is heated. The dust can ignite either on the plate or in the volume of the furnace. The furnace tube is closed with a flap, which means that the dust cloud is likely to stay at an elevated temperature more readily than in the Godbert–Greenwald furnace. The BAM furnace tube is horizontal, and particles falling onto the tube floor can smoulder to give off flammable gases that could ignite at lower temperatures than would ignite the dust cloud. If any ignition, no matter how produced, is noted within 10 s of dust injection, it is taken to be an ignition of the dust cloud.

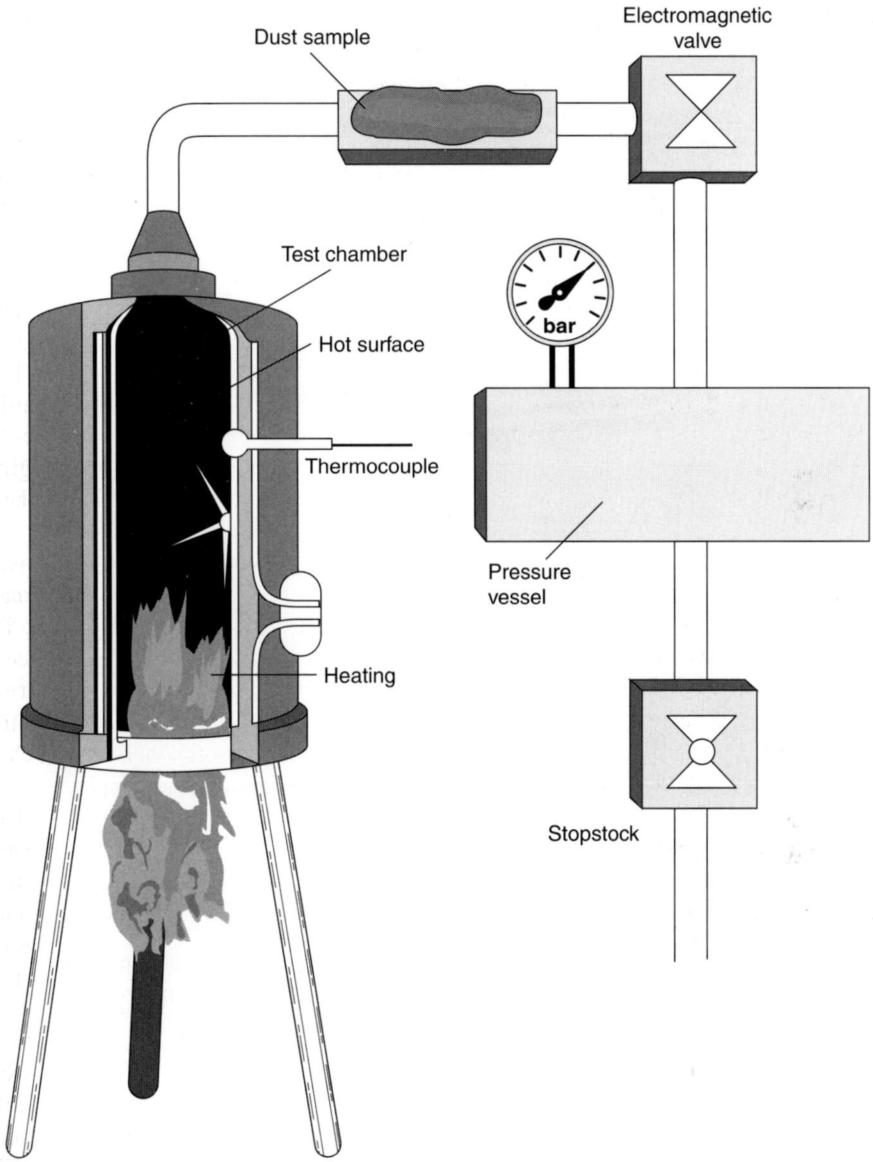

Figure 6.3 Godbert–Greenwald furnace apparatus for measuring the Minimum Ignition Temperature of dust clouds.

The Minimum Ignition Temperature specifies the temperature that must not be exceeded by any hot surface with which the cloud may come into contact. In the powder handling plant, it is used to assess the possibility of autoignition in the plant process. Few powder-handling plants, however, operate under conditions that could lead to autoignition of the dust cloud. The major use of the Minimum Ignition Temperature is in the selection of either electrical or mechanical equipment for use where potentially explosive atmospheres are a possibility.

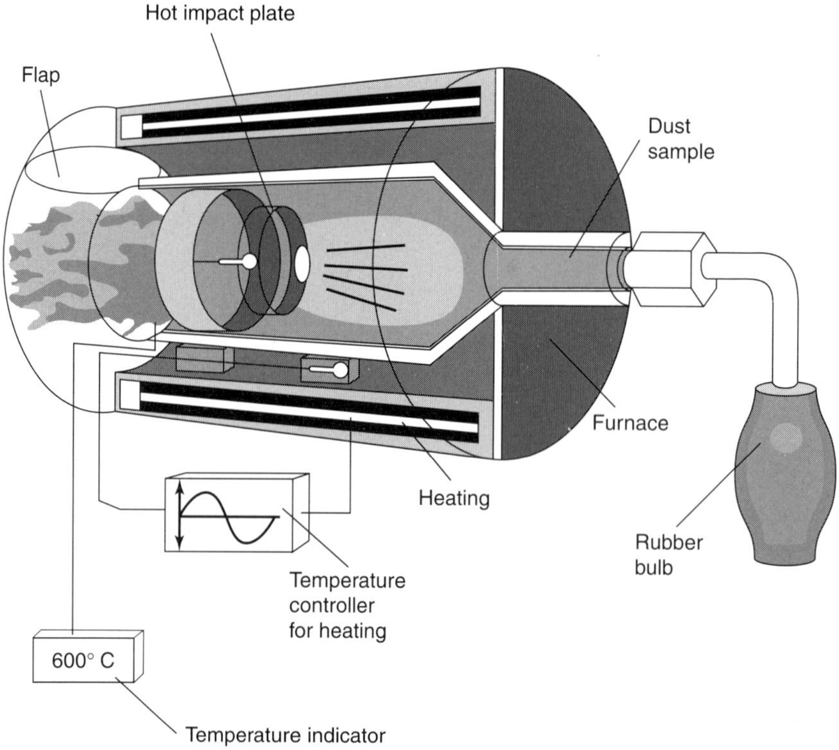

Figure 6.4 BAM furnace apparatus for measuring the Minimum Ignition Temperature of dust clouds. Reproduced with permission.

Equipment intended for use in explosive environments is allotted to a temperature class depending on the maximum surface temperature that will develop in operation. The Temperature Classification is given in Table 6.1. There is a temperature limitation when dust clouds are likely to be present (Standard BSEN 50281-1-2:1999). The maximum surface temperature of equipment exposed to a dust cloud must not exceed two-thirds of the Minimum Ignition Temperature measured in the Godbert–Greenwald furnace.

Table 6.1 Temperature class: maximum surface temperatures.

Temperature class	T1	T2	T3	T4	T5	T6
Maximum surface temperature (°C)	450	300	200	135	100	85

In practical circumstances, conditions are likely to be different from those in the furnace tests. Inside a dryer or other plant where the contact time can be long, the potential for a lower ignition temperature than the Minimum Ignition Temperature is present. Griesche and Brandt (1976) have shown that the ignition temperature of a dust-cloud decreases substantially when the residence time of a dust in the Godbert–Greenwald furnace increases.

Conversely, however, in practical conditions where the dust cloud is in contact with a hot surface less confining and smaller than that in the furnace test, the ignition temperature may be above the Minimum Ignition Temperature. Measurements by Pinkwasser (1986) show that dusts with Minimum Ignition Temperatures of approximately 400°C require temperatures of 1000–1200°C to ignite if the area of the hot surface is in the range 12–20 mm^2. The configuration of the hot surface is also a factor. The Minimum Ignition Temperature measurements from the BAM oven are lower than expected from an extrapolation of Pinkwasser's results because the hot surface is a relatively large one compared with the size of the cloud, and surrounds it as a cylinder (Pinkwasser's areas were made up of hot wires and coils). In practice, therefore, the plant design and conditions may significantly alter the potential ignition temperature of an explosive dust atmosphere, but any departures from use of the Minimum Ignition Temperature must be based on reliable evidence.

6.3.3.2 *Ignition of dust layers by hot surfaces* Layers or accumulations of a flammable dust are prone to smoulder or combust with flame if the surface that they cover is at a sufficiently high temperature. If an explosive atmosphere of dust is present, then burning deposits are a potential ignition source, either at the point of the deposit or if parts of the deposit break away and are carried as burning clumps into a dust cloud in a downstream process vessel. The temperatures at which dust deposits ignite on hot surfaces are much lower than the Minimum Ignition Temperatures for a dust cloud.

The Minimum Ignition Temperature of a dust layer is defined as (BSEN 1127-1:1998) 'the lowest temperature of a hot surface on which a dust layer will ignite under specified test conditions'. The Layer Ignition Temperature is measured in a standard test using a layer of specified thickness. European Standards specify a 5 mm layer (BSEN 50281-2-1:1999); standards in the USA specify 12.5 mm.

The apparatus, shown in Figure 6.5, consists of an electrically heated round metal plate with a diameter of 200 mm and a thickness of 20 mm (BSEN 50281-2-1:1999). Once the plate has reached a steady state, the temperature across the plate must be uniform to within ±5 K, not change by more than ±5 K when the dust layer is placed on the plate, and return to within ±2 K of the original value within 5 min. The thermocouple for the temperature control is placed in the middle of the plate (1 ± 0.5) mm below the upper surface. A similar thermocouple is placed close to the control thermocouple and is connected to a recorder. The temperature of the heating plate is recorded during the experiment.

A metal ring with an inner diameter of 100 mm and height of 5 mm is placed in the middle of the plate. The metal ring is filled with the dust sample when the chosen temperature is reached. The dust sample remains on the hot surface until ignition or self-heating without ignition occurs. If there is no ignition or self-heating within 30 min, the experiment is stopped. The experiment is then repeated at a higher temperature and with a new sample. If the dust layer ignites, the experiment is repeated at a lower temperature. The experiments are continued until the ignition temperature is found.

The lowest temperature for an ignition of the dust layer occurs is recorded as the Layer Ignition Temperature. The highest temperature at which ignition does not occur has to be verified by three measurements. If the dust is not ignited by a hot surface under 400°C, the experiments are stopped. The time taken for ignition or the time taken to reach the maximum temperature is measured and recorded. If no ignition occurs, the time until the end of an experiment is also recorded.

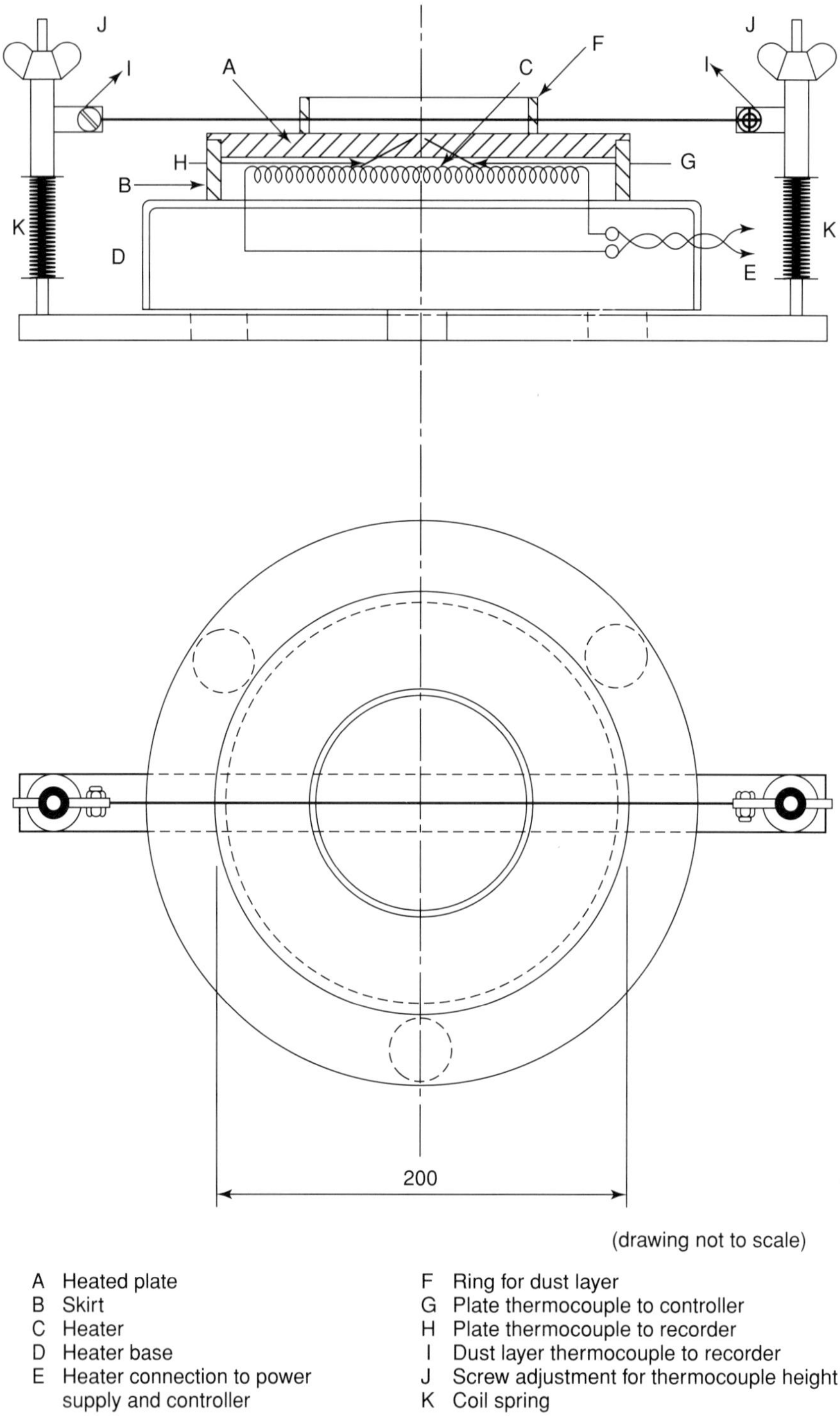

Figure 6.5 Test apparatus for measuring the Minimum Ignition Temperature of dust layers. Reproduced with permission.

The Minimum Ignition Temperature of a dust layer gives an indication of the temperature that a hot surface should not exceed if it is likely to be covered by a layer of a flammable dust. A temperature limitation is imposed for electrical or mechanical equipment used in dusty environments. The maximum surface temperature should not exceed a temperature 75 K below the measured Minimum Ignition Temperature of a 5-mm-thick layer of the dust (Standard BSEN 50281-1-2:1999).

When considering the test results, it has to be borne in mind that as the layer thickness increases, the insulating effect increases, and the rate of dispersion of heat from the layer decreases. The ignition temperature consequently falls. Second, the dust layer is placed on the hot surface once the test temperature has been reached. In practice, surfaces may heat up beneath the dust layer. Third, the hot surface is flat and has a specified size. Ignition temperatures vary, depending on the size and configuration of the hot surface. Fourth, the hot plate is temperature-controlled. In practice, the surface can be heated by a source of a constant power, and this can lead to lower layer ignition temperatures. The length of time taken for a constant power test, however, militates against it as an effective practical test. It takes several hours for the hot surface to reach a relatively constant temperature each time the power is adjusted.

The influence of dust layer thickness on the Layer Ignition Temperature can be estimated by methods given in standards (Standard BSEN 50281-1-2:1999; BSEN 50281-2-1:1999). In BSEN 50281-2-1:1999, the method is to draw a graph of the logarithm of the layer thickness vs. the reciprocal of the Ignition Temperature (absolute temperature in K) and estimate a value by linear interpolation or extrapolation. For this calculation, at least two values of layer ignition temperature have to be determined experimentally. One should be the value for the 5 mm layer, and the other the value of the 20 mm or 30 mm layer.

In Standard BSEN 50281-1-2:1999, a maximum permissible surface temperature can be estimated by a graphical method. The maximum permissible surface temperature is always much lower than the experimental ignition temperature, and thus there is usually a sizeable safety margin. Measurements of layer ignition temperatures at different thicknesses (Testing Methods for Electrical Apparatus Installed in a Dusty Environment with a Potential Risk of Explosion) show that the method in BSEN 50281-2-1:1999 provides satisfactory estimates.

In practice, it is recommended that layer ignition temperatures be determined experimentally. When thick dust accumulations cover a hot surface, the ignition temperature can decrease even further. Tests have been carried out with conical accumulations of dust covering heated surfaces (Testing Methods for Electrical Apparatus Installed in a Dusty Environment with a Potential Risk of Explosion).

In practice, the critical conditions that lead to layer ignition can vary widely, depending on the size of the heated surface, the size of the dust accumulation, the thermal conductivity of the dust layer, the power being fed into the hot surface and the time available for the dust to undergo heating. A power input of less than 5 W can produce temperatures in a hot spot of 10 mm diameter sufficient to ignite deposits of wheat flour (Nelson 1995). More extensive surfaces require much more power to reach ignition temperatures and much more time (Bennett *et al.* 2002).

If the air above the layer is at a temperature higher than normal room temperature, the layer ignition temperature will fall. Layer ignition temperatures typically decrease by 40–60°C at an air temperature of 100°C. It is important, therefore, that ignition temperatures are measured under conditions that reflect as closely as possible those in the intended application.

Standard tests are available for assessing the ignition behaviour of bulk dust samples. These tests assess the temperature at which accumulations will self-ignite by reaction with atmospheric oxygen or will undergo exothermic decomposition in the absence of air. Furnace tests are used to measure the temperature at which a sample of bulk material will reach, through a reaction, a temperature in excess of either the furnace or an inert reference sample. Hot storage tests, differential thermal analysis and tests for storage life estimate the temperature at which a bulk material can be stored. These tests are outlined in International Social Security Association (1996). Abbot (1990) describes tests developed to simulate various conditions and obtain measurements of the temperature at which exothermic reaction begins.

When a dust accumulation is ignited, it can burn in a variety of ways. Harper *et al.* (1997) have discussed the burning behaviour of powder deposits on hot surfaces. The stages of ignition and the form of the combustion zone can be complex. Some powders burn directly in the solid phase either with a flame or by smouldering, others melt and burn as a liquid, while some burn with a large amount of flame. Some dusts can evolve large amounts of flammable gas when subjected to heat. A change from solid to liquid or agglomeration/expansion of dust particles to give an extended mass of material can block burning if diffusion of oxygen to the seat of burning is prevented.

A measure of the ignitability of a dust layer and the intensity of burning of a dust layer is the Combustion Class (CC) (International Social Security Association 1996a). This classification is based on the behaviour of a defined dust heap when subjected to a gas flame or hot platinum wire:

(1) CC1 No ignition; no self-sustained combustion,
(2) CC2 Short ignition and quick extinguishing; local combustion of short duration,
(3) CC3 Local burning or glowing without spreading; local sustained combustion but no propagation,
(4) CC4 Spreading of a glowing fire; propagating smouldering combustion,
(5) CC5 Spreading of an open fire; propagating open flame,
(6) CC6 Explosible burning; explosive combustion.

The train-firing test assesses flammability with reference to different ignition sources, and allots the Combustion Class.

6.4 Explosion protection

Protection Techniques are used either to halt an explosion in its very early stages or to ensure that destructive pressures do not occur. The methods by which this is achieved are, venting, suppression, containment and isolation.

6.4.1 *Measuring explosibility characteristics*

Proper application of protection techniques requires measurements of the maximum explosion pressure, $P_{\max}$, and the maximum rate of pressure rise, $(\mathrm{d}P/\mathrm{d}t)_{\max}$.

6.4.1.1 *The Maximum Explosion Pressure, P_{max}* The Maximum Explosion Pressure is defined as (BSEN 1127-1:1998) 'maximum pressure occurring in a closed vessel during an explosion of an explosive atmosphere determined under specified test conditions'. The test is carried out across the explosion range of the combustible material and the highest value of P_{max} taken as the explosion characteristic of the fuel.

The recommended test vessel is either a sphere or cylinder of aspect ratio 1:1 ± 10% with a volume of 1 m^3 (International Social Security Association 1996a). The dust is injected under pressure from external reservoirs so that a well-dispersed dust cloud is produced with a turbulence level that exceeds that likely to be met with in most industrial plants. The ignition source is positioned at the centre of the vessel and consists of two chemical igniters of specified composition, each with a total energy of 5 kJ. Ignition occurs 0.6 s after dust injection.

Alternative test methods are allowed so long as conformity with the 1 m^3 test can be demonstrated, as described in the standard. The 20 litre test vessel and procedure (Cesana & Siwek 2001b) can be taken as being in conformity with the 1 m^3 test. The 20 litre sphere is shown in Figure 6.6. The Maximum Explosion Pressure can be as high as 13 bar, but is usually about 10 bar.

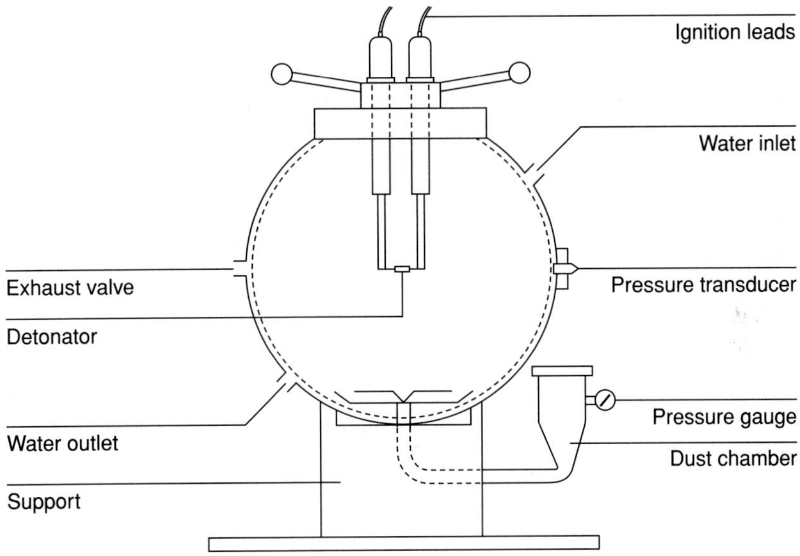

Figure 6.6 Twenty-litre sphere test apparatus for measuring the explosibility characteristics of a dust. Reproduced with permission.

If the basis of safety relies on either explosion suppression or explosion venting, knowledge of the Maximum Explosion Pressure is required to estimate the degree of protection, e.g. the amount of suppressant or the size of the vent area. When the basis of safety relies on explosion containment, the Maximum Explosion Pressure signifies the pressure that a single contained vessel, isolated from any other vessels and without any internal compartments, must withstand. If pressure piling is a possibility (Holbrow *et al.* 1999), the data must be corrected for pressure piling and the vessel designed to withstand the higher explosion pressures that may result.

The Maximum Explosion Pressure decreases as temperature increases. This is because there is less reacting material available as the initial temperature increases. The trend is linear if the absolute temperature is expressed as $(1/T)$ (Bartknecht 1981). Wiemann (1987) has normalised this effect across a number of dusts, plotting the ratio of the explosion pressure against the inverse ratio of the initial temperatures. The Maximum Explosion Pressure increases linearly as the initial pressure increases (Bartknecht 1981; Wilen & Rautalin 1999).

6.4.1.2 *The Maximum Rate of Pressure Rise, $(dP/dt)_{max}$* The Maximum Rate of Pressure Rise is a measure of the speed of the explosion in a confined vessel and essentially quantifies the rate of expansion due to the combustion. If a vessel is protected by explosion venting, the expansion has to be relieved if damage is to be avoided, and the size of the venting area depends on the rate of pressure rise the combustion of the fuel generates. Likewise, in effective explosion suppression, the amount of suppressant required and its speed of delivery depend on knowing the rate of combustion.

The maximum rate of explosion pressure rise is defined as (BSEN 1127-1:1998) the 'maximum value of the pressure rise per unit time during explosions of all explosive atmospheres in the explosion range of a combustible substance in an enclosed vessel under specified test conditions'. The maximum rate of pressure rise is measured in the same test vessel and by the same procedure as for the Maximum Explosion Pressure.

Dusts have their explosibility ranked according to the rate of pressure rise they generate in an enclosed explosion. A method for normalising this rate so that it is independent of volume exists, and it is this explosibility characteristic, i.e K_{St} value, that is used in designing the degree of explosion protection.

The K_{St} value is calculated by:

$$K_{st} = (\mathrm{d}P/\mathrm{d}t)_{max} V^{1/3},$$

where $(\mathrm{d}P/\mathrm{d}t)_{max}$ is the maximum rate of pressure rise (bar/s) at the optimum dust concentration; V is the test vessel volume (m^3); K_{St} is the dust explosibility characteristic (bar m/s).

Flammable dusts are separated into Groups depending on the K_{St} value:

(1) Group St1: 0 bar m/s $< K_{St} \leq$ 200 bar m/s
(2) Group St2: 200 bar m/s $< K_{St} \leq$ 300 bar m/s
(3) Group St3: 300 bar m/s $< K_{St} \leq$ 800 bar m/s

The K_{St} is essentially the maximum rate of pressure rise, at the optimum dust concentration, in a 1 m^3 vessel under standard conditions. It is taken as a measure of the degree of explosibility of the dust, is assumed to be independent of vessel volume and is used in estimations of explosion protection requirements. The K_{St} is not a fundamental property of the dust but an artificial parameter that serves a very useful purpose in the design of explosion protection for the powder-handling plant. The value of K_{St} can be taken to be invariant with volume because dust-explosion protection guidelines are based on tests in which the dust explosions have been arranged experimentally to conform to the measured K_{St} value. Changes to the dust-injection conditions and the ignition delay following the injection alter the state of the relatively high turbulence used in dust explosion testing so that the rate of pressure rise in a test vessel can be varied to conform to the K_{St} value as vessel volume increases. It is important that the correct value of K_{St} is used when designing explosion protection measures. Incorrect values will lead to either overprotection, which is unnecessarily expensive, or to underprotection, which leads to the risk of a damaging explosion.

The K_{St} value increases as pressure increases (Bartknecht 1981; Wilen & Rautalin 1999). The effect of temperature on the K_{St} value is dust-dependent: it may increase as temperature increases, decrease or stay approximately the same (Bartknecht 1981). This unpredictability is due to two conflicting influences. As temperature increases, the rate of reaction increases, tending to increase the rate of pressure rise, but, at a given pressure, the amount of reacting material is less, tending to decrease the rate of pressure rise.

Although guidelines for explosion protection are based on the K_{St} value, in an item of dust-handling plant the dust cloud may differ in terms of concentration, dust homogeneity and turbulence from that generated in the standard test vessel. Thus, the rate of pressure rise in an explosion in the equipment may either be less or greater than that presumed by use of the guidelines. The guidelines may, therefore, prove either too conservative or insufficient, respectively. Draft European standards take this into account by allowing departures from the guidelines if they can be supported by experimental evidence.

6.4.2 *Explosion-protection techniques*

6.4.2.1 *Explosion containment* Containment is achieved by constructing process vessels with sufficient strength to withstand the full explosion pressure likely to be generated. There are two options: explosion-pressure-resistant vessels are designed to contain the explosion without rupture or deformation; explosion-pressure-shock-resistant vessels must not rupture, but some permanent deformation is allowed.

The Maximum Explosion Pressure of a dust is measured in a single compact vessel that contains no internal obstructions, but in practice, the dust-handling plant may differ from this ideal situation, and a straightforward application of the Maximum Explosion Pressure to the plant designed for containment is not always an option.

If an explosion propagates through a connecting pipeline between vessels, pressures higher than the Maximum Explosion Pressure can be generated (Holbrow *et al.* 1999). This is due to a process called ‘pressure piling’ and arises when an explosion in one vessel propagates into a smaller one. The expansion due to the oncoming explosion compresses the dust cloud so that the flame entering the smaller vessel from the pipeline ignites an explosive atmosphere that is at a pressure above atmospheric. As a result, the explosion pressure generated in the second vessel is correspondingly higher than the Maximum Explosion Pressure. Some guidance is available for estimating the explosion pressures generated in systems of two connected, enclosed vessels (Holbrow *et al.* 1999). Similarly, obstructions inside a vessel that divide the internal volume into connected compartments may produce pressure piling effects. Transmission of an explosion from one vessel to another can be stopped by isolation methods.

6.4.2.2 *Explosion venting* Explosion relief venting involves fitting into the walls of a vessel areas of weakness that open up early in the explosion and allow unburned dust cloud and combustion products to vent into the open air. Most of the explosion takes place outside the vessel, and as a result, the pressure rise inside is considerably reduced. Figure 6.1 shows a vented explosion. Devices that cover vent areas are described in Barton (2002).

If the venting design has been done properly, damage to the vessel is avoided. The size of the vent and its failure pressure, P_{stat}, have to be chosen so that the highest internal pressure, known as the reduced explosion pressure ($P_{red,max}$), is below that which would cause damage. Incorrectly sized relief vents will result in either the plant not being properly protected or impracticable or unnecessarily expensive vents being fitted.

A method for sizing the area of a dust explosion vents is (VDI 3673 2002):
For 0.1 bar overpressure $\leq p_{red,max} < 1.5$ bar overpressure

$$A_e = B(1 + C \log L/D) \text{ in m}^2 \tag{6.1}$$

with

$$B = \left(3.264 \times 10^{-5} p_{max} K_{St} p_{red,max}^{-0.569} + 0.27(p_{stat} - 0.1) p_{red,max}^{-0.5}\right) V^{0.753} \tag{6.2}$$

$$C = (-4.305 \log p_{red,max} + 0.758). \tag{6.3}$$

For 1.5 bar overpressure $\leq p_{red,max} \leq 2.0$ bar overpressure

$$A_e = B. \tag{6.4}$$

The equations are valid for:

(1) enclosure volume $0.1 \text{ m}^3 \leq V \leq 10\,000 \text{ m}^3$
(2) static activation overpressure of the venting device 0.1 bar $\leq p_{stat} \leq 1$ bar
(3) maximum reduced explosion overpressure $p_{stat} < p_{red,max} < 2$ bar
(4) it is recommended that $p_{red,max}$ should exceed p_{stat} by at least 0.02 bar
(5) maximum explosion overpressure 5 bar $\leq p_{max} \leq 10$ bar for: 10 bar m/s $\leq K_{St} \leq$ 300 bar m/s
(6) maximum explosion overpressure 5 bar $\leq p_{max} \leq 12$ bar for: 300 bar m/s $< K_{St} \leq$ 800 bar m/s
(7) dust explosibility characteristic 10 bar m/s $\leq K_{St} \leq$ 800 bar m/s
(8) length:diameter ratio of the vessel $1 \leq L/D \leq 20$

In practice, the dust flame outside a vent cannot be ignored because inside the plant the dust concentration is usually high, well above the concentration that will react with the available oxygen. When expelled into the open, this dust is able to react, and the resulting flame can be extensive and develop appreciable pressures. Methods are available for estimating the size of the external fireball, as are techniques for limiting its size and transporting the vented cloud to a safe place (Barton 2002).

It is inadvisable to place vented dust plant inside a work place, and vent ducts are used to take the vented dust outside the building. These, however, cause a restriction to the venting process that can produce high pressures inside the plant. Methods are available to take this effect into account (Barton 2002).

6.4.2.3 *Explosion suppression* Explosion suppression is a technique whereby an explosion is extinguished in its very early stages by the injection of a suppressant into the process vessel. The incipient explosion is detected shortly after ignition by detectors fitted in the vessel walls. The pressure rise caused by the explosion travels at a much higher speed than the dust flame and is thus detected before the flame has travelled far from where the ignition occurs. Once the flame has been detected, a signal is sent to a set of pressure vessels that contain the suppressant material, usually under pressure. The suppressant is released into the process vessel and extinguishes the explosion before a high overpressure is reached. Effective suppressants are inert powders and water.

The explosion is only extinguished if sufficient suppressant is injected rapidly enough and dispersed at a sufficiently high concentration throughout the vessel. Otherwise, the explosion can escape effective suppression and generate overpressures that may cause damage. In practice, suppression systems are designed by specialist firms. A suppression system is shown diagrammatically in Figure 6.7.

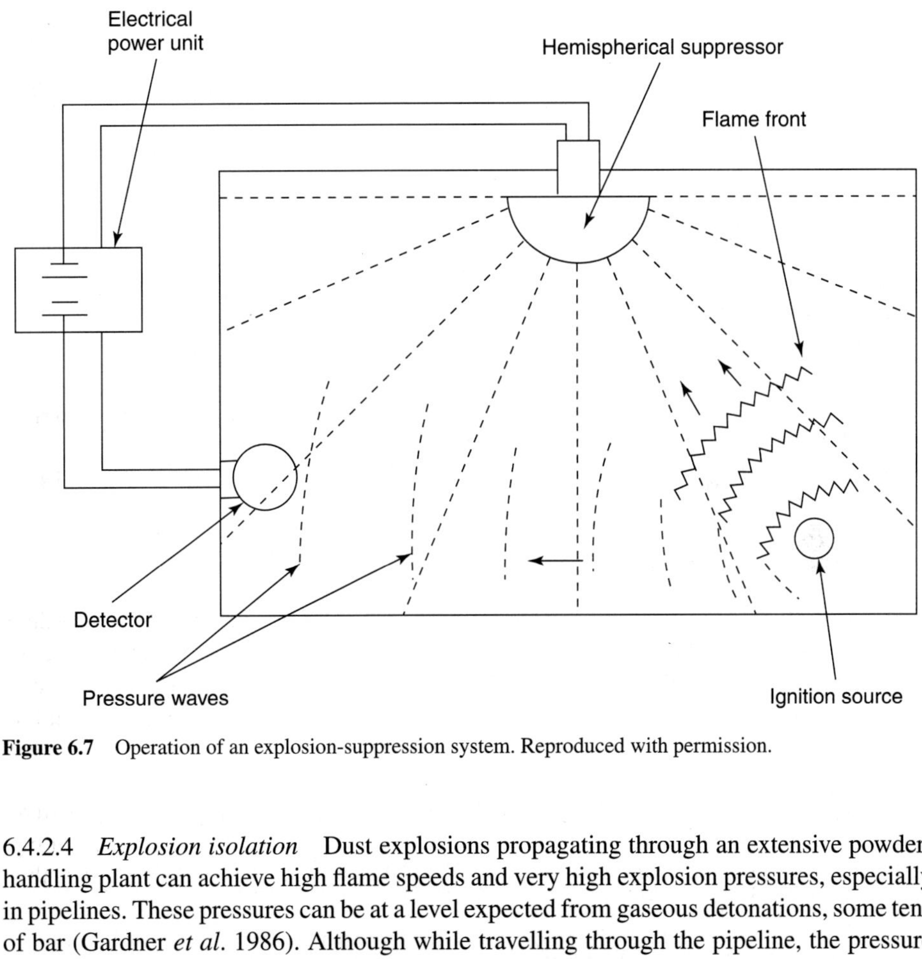

Figure 6.7 Operation of an explosion-suppression system. Reproduced with permission.

6.4.2.4 *Explosion isolation* Dust explosions propagating through an extensive powder-handling plant can achieve high flame speeds and very high explosion pressures, especially in pipelines. These pressures can be at a level expected from gaseous detonations, some tens of bar (Gardner *et al.* 1986). Although while travelling through the pipeline, the pressure spikes are too rapid and are over too quickly at any given location to cause damage, if the explosion hits an obstruction squarely on, then extensive damage can result. Transmission of an explosion through a pipeline from one enclosure to another can lead to a violent secondary explosion (Holbrow *et al.* 1999). Flame transmission can be prevented by isolation techniques that include triggered suppression barriers and fast-acting valves (Barton 2002).

References

Abbot, J. (1990) *Prevention of Fires and Explosions in Dryers – A User Guide*, 2nd edn. The Institution of Chemical Engineers. Rugby, UK.

American Society for Testing Materials (1997) *ASTM E1491-97 – Standard Test Method for Minimum Autoignition Temperature of Dust Clouds*. ASTM International, West Conshohocken, PA.

American Society for Testing Materials (2000) *ASTM E1515-00 – Standard Test Method for Minimum Explosible Concentration of Combustible Dusts*. ASTM International, West Conshohocken, PA.

Bailey, M., Hooker, P., Caine, N. & Gibson, N. (2001) Incendivity of electrostatic discharges in dust clouds; the minimum ignition energy problem. *J. Loss Prev. Proc. Ind.*, **14**(2), 99.

Bartknecht, W. (1981) *Explosions Course Prevention Protection*. Springer, Berlin.

Bartknecht, W. (1993) *Explosion Protection – Fundamentals and Applications*. Springer, Berlin (in German).

Barton, J. (ed.) (2002) *Dust Explosion Prevention and Protection. A Practical Guide*. Institution of Chemical Engineers, Rugby, UK.

Beck, H., Glienke, N. & Mohlmann, C. (1997) *Combustion and Explosion Characteristics of Dusts*. Berufgenossenschaftliches Institut fur Arbeitssicherheit, BIA Report 13/97.

Bennett, D., Lunn, G.A., Torrent, J.V., Querol, E. & Fritze, J-P. (2002) *Hot Surface Ignition of Flammable Dust Deposits*. Paper to the ESMG Conference on Explosion Protection. ESMG, Nuremberg.

Boyle, A.R. & Llewellyn, F.J. (1950) The electrostatic ignitability of dust clouds and powders. *J. Soc. Chem. Ind. Trans.*, **69**, 73–181.

BS 5958: 1991 (1991) *Code of Practice for Control of Undesirable Static Electricity. Parts 1 and 2*. British Standards Institution, London.

BSEN 1127-1:1998 (1998) *Explosive Atmospheres – Explosion Prevention and Protection. Part 1. Basic Concepts and Methodology*. British Standards Institution, London.

BSEN 13821:2002 (2002) *Potentially Explosive Atmospheres – Explosion Prevention and Protection – Determination of Minimum Ignition Energy of Dust/Air Mixtures*. British Standards Institution, London.

BSEN 50 020:2002 (2002) *Electrical Apparatus for Potentially Explosive Atmospheres – Intrinsic Safety 'i'*. British Standards Institution, London.

BSEN 50281-2-1:1999 (1999) *Electrical Apparatus for Use in the Presence of Combustible Dust – Part 2-1: Test Methods – Methods of Determining Minimum Ignition Temperatures*. British Standards Institution, London.

Cesana, Ch. & Siwek, R. (2001a) *Mike 3. Minimum Ignition Energy. 3.3*. Kuhner AG, Birsfelden, Switzerland.

Cesana, Ch. & Siwek, R. (2001b) *Operating Instructions 20-l-Apparatus 6.0*. Kuhner AG, Birsfelden, Switzerland.

Dangerous Substances and Explosive Atmosphere Regulations (DSEAR) (2002).

Directive 1999/92/EC of the European Parliament and of the Council of 16 December 1999 on Minimum Requirements for Improving the Safety and Health Protection of Workers Potentially at Risk from Explosive Atmospheres (2000). *Off. J. Eur. Commun.*, **L23**, 57–63.

Dust Explosion Test: Modified Hartmann Apparatus. http:/www/kuhner.com/DOCUMENT/b035_028.pdf

Eckhoff, R.K. (1997) *Dust Explosions in the Process Industries*, 2nd edn. Butterworth-Heinemann, London.

EU Funded Project SMT4-CT98-2217 (1999) *Gas Ignition Test for Heavy Current Low Voltage Circuits*. Final Report.

Gardner, B.R. Winter, R.J. & Moore, M.J. (1986) *Explosion Development and Deflagration-to-Detonation Transition in Coal Dust/Air Suspensions. Twenty-First Symposium (International) on Combustion*, p. 335. The Combustion Institute.

Gibson, N. & Lunn, G.A. (2003) *The Interpretation of Flammability and Ignitability Test Data Determined in Standard Tests*. Health and Safety Laboratory Report. TD/03/01.

Griesche, G. & Brandt, D. (1976) Einflussfaktoren auf die Zundtemperatur von Staub-Luft-Gemischen. *Die Technik.*, **31**, 504.

Harper, D.J., Plain, K.E., Wilton, J.S. & Gibson, N. (1997) *Use of Intrinsically Safe Circuits and Enclosures to Control Ignition Risk from Equipment in Powder Handling Plant. IchemE. Symposium Series No. 141. Hazards XIII. Process Safety. The Future*, p. 463.

Hertzberg, M. & Cashdollar, K.L. (1987) Introduction to Dust Explosions, in Cashdollar K.L. & Hertzberg, M. (eds) *Industrial Dust Explosions*. ASTM International, West Conshohocken, PA, p. 5.

Holbrow, P., Lunn, G.A. & Tyldesley, A. (1999) Dust explosion protection in linked vessels: containment and venting. *J. Loss Prevent. Proc. Ind.*, **12**, 227.

International Social Security Association (ISSA) (1996a) *Determination of the Combustion and Explosion Characteristics of Dusts. ISSA Prevention Series No. 2018*. The International Section for the Prevention of Occupational Risks in the Chemical Industry. The International Social Security Association (ISSA), Heidelberg, Germany.

International Social Security Association (ISSA) (1996b) *Static Electricity. Ignition Hazards and Protection Methods; a Practical Guide. ISSA Prevention Series No. 2017*. The International Section for the Prevention of Occupational Risks in the Chemical Industry, the International Social Security Association (ISSA), Heidelberg, Germany.

Line, L.E., Rhodes, H.A. & Gilmer, T.E. (1959) The spark ignition of dust clouds. *J. Phys. Chem.*, **63**, 290.

Nelson, M. (1995) Detection and extinction of fire and smouldering in bulk powder. Paper presented at *Dust Explosions. Protecting People, Equipment, Buildings and Environment*. IBC Technical Services, London, 11/12 October.

PD CLC/TR SO4O4 (2003) *Electrostatics. Code of practice for the avoidance of hazards due to static electricity*. British Standards Institution, London.

Pinkwasser, Th. (1986) *On the Ignition Capacity of Free-Falling Smouldering Fires*. Euromech Colloquium 208, Explosions in Industry.

Smielkow, G.I. & Rutkowski, J.D. (1971) Badania Zjawiska Zaplonu Mieszanin Pylowopowietrznych wywolanego wyladonaniami iskrowymi. *Chemia Stosowana*, **XV3**, 283.

Standard BSEN 50281-1-2:1999 (1999) *Electrical Apparatus for Use in the Presence of Combustible Dust – Part 1-2: Electrical Apparatus Protected by Enclosures – Selection, Installation and Maintenance*. British Standards Institution, London.

Testing Methods for Electrical Apparatus Installed in a Dusty Environment with a Potential Risk of Explosion (EADE). EU funded project Contract No. SMT-CT98-2273. Synthesis report available on the SAFETYNET website: www.safetynet.de.

The Machinery Directive (1998) Directive 89/392/EEC, as subsequently amended, of the European Parliament and the Council of 22 June 1998 on the approximation of the laws of the Member States relating to machinery. *Off. J. Eur. Commun.*, **L207**, 1–46.

The Potentially Explosive Atmospheres Directive (1994) Directive 94/9/EC (ATEX 100A) of the European Parliament and the Council of 23 March 1994 on the approximation of the laws of the member States concerning equipment and protective systems intended for use in potentially explosive atmospheres. *Off. J. Eur. Commun.*, **L100**, 1–29.

Uth, H-J. (1999) *Prevention of Dust Explosion Hazards in Germany – a Regulator's View*. In: European Seminar on 'Dust Explosions', 13–15 April 1999, Metz, France.

van Larr, G.F.M. (1996) Interpretation of the standard gas and dust explosion characteristics based upon process knowledge: the key to realistic safety in real industrial plant. In: *2nd World Seminar on the Explosion Phenomenon and on the Application of Explosion Protection Techniques in Practice (EUROPEX)*, 4–8 March, 1996. Gent, Belgium.

VDI 3673 (2002) *Part 1. Pressure Venting of Dust Explosions. Verein Deutscher Ingenieure*. November, 2002.

Wiemann, W. (1987) Influence of temperature and pressure on the explosion characteristics of dust/air and dust/air/inert gas mixtures. In: *Industrial Dust Explosions* (eds K.L. Cashdollar & M. Hertzberg), p. 33. ASTM International, West Conshohocken, PA.

Wilen, C. & Rautalin, A. (1999) Safety aspects of fuel handling in IGCC and PFBC plants. In: *15th International Conference on Fluidized Bed Combustion*, Savannah, USA, 13–16 May 1999.

Zeeuwen, J.P. & van Larr, G.F.M. (1984) An overview of the research at TNO in the dust explosion field over the last decade. *VDI-Berichte*, No. 494, Seite 11, pp. 109–18.

7 Characterisation of bulk materials – industrial practice

LYN BATES

7.1 Introduction

Loose solids are stored, handled and processed in prodigious quantities and diverse manners throughout industry. Many factors influence the behaviour of such materials, such as the intrinsic nature of the basic material of composition and the ambient conditions and operation circumstances of the equipment in use. In order to secure reliable performance, it is essential to be able to predict the behaviour of the material and take account of these characteristics in the equipment design. The failure to achieve plant designs that provide trouble-free operation when handling bulk materials is a major source of industrial inefficiency.

Even apparently minor flow difficulties often give rise to serious operating problems and prove intractable to cure. Severe flow problems cause many major production losses, shortfalls in production, product losses, rectification disruption, and even catastrophic failures. The diversion of effort by engineers and management incur both direct and indirect costs unrelated to the capital value of the equipment involved. Consideration of the potential consequential aspects of a failure to provide the required performance would frequently introduce a different perception to cost differences between suppliers for crucial items of equipment for storage and handling duties. The significance of poor plant performance resulting from powder flow problems is rarely appreciated until it has been experienced.

Analysis of the behavioural characteristics of materials in granular powder form is severely limited by the nature of particulate systems. Fundamental difficulties arise from the great number of variables involved, such as the basic properties of the constituent particles, their structure in the mass, and the various systems and sequences of stress to which the bulk may be subjected.

Further complications arise from exposure to differing ambient and process conditions and their variations, second phase effects of the void content and from any changes which may take place due to chemical, thermal, bacterial, ageing, electrostatic, attrition or other factors which effect the condition, consistency, uniformity, stability or other characteristic of the bulk material. These factors also have to be related to the equipment scale, geometry and constructional features of the installation, and particularly to the containing contact surfaces against which the material is expected to slip.

Features of interest similarly vary. These range from subjective judgements of appearance and texture to quantitative measurements of physical characteristics. For such reasons, the comprehensive study of bulk material properties is an involved subject having many interacting variables. The detailed influence of measured powder properties on equipment design is usually the domain of the specialist. Nevertheless, a systematic approach provides a framework of understanding which will indicate which measurements should be taken for a guide to key aspects of performance.

There are three broad areas of industrial interest in bulk characterisation, health and safety, plant performance and product quality. A primary focus is essentially centred on health and safety issues. These have a major influence in the degree of containment necessary in respect of loading, operation and unloading, but they also bear strongly on aspects of performance reliability, cleaning, residue, product changes, sampling and other issues that may involve manual intervention. The integrity of seals and maintenance demands also assumes significant design importance when the product properties constitute a hazard to personnel. Account may be taken of the potential degree of exposure in terms of magnitude and frequency, according to the nature of the bulk materials.

Generally, inherent product hazards in relation to personnel are well known and documented in relation to their chemical and physical composition by way of identification under 'risk phrases' as shown in Box 7.1. Such properties are inherent to the basic material, but their influence may be reviewed by considering the form of flow route undertaken in the project. Dust generating, for example, assumes more prominence when products containing fines have adverse ingestion effects. In many cases, the tolerable levels of exposure are to be determined, and containment procedures and techniques are well established. Some industries, as with lead and radioactive substances, have regular monitoring and personnel health checklists ensure that exposure limits are not exceeded.

It is useful to examine the relationship between the primary properties of a particle and how these form characteristics of the bulk material, which in turn influences the way in which the product will behave in given circumstances. The fundamental particle properties, although of much interest for a variety of reasons, do not allow accurate predictions to be made of the behaviour of the mass of particles in bulk form because of many interactive and external influences (Figure 7.1). Measured physical properties of the bulk material can be used for design purposes, but these do not translate directly to dynamic behavioural characteristics, that have to be characterised and quantified by other means.

A crude categorisation grouping can be made according the nature of changes that may take place in the bulk material. This will highlight the type of precautions and tests that should be made. Note that a bulk material will have a response to all queries, so the general nature of the material may be represented by four digits.

	Yes	**No**
Type 1	Products gain strength with compaction	
Examples	Flour, cement, coal fines	Dry sand, granular sugar
Type 2	Products that physically alter in composition (segregation, attrition)	
Examples	Coarse coal, tea, coffee granules	Titanium dioxide powder
Type 3	Products that form bonds (cake, sinter, fuse, chemically active, melt)	
Examples	Salt, soluble crystals, warm plastic, phosphates	Dry, inert minerals
Type 4	Products that change with time (degrade, decompose, age)	
	Agricultural products, cement	

Box 7.1 Risk phrases used in the countries of EU (phrases in parentheses are no longer in use): nature of special risks attributed to dangerous substances and preparations.

R1	Explosive when dry
R2	Risk of explosion by shock, friction, fire or other sources of ignition
R3	Extreme risk of explosion by shock, friction, fire or other sources of ignition
R4	Forms very sensitive explosive metallic compounds
R5	Heating may cause an explosion
R6	Explosive with or without contact with air
R7	May cause fire
R8	Contact with combustible material may cause fire
R9	Explosive when mixed with combustible material
R10	Flammable
R11	Highly flammable
R12	Extremely flammable
R13	(Extremely flammable liquified gas)
R14	Reacts violently with water
R15	Contact with water liberates highly flammable gases
R16	Explosive when mixed with oxidizing substances
R17	Spontaneously flammable in air
R18	In use may form flammable/explosive vapour–air mixture
R19	May form explosive peroxides
R20	Harmful by inhalation
R21	Harmful in contact with skin
R22	Harmful if swallowed
R23	Toxic by inhalation
R24	Toxic in contact with skin
R25	Toxic if swallowed
R26	Very toxic by inhalation
R27	Very toxic in contact with skin
R28	Very toxic if swallowed
R29	Contact with water liberates toxic gases
R30	Can become highly flammable in use
R31	Contact with acids liberates toxic gas
R32	Contact with acids liberates very toxic gas
R33	Danger of cumulative effects
R34	Causes burns
R35	Causes severe burns
R36	Irritating to eyes
R37	Irritating to respiratory system
R38	Irritating to skin
R39	Danger of very serious irreversible effects
R40	Possible risks of irreversible effects
R41	Risk of serious damage to eyes
R42	May cause sensitization by inhalation
R43	May cause sensitization by skin contact
R44	Risk of explosion if heated under confinement
R45	May cause cancer
R46	May cause heritable genetic damage
R47	(May cause birth defects)
R48	Danger of serious damage to health by prolonged exposure
R49	May cause cancer by inhalation
R50	Very toxic to aquatic organisms
R51	Toxic to aquatic organisms
R52	Harmful to aquatic organisms
R53	May cause long-term adverse effects in the aquatic environment
R54	Toxic to flora
R55	Toxic to fauna
R56	Toxic to soil organisms
R57	Toxic to bees
R58	May cause long-term adverse effects in the environment

R59	Dangerous for the ozone layer
R60	May impair fertility
R61	May cause harm to the unborn child
R62	Possible risk of impaired fertility
R63	Possible risk of harm to the unborn child
R64	May cause harm to breastfed babies
Combination of R-phrases	
R14/15	Reacts violently with water liberating highly flammable gases
R15/29	Contact with water liberates toxic, highly flammable gas
R20/21	Harmful by inhalation and in contact with skin
R20/22	Harmful by inhalation and if swallowed
R20/21/22	Harmful by inhalation, in contact with skin and if swallowed
R21/22	Harmful in contact with skin and if swallowed
R23/24	Toxic by inhalation and in contact with skin
R23/25	Toxic by inhalation and if swallowed
R23/24/25	Toxic by inhalation, in contact with skin and if swallowed
R24/25	Toxic in contact with skin and if swallowed
R26/27	Very toxic by inhalation and in contact with skin
R26/28	Very toxic by inhalation and if swallowed
R26/27/28	Very toxic by inhalation, in contact with skin and if swallowed
R27/28	Very toxic in contact with skin and if swallowed
R36/37	Irritating to eyes and respiratory system
R36/38	Irritating to eyes and skin
R36/37/38	Irritating to eyes, respiratory system and skin
R37/38	Irritating to respiratory system and skin
R39/23	Toxic: danger of very serious irreversible effects through inhalation
R39/24	Toxic: danger of very serious irreversible effects in contact with skin
R39/25	Toxic: danger of very serious irreversible effects if swallowed
R39/23/24	Toxic: danger of very serious irreversible effects through inhalation and in contact with skin
R39/23/25	Toxic: danger of very serious irreversible effects through inhalation and if swallowed
R39/24/25	Toxic: danger of very serious irreversible effects in contact with skin and if swallowed
R39/23/24/25	Toxic: danger of very serious irreversible effects through inhalation, in contact with skin and if swallowed
R39/26	Very toxic: danger of very serious irreversible effects through inhalation
R39/27	Very toxic: danger of very serious irreversible effects in contact with skin
R39/28	Very toxic: danger of very serious irreversible effects if swallowed
R39/26/27	Very toxic: danger of very serious irreversible effects through inhalation and in contact with skin
R39/26/28	Very toxic: danger of very serious irreversible effects through inhalation and if swallowed
R39/27/28	Very toxic: danger of very serious irreversible effects in contact with skin and if swallowed
R39/26/27/28	Very toxic: danger of very serious irreversible effects through inhalation, in contact with skin and if swallowed
R40/20	Harmful: possible risk of irreversible effects through inhalation
R40/21	Harmful: possible risk of irreversible effects in contact with skin
R40/22	Harmful: possible risk of irreversible effects if swallowed
R40/20/21	Harmful: possible risk of irreversible effects through inhalation and in contact with skin
R40/20/22	Harmful: possible risk of irreversible effects through inhalation and if swallowed
R40/21/22	Harmful: possible risk of irreversible effects in contact with skin and if swallowed
R40/20/21/22	Harmful: possible risk of irreversible effects through inhalation, in contact with skin and if swallowed
R42/43	May cause sensitization by inhalation and skin contact
R48/20	Harmful: danger of serious damage to health by prolonged exposure through inhalation
R48/21	Harmful: danger of serious damage to health by prolonged exposure in contact with skin
R48/22	Harmful: danger of serious damage to health by prolonged exposure if swallowed

R48/20/21	Harmful: danger of serious damage to health by prolonged exposure through inhalation and in contact with skin
R48/20/22	Harmful: danger of serious damage to health by prolonged exposure through inhalation and if swallowed
R48/21/22	Harmful: danger of serious damage to health by prolonged exposure in contact with skin and if swallowed
R48/20/21/22	Harmful: danger of serious damage to health by prolonged exposure through inhalation, in contact with skin and if swallowed
R48/23	Toxic: danger of serious damage to health by prolonged exposure through inhalation
R48/24	Toxic: danger of serious damage to health by prolonged exposure in contact with skin
R48/25	Toxic: danger of serious damage to health by prolonged exposure if swallowed
R48/23/24	Toxic: danger of serious damage to health by prolonged exposure through inhalation and in contact with skin
R48/23/25	Toxic: danger of serious damage to health by prolonged exposure through inhalation and if swallowed
R48/24/25	Toxic: danger of serious damage to health by prolonged exposure in contact with skin and if swallowed
R48/23/24/25	Toxic: danger of serious damage to health by prolonged exposure through inhalation, in contact with skin and if swallowed
R50/53	Very toxic to aquatic organisms, may cause long-term adverse effects in the aquatic environment
R51/53	Toxic to aquatic organisms, may cause long-term adverse effects in the aquatic environment
R52/53	Harmful to aquatic organisms, may cause long-term adverse effects in the aquatic environment

It is important to recognise that detailed material characterisation is a comprehensive and interactive exercise. It embraces the total life cycle of the project, the product lifecycle in the plant and any exceptional operating conditions that may at some stage be expected to occur within the total history of the plant operation. The significance of plant shutdowns, whether planned or unexpected, machine breakdowns and maintenance needs, changes of process, accidents or outside influences may be categorised as falling within circumstances that may have planned accommodation procedures so rare or exceptional that emergency measures should be undertaken if the need arises. These features are in association with working conditions that influence how the material will behave.

While such health and safety concerns have an overriding influence on plant design, they tend to exert boundary conditions on the equipment selected rather than dictate the forms. Equipment form is determined by the process function of the plant, and it is in this area that the two main operational interests of product categorisation come into prominence. Characteristics relating to the smooth and efficient operation of the plant have a vital role in the design process because reliable and economic performance is crucial to the viability of the operation. Great interest is also attached to factors that influence the quality of the product.

A distinguishing feature between these two interests is that performance-related factors are almost invariably connected to the mechanics and physics of particulate solids, and bear on the engineering task in a predictable way. By contrast, quality issues are almost invariably related to user preference and judgement, generally as dictated by the perceived utilities of the product. Whereas the nature and condition of the constituent particles have strong engineering implications, it is difficult in some cases to define their influence on product quality with precision. Where specific characteristics are the essence of a process, bounding examples may be the only way to set a standard of quality.

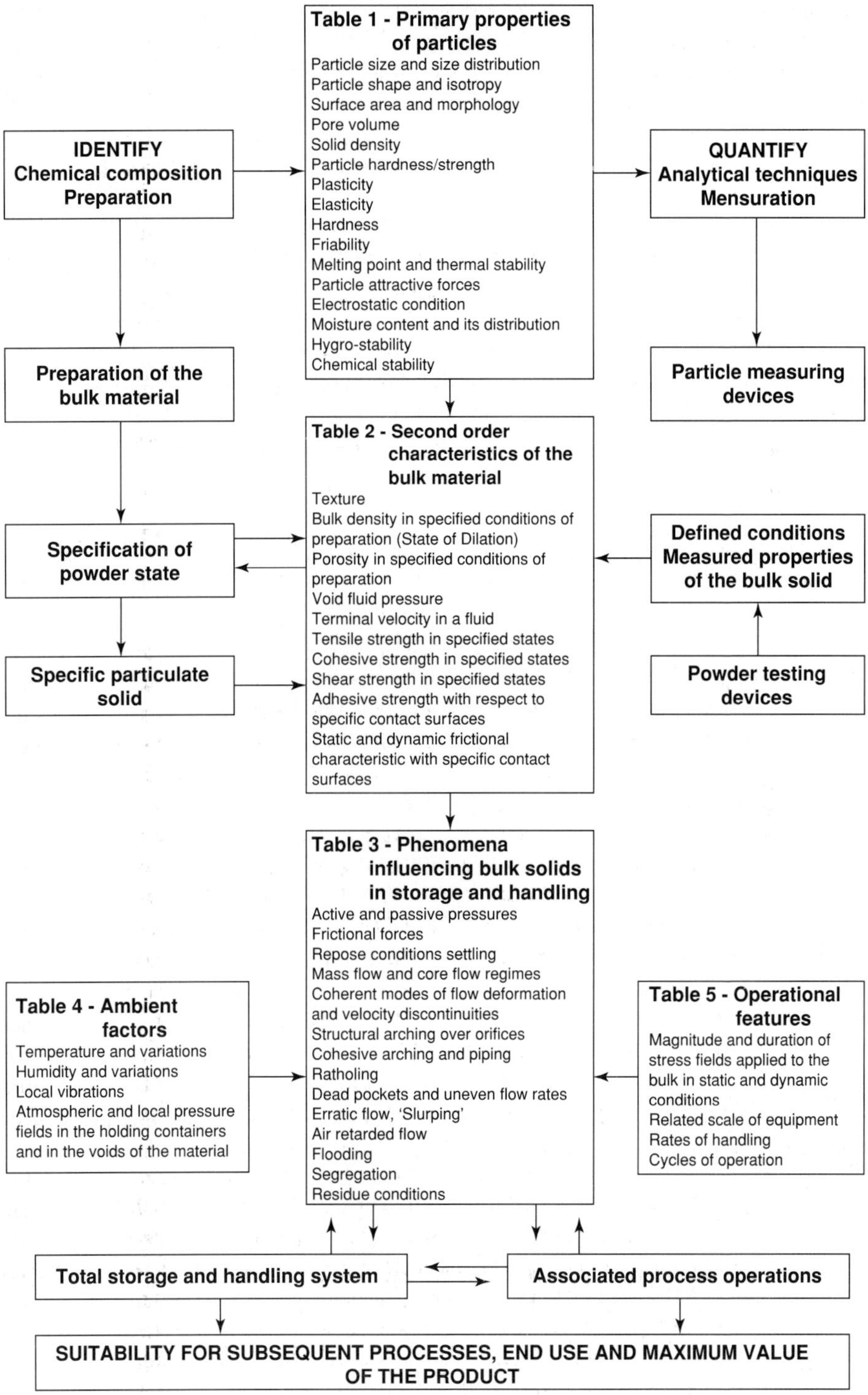

Figure 7.1 Aspects of bulk solids relating to storage and handling.

A feature of great importance is that, by design or otherwise, the product changes during its process through the plant. Quality may vary during this path, but while progressive changes may be accumulative, main product concern relates to the ultimate condition in which the ultimate user receives the material. By contrast, the functional operation requires that at each stage of the process, the condition of the material is suitable for the operation in hand and that it is delivered forward in a condition for this next operation to be reliably and efficiently performed.

Taking operational issues, there are a host of behavioural conditions of bulk solids that may occur. It is good industrial practice to conduct a thorough review of their likelihood in a sequential flow audit. This should take account of the nature of the particulate solid in relation to the scale and form of the equipment to be used and the bounds of ambient condition that may prevail within the sequence and timescale that can apply. This is a demanding exercise, but neglect of detail is the source of so many emergency failures that have endowed solids handling plants with their unenvied reputation for poor performance.

For a plant flow audit, it is useful to classify the range of operating conditions to which the bulk material will be exposed:

A: The degree of confinement, freedom of the material to dilate according to conditions
Examples

A.1	Totally unconfined – Trajectories and free fall
A.2	Partially confined – Transfer chutes, belt and screw conveyors
A.3	Totally confined – Flooded chutes, storage bins and hoppers
A.4	Flow boundary – Contact surface, shear plane

B: The effective stress conditions acting on the material (note degree of confinement)
Examples

B.1	Incipient failure, initial deformation (stress history and prevailing stresses pertinent)
B.2	Transitional state, changing stress conditions
B.3	Steady state flow, 'critical state' conditions prevail (stress/density relationship)

C: Ambient conditions
Examples

C.1	Steady or fluctuating (timescale and consistency)
C.2	Range and form of variation

Behavioural characteristics of bulk solids are best thought of as 'attributes' that are a consequent of a specific combination of physical, chemical, electrostatic, thermal and other fundamental properties. In this particular condition, a mass of bulk material will exhibit secondary aspects of behaviour such as bulk strength, cohesion, surface friction, bulk density variations, reactivity in given environmental conditions and conditions that give rise to a range of behavioural responses. Some consequences may occur that are not significant to the operation because of their limited nature or inconsequential effect. Other effects may be slight but may be highly significant because of the sensitive nature of the attribute. However, it is good practice to consider all prospects and their accumulative and future consequences to cover all eventualities, rather than overlook a potential problem by default.

A number of bulk attributes influence the 'handleability' of a bulk solid. These properties may affect the behaviour of the material and/or the type of equipment used. The legends are as follows:

- A: sensitising
- C: corrosive
- E: explosive
- Fo: flammable
- Fx: extremely flammable
- H: skin absorption
- K: carcinogenic
- Lk: no classification required
- N: danger to the environment
- M: toxic
- O: oxidising
- R: causes birth defects
- Tx: very toxic
- Xi: irritant
- Xn: harmful

These attributes may be structured according to the nature of the property and its effect on the specification of the equipment, as follows:

Property	**Effect**	**Typical material**
Packs under pressure	More difficult to flow	Hydrated lime, pigments
Cohesive	Forms poor flow mass	Flour, fly ash, titanium dioxide
Fibrous	Interlocks to resist shear	Hair, wood shavings
	Gains strength with compaction	Asbestos
Fluidises	Difficult to contain	Fly ash, talcum powder
Fatty	Sticks together to resist flow	High-fat mixes, waxes
Elastic	Deforms at contact points to resist flow	Cork, rubber granules, plastic pellets
Plastic	Deforms at contact points to resist flow	Plastics
Chemically active	Forms solid mass Must prevent	Ground phosphate
Melts	Can fuse to solid	Plastic, ice
Sinters	Fuses to solid mass	Warm plastics
Fuses	Forms solid mass Must prevent	Raw rubber
Cakes	Forms solid mass Must prevent	Salt, sugar, crystals
Wet	Sticky, may dry to 'cake'	Filter and centrifuge cake
Sticky	Adhesive to surfaces	Damp and fatty products
High friction	Resists slip on contact surface	Titanium dioxide
Abrasive	Wears plant	Sand, aggregate, crystals
Corrosive	Attacks surfaces	Salt, acidic chemicals

Friable	Delicate handling needed	Tea, coffee granules, flakes
Explosive	Must contain, inert, suppress or vent	Coal dust, aluminium powder, flour
Flammable	Must prevent or protect equipment	Wood shavings
Dusty	Hazard and objectionable Should contain, collect or suppress	Cement, fly ash
Hygroscopic	Becomes sticky	Sugar, soda ash
Deliquescent	Dissolves	
Noxious	Offensive to operatives Must contain	Sewage sludge, waste
Toxic	Dangerous to operatives Must contain	Arsenic powder, active drugs
Irritant	Hazard to operatives Must contain	
Sensitiser	Hazard to operatives Must contain	Penicillin intermediates
Degradable	Cleanability	Organic products
Hot	Hazard to operatives,	Kilned powder
Sharp	Hazard to operatives	Cullet, metal cuttings
Ultra-pure	Sanitary	Drugs, meat, fish

This list is not exhaustive but illustrates that interactive factors may determine the overall specification of equipment. Fundamental to the performance of the plant, however, is that the flow behaviour must be reliable and safe, which means that it must be predictable and the potential hazards recognised and accommodated.

Some of these attributes are connected and/or interactive. For example, particle attrition leads to the formation of fines, which are then susceptible to segregation. Numerous other consequences can then develop, such as vulnerability to dust explosions or personnel hazards because of the local accumulation of fines or coarse fractions. Flow mechanisms also induce various forms of segregation that can have variable effects on further behaviour and dust-related hazards.

The initial evaluation stage does not necessarily require specialised skills or expensive equipment. Simple tests can often be utilised to indicate whether a problem is likely or not, and to indicate which aspect requires further and more detailed attention. A common industrial problem is, from the morass of interacting variables, deciding upon the form and degree of testing that should be conducted.

7.2 Aspects of powder behaviour

With the exception of simple cases, the behaviour of a particulate material cannot be reliably predicted from knowledge of the primary properties of the particles of which it is composed. Various secondary properties of the bulk are therefore measured directly to evaluate the mechanical characteristics of the mass. Measured values of how the material deforms under stress and knowledge of the slip properties on a contact surface are used as the basis of design for flow of a bulk solid in hoppers and chutes. Assessments can also be made of power requirements for feeders and conveyors, and also for various operations such as blending and mixing. These values also form the basis by which a material is specified or categorised for storage and handling applications and for some process operations. To secure quantified values for a bulk material characteristic, it is necessary to conduct controlled tests.

Testing in appropriate circumstances also reveals how these properties, and the corresponding behaviour of the material in specific circumstances, are influenced by changes in the bulk condition of the material. The effects of the interaction of differing variables, when reproduced in laboratory conditions, offer the prospect of determining how a material will behave in a range of operational conditions.

7.2.1 *The selection of a sample for testing*

A simple initial classification of the bulk material may be accomplished by considering its nature under the various risk phrases (Box 7.1) This simple concept approach draws attention to the precautions necessary when planning and evaluating sample selections and when choosing the test method to be adopted to secure meaningful measured values from a test. The sample must be fully representative of the conditions of interest. That is, the sample and the test process followed must reflect the worst conditions to be encountered in the equipment and process under consideration for the specific design feature being considered.

The first step in this sample-selection process is to determine what forms of variation may prevail in the material in the circumstances and at the scale of scrutiny significant to the application. Variations in the material's condition may be considered under three headings:

(1) Uniformity: Is the material homogeneous, i.e. of even composition throughout the mass at all times, not subject to local variations of particle size, isotropy, moisture content, temperature or any other feature that may affect its condition or behaviour?
(2) Consistency: Is the material always of a similar nature on a day-to-day, year-to-year or source-to-source basis? Naturally derived products, whether organic or mineral, may be expected to vary to some degree according to their origin and history, and some process operations are also prone to have variable product condition outputs.
(3) Stability: Over the relevant plant life of the material, will changes take place in its condition due to age, chemical or thermal action or other internal or external variants to adversely change the nature of the material?

It must be appreciated that, so far as powder testing is concerned, any variation of a bulk material composition or condition essentially represents a different 'material'. The behaviour of a bulk material in one condition cannot be predicted or reliably inferred from that in another. When two or more variables interact, the consequences are even less related. For this reason, it is essential that the quality of material secured for a test is appropriately representative and that the testing circumstances reflect the relevant application conditions.

The second step of the sample selection process is to establish the particular condition(s) which give the 'worst' product properties to be accommodated in the equipment or the process under review. A simple review of the product with respect to the various operations throughout the material life in the plant may reveal obvious states that exhibit such 'worst' behavioural qualities. Samples in these appropriate conditions may then be tested.

In other circumstances, a more systematic study of the flow route and the product life cycle is required. This is conducted to determine whether such features as ambient changes, preferential condensation, degradation, segregation or uneven residence periods in the equipment give rise to inconsistencies of the product nature that are not present in the original substances. Account should also be taken of relatively abnormal, yet possible, circumstances, such as freak ambient conditions, minor mishaps, extended stoppages, rework handling and the like.

The third stage of the sample selection process is to secure or prepare a sample in the condition required. If there is any doubt about what sample condition should be tested, a two-stage programme should be conducted. First, pre-testing a range of samples to establish the range of variation that the equipment has to accommodate. The final tests are then undertaken with material conditions representing the 'worst' properties for the particular feature under consideration.

It should be noted that the 'worst' condition of the material for some sections of handling plant may be favourable for other sections. For example, the opening size of a hopper outlet to provide reliable flow of a material in a difficult flow condition will lead to higher overpressures on a feeder if a more free-flowing condition of material is then handled. Likewise, high friction on the upper region of hopper walls will reduce compacting pressures in the lower region of the hopper but require steeper wall inclination to give slip on the converging surfaces, and vice versa. The density values for settled and dynamic conditions must also be determined and used to reflect the appropriate volumes for storage and handling rates.

Great care is needed in the selection of material samples and choosing test conditions to avoid deriving useless or misleading data. Prior published data must also be used with great discretion.

An I.Mech.E (1994) publication outlines a basis of specification of a bulk material for storage and handling purposes. The material's range of conditions and the nature of its use must be established before testing. It is important to cover acceptable extremes of product and operating conditions, not only 'design values' that represent 'normal' conditions of operation. While quantified values should be used wherever possible, some characteristics, such as appearance or value, are subjective. It is vital that authoritative definitions of important features are identified, specified and agreed, for contractual purposes. Where numerical values cannot be recorded, there should be definitive samples or conditions specified as permissible bounds of variation.

As an aide-mémoire to the differing considerations affected by bulk material properties, features of interest may be considered under the following 'P' headings:

- Production: periods of run and standing, frequency of use, stock
- Performance: quantities, rates, reliability, accuracy
- Product: quality, segregation, appearance, texture, degradation, hygiene
- Personnel: health and safety, working conditions
- Plant: durability, maintenance, corrosion, containment, power
- Pollution: local and broad environment, wastage, dirt and spillage

- Permanence: the working life of the equipment
- Position: considerations of space and site location
- Panic: what is at stake if the equipment does not work properly in terms of operating losses and retrofit requirements?
- Price: the implications of the above features to the plant capital and revenue costs over the working life of the system
- Profit: how dependent is the return on capital to the reliability of the equipment?

The value of securing valid test results can then be assessed against these various factors to indicate what testing costs are justified. A way to review the project is by way of a flow audit that systematically follows the product route through its plant life and considers the current and accumulated influence of the system dynamics on the state and condition of the product.

7.2.2 *Reasons for testing*

Interest in the properties of powders arises for many reasons. The objectives of the test influence both the selection of the testing methods and the amount of testing undertaken.

Typical reasons for powder testing include:

- Registration: of a specific property for identification or reference purposes
- Value evaluation: determination of selling quality
- Contractual specification: e.g. defining bulk material properties for a storage or handling project or the supply of bulk processing equipment with a guaranteed performance
- Behavioural evaluation: the assessment or prediction of some behavioural aspect in a bulk storage, handling or process operation
- Comparison: of some properties with a similar or different material, or against a reference source to determine if a better or worse scenario exists for a specific function; systematic tests may be used to give an order of ranking for these values and products
- Categorisation: so that the prospective behaviour in a production or use operation may be assessed relative to known products or predicted within acceptable bounds
- Equipment manufacture: for the design of hoppers, chutes, handling and process equipment such as pelletising, tableting, roll pressing and the like
- Identify phenomena: to examine tendencies to arch, flush, settle to differing densities, adhere to contact surfaces, etc.
- Indicate the effect of product and process variations: allow sensitivity assessments and set bounds of acceptability for the product condition
- Quality control: to establish standards, monitor production processes, quality limits, formation of data bank, etc.
- Investigatory: for the examination of production problems and process variations, trouble shooting and failure analysis. To provide expert based data for litigation
- Product research and development
- Equipment research and development
- Fundamental research: including the development of powder testing devices and procedures

- Specification: of a bulk material for purchase or sale, multi-sourcing of raw materials, assessing products with natural variations, monitoring multiple production facilities, imparting long term standards and specifying particular characteristics.

7.3 The basis of test methods

These notes relate to flow-related properties of materials in bulk form, i.e. how the material deforms under stress and how it will slip on contact surfaces. Bulk strength and wall friction properties are therefore key performance parameters for flow in gravity flow and all devices that promote the movement of bulk materials. These features apply to chutes and hoppers, as well as mechanical conveyors, such as screw feeders, screw and scraper conveyors, mixers and the like and wherever bulk material flows or is caused to change shape or volume. They also apply to activities as diverse as roll pressing, tableting and compacting, sack filling, stockpiling and the behaviour of bulk cargoes at sea.

For some purposes, the type of test to be employed is well established by industrial practice and may be supported by the existence of a related design procedure, as with shear cell tests for hopper design. There is, however, a common requirement for simple, quick tests to evaluate the material and identify whether it is likely to give handling problems, or otherwise show a need for further investigation. The range of interest varies according to the purpose for which the test is to be carried out and the state of the art is not sufficiently developed to have established the adoption of standards or universally adopted methods for all these types of test. In these circumstances, there are two basic options: a distinction should be made between once-only tests, at the initial or design stage of a project, and regular or routine tests that are carried out as part of the ongoing operation.

7.3.1 *Empirical tests*

Many forms of empirical tests have been adopted; some particular to a specific bulk material or process and others for general use. They may be of qualitative form or result in a specific measured value. Such tests rely upon the evaluation of some phenomena-based condition that can be related to the circumstances of interest of the bulk solid. Where these closely simulate operational or production events, there is a direct indication of performance. Where such tests lack a credible scientific base, there is a danger that the method may be suspect, not fully representative or be subject to operator technique or some other variable.

Particular care should be taken to ensure that the measurements are not influenced by the manner of formation of the sample *prior* to the preparation and conduction of the test. If the objective of the test is to correlate behaviour in a less direct manner, a clear grasp of the fundamentals of powder technology is required to appreciate the implications of slight differences in relation to sensitive applications or in translating the results to other applications.

7.3.2 *Tests founded upon fundamental principles*

These are appropriate when the features of interest can be related to a property of the bulk material that can be measured in a direct manner. Some measurements, such as bulk density, require little in the way of equipment or skills. Other tests, such as shear cell work, demand considerable experience in both the procedure and the interpretation of results. Some proprietary test devices are available to measure specific data and provide direct measurements for design or evaluation purposes.

In all cases, a basic understanding of the principles of powder technology will give insight into the significance of the various measurements that are taken and also provide a guide to the degree of sensitivity of results that may be expected.

7.4 Main feature of interest – 'flow'

A powder mass is fundamentally different to the behaviour of a liquid, as powders have the ability to sustain internal stresses. Bulk materials deform internally, and the boundary surface of the flow channel may be on the contact surfaces of the containing vessel or within a static region of the bulk material itself. When a bulk material moves en masse in a coherent form or along a parallel channel, the bulk is not 'flowing' in the conventional sense as applicable to a bulk solid.

A distinction has to be made between two different circumstances. Flow that takes place under *unconfined conditions* allows deforming material to dilate or contract according to internal stress conditions. In *confined flow,* the total cross-section of the flow channel is bounded by a confining surface of a container wall or a static bed of the material itself, and the confining boundaries exert confining stresses on the moving media. In unconfined conditions, the slope of the contact surface needs only to exceed the angle of contact friction for the material to run clear of the walls. The flow of confined products is often inhibited by the demand of internal work to deform the mass to suit changes of cross-sectional area or direction.

Unconfined flow takes place on the draining surface of a non-mass flow hopper as it empties towards the central core in a characteristic 'funnel' shape. It also relates to the final slip from the walls when a hopper is empty in the central regions. In the first instance, the surface slope formed by the 'drained angle of repose' reflects the conditions of formation of the bulk material. The effect of being at rest for some time under load may create bulk strength sufficient to support a 'cliff' or a steep wall. When this happens, the term 'angle of repose' has no real meaning. Likewise, the material that has been held against the wall surface for some time under load may tend to stick due to the formation of surface adhesion or bonds. The 'angle of friction' can then be misleading, as the plot of force to promote slip against wall contact pressure is not a straight line in these circumstances. Testing of the material under appropriate conditions of time and contact pressure for surface effects or unconfined failure will highlight the extent to which these features may give rise to problems.

Similarly, for flow in chutes where the material does not fill the whole cross-section, static friction is measured to assess the conditions for initiating flow if the material has to be started from rest, or dynamic conditions of wall friction if the material is always sliding on the chute surface. If the bulk material fills the whole cross-section of the chute, it is termed a 'flooded' chute and has to considered as a form of hopper. The transfer capacity of a full chute is less than when the chute is not totally filled because unconfined material can flow much more freely than when is a state of confinement.

With regard to the behaviour of a bulk material which moves under various conditions of loading in conditions that fully confine the cross-section of flow channel, whether in a converging, parallel or diverging channel, there are two principal areas of interest:

(1) The conditions under which a material slips on the contact surfaces, i.e. the wall friction and any surface cohesion developed under differing contact pressures and duration periods, as for unconfined flow.
(2) The conditions under which the material shears, i.e. the internal strength of the bulk material in differing states of compaction. This is a more complicated situation than unconfined deformation because confining stresses, as well as the 'state' of the material, as characterised by its condition of density or compaction, influences its behaviour under the applied stresses.

7.5 Selection of appropriate test

For any new investigation, an initial requirement is to assess whether problems are likely to arise and if these should be more thoroughly evaluated. To do this quickly and economically, a two-stage approach is suggested.

(a) 'quick', simple tests (for initial evaluation)
(b) measured value tests (if then merited)

7.5.1 *'Quick tests'*

These 'quick tests' are put forward as virtually instantaneous, crude, low-cost tests to give an initial 'feel' to the nature of a new material, guidance as to potential hazards and show whether further testing is required. For hazardous materials, tests a4 and a5 may be appropriate. These tests are no substitute for properly conducted tests of measured value, but with only a little experience and practice, they can serve to provide a preliminary, but valuable, indication as to the behavioural characteristics of a product.

Further testing then depends upon the interpretation and the application interest.

	Tests	Indication
a1	'Snowball' test	Shear strength
a2	Wall 'stiction'	Wall adhesion
a3	'Cliff' formation	Cohesion
a4	'Agitated' bottle	A tendency to 'flush'
a5	'Flung' bottle	A tendency to compact
a6	Poured repose	Dilate shear strength
a7	Tapped test	'Compactability'
a8	Hausner ratio	'Compressability'

7.5.1.1 *Snowball test* Provided that the material can be handled without hazard, a quick way of finding whether the material is likely to present flow problems is to pick up a handful of the product and squeeze firmly. If the resultant 'ball' hangs together in a firm shape when the material is released, the material will not flow easily when subjected to pressure, and further examination is necessary. Should the material feel light and 'fluffy' during the test and squeeze out between the fingers, repeat the test very slowly with material that has been allowed to settle for some time. Materials that behave in a fluid-like manner when dilated tend to compact into a poor flow condition. If the material is apparently non-cohesive but, when gripped firmly with a slightly open hold, tends to 'lock up' and will not fall out of the packed mass freely when the hand is inverted, the material is likely to present flow problems under deep bed conditions.

Flaky, fibrous and elastic materials, such as plastic flakes, sawdust, maize meal and ground cork, will freely separate, even after firm compaction, but strongly resist shear in a confined, compacted condition due to the mechanical interference of the particles. This effect can be detected in a 'Snowball' test by the feeling generated that the bulk product 'locks together' when gripped firmly, but will dribble away through the fingers if pressure is relaxed. Such materials can arch over large outlets in storage hoppers. Further testing or experienced design attention is required to avoid or deal with overpressures acting on such products. If the material separates easily during and after the application of pressure, with no particles adhering to each other, it is unlikely that its condition changes in service, but expert evaluation is required for storage hopper design.

7.5.1.2 *Wall 'stiction' – for damp or cohesive materials* Form a snowball of the material as above, and press against a face of a material that will be used as a contact surface. If some material sticks, be wary of wall adhesion and progressive buildup of product on corners and surfaces exposed to high contact pressure. Carry out wall friction tests with force transducers at appropriate pressures. Alternatively, press the sample between two plates at pressures reflecting the conditions of application. If the material forms firm cakes or flakes, or sticks to the surface when separated, the need for further testing is indicated.

7.5.1.3 *Cliff test' – for fine, uniform materials* This is used for fine, uniform materials. Pour the material gently through a funnel onto a horizontal surface, noting the repose condition. Cut down the centre of the pile with a dividing knife, and move one section sideways. If the remaining pile holds a 'cliff' face at the division, poor flow conditions can be expected. If the face slumps to a similar repose angle as the original surface, it should present no serious flow difficulties.

7.5.1.4 *Agitated bottle* Vigorously shake a sealed glass jar approximately 75% full of the sample material. Observe if the material remains fluid after holding still a few seconds and how quickly the surface falls to a stable level. If it settles very quickly and the surface consistently forms a similar repose angle each time the surface is tilted slowly back and forth, it is non-cohesive and should flow well in light compact conditions. If the material takes some time to settle and inverting the container quickly shows the material to hold as a 'piston' in the jar or 'cracks' appear in the bulk, then expect poor flow conditions after a long period of standing but also consider the prospects of the material 'flushing' following dilate handling, such as loading by means of pneumatic conveyor or following free fall into a hopper, or from a chute, unless restrained. The collapse of arches in or from a hopper can result in such material behaving in a fluid manner.

7.5.1.5 *'Flung' bottle* Agitate and settle material in a jar as above. Swing the jar vigorously a few times to cause the contents to be accelerated against the bottom of the container by centrifugal action. Invert the jar to see if the material holds together or collapses freely. The former indicates that poor flow is likely.

7.5.1.6 *Poured repose* The prime value of such a test is to determine the form of the surface profile that will be built up during the filling of a hopper. The inclination of the surface shows the allowance that must be made for 'ullage', which represents the unfilled region of a hopper around a material inlet point. It does not indicate how the surface will 'drain' as the hopper empties because the pouring condition is essentially loose while the emptying condition is invariably settled and compacted by overpressures on the regions below the surface of the original hopper contents.

At a large scale, away from the actual impact point of the filling stream, the inclination of the poured repose surface tends to increase with distance from the central point of pouring and stabilise at a slightly higher angle of slope as the diameter of the repose cone increases. This is because the rate of divergence of the particles sliding down the cone of repose decreases as the surface becomes less curved. Conversely, with free-flowing products, drained repose angles are steeper in the focus of the material's convergence than the angle in regions further from the centre of flow. In essence, the effect is caused by particle interference in converging flow and the freedom from restraint given by divergence. Ultimately, both angles coincide at the repose conditions of a plane, provided the material has no cohesive properties.

This test can be used to measure the efficiency of flow aid additives or 'lubricants'. The material is fed through a coarse screen to drop onto the flat end of a 50-mm-diameter bar until a cone is formed. The angle of the surface to the horizontal is taken as a measure of the dilate shear characteristics of the material, hence its flow when not strongly compacted.

7.5.1.7 *Tapped density test* A quantity of material is gently poured into a tall measuring vessel or jar and allowed to settle for approximately 1 min. It is then subjected to repetitive light tapping or lifting and dropping approximately 25 mm onto a hard surface, say 20 times. The change in level is noted between its initial fill condition, the level after gravity settling and that after the application of tapping. Materials that significantly alter in density at these stages are likely to be variable in condition from behaving as a fluid when dilated to offering poor flow properties when settled under consolidating conditions. Note that the test must be performed at the temperature which the material will experience in service, as hot air in the voids of a fine powder will give very different results than when cold air is present due to the higher viscosity of air at elevated temperature inhibiting the escape of air from the voids during settling.

7.5.2 *Measured value tests*

b1	Wall friction	ϕ_w
b2	(i) Tensile strength	T_1
	(ii) Tensile strength	T_2
b3	Cohesion	C
b4	Unconfined failure strength	F_c
b5	Maximum principle stress/density	σ_1/γ
b6	Internal angle of friction	δ
b7	Permeability at given densities	ψ
b8	Shear strength	f

7.5.2.1 *Wall friction* – ϕ_w This is fundamental to hopper and chute design (Code of Practice for Hopper Design; The Design of Chutes and Conveyor Transfer Points) and relevant to wall contact in transport containers, process vessels and the like, mixer blades or any surface at which sliding, wiping or pressure contact takes place in all handling and process applications, including power and handling capacities for screw conveyors, scraper, vibratory, en masse and pneumatic conveyors. It is relevant to wall contact in transport containers, process vessels and the like, mixer blades or any surface at which sliding, wiping or pressure contact takes place, e.g. heat exchangers for solids.

7.5.2.2 *Tensile strength* The reason two tensile strengths are cited is that the failure strength of a powder compact in tension depends upon the relationship of the direction of the compacting forces to those forces applied to pull the specimen apart. The maximum tensile strength is developed in contact surfaces in line with the maximum principle stresses of the compacting force, as this is a direct action. The bulk will fail in tension at a lower value if the separating forces are applied at 90° to the axis of the compacting force. The particle-to-particle adhesive forces developed in this orientation are as a result of the lower influence of the minimum principle stress developed in the sample preparation. Industrial interest generally centres on the minimum strength at which a material will fail. This form of loading and stressing is also much easier to undertake without recourse to such complications as the use of adhesives. This method of measuring tensile strength is the only one served by a commercial instrument.

There is also a third tensile strength for a sample prepared by tri-axial compaction. In such a case, the orientation of failure plane is irrelevant. This stress system rarely occurs in industrial processes; hence this value is not considered in this review.

(1) Maximum tensile strength developed in a specific compact, T_1: This is a complementary measurement to Cohesion and Shear strength yield loci, a fundamental property of particle-to-particle strength, virtually independent of mechanical interference considerations (except in the case of fibrous, elongated or dendritic products). The value is useful for research in that the measurements indicate fundamental particle properties and surface effects. It is a relatively quick and inexpensive test, but as the forces involved are small, some care is needed in both equipment design and the procedure for carrying out the tests to secure the delicate measurements without introducing spurious internal or external forces. It may be useful for quality control, comparison tests measuring the effects of additives or moisture content, for flow assessment, measuring the strength of 'green' compacts, pills, roll pressings, extrudates or similar.

(2) Minimum tensile strength developed in a specific compact, T_2: This is a basic measurement of the lowest stress condition under which a material will fail under applied loads, a fundamental property of the bulk related to particle surface effects and principle stress ratios. This value complements the measurement T_1 but can be used directly to quantify the relative or absolute strengths of pressed cakes, tablets, formed compacts and the like. The measurement is useful for research or as a quick industrial test for assessment/comparison purposes as a measure of 'flowability' or other strength feature. A material with a high tensile strength is invariably cohesive. A tensile force acts on a failure surface as equivalent to an external normal force of

equal value. This force multiplied by the internal frictional results in a resistance to internal slip in the absence of external forces, i.e. is a component of the material's cohesive strength. However, Cohesion relates to the relative motion of particles along a slip plane, and therefore inter-particle interferences affect the slip resistance. This complicates the relationship of tensile strength to cohesive strength but indicates the more simple nature of tensile strength, compared with the compound characteristics of Cohesion.

7.5.2.3 *Cohesion, C* This is a fundamental characteristic of a bulk material reflecting both the particle adherence tendency and the structural composition of the mass as mechanical overlapping of the particles in the shearing layer obstructs initial yield of the bulk. This measurement of shear strength of a pre-compacted material no longer subject to a normal load on the failure surface indicates such behaviour as the strength of bag set materials or starting loads in equipment subjected to historical compacting loads only. Cohesive strength complements shear cell yield loci by providing the intercept on the yield loci diagram at the point of zero external load in the failure plane. The measurement may be employed as a quick measurement to assess flow prospects or comparative behavioural values of bulk materials in various states of pre-compaction and the absence of normal loading at the time of applied shear stress, i.e. unconfined failure conditions.

7.5.2.4 *Unconfined failure strength, F_c* This is the strength in the outer layers of a bulk material not constrained by a contact surface as on the surface of an arch in a hopper, the repose face of a cliff in a stockpile or 'pipe' in a flow channel. The measurement is very basic to a yield locus and directly indicates the interesting lower stressed region of the curve for a particular consolidating load, σ_1. Its direct measurement has generally been neglected or not pursued due to various practical difficulties of securing values not affected by other factors, such as wall friction in the apparatus and the need for the smooth application of the preparation and failure loads. More recent interest in this device for assessing the strength of coal samples and 'caked' products indicates that there is promise of further work in this area.

7.5.2.5 *Maximum principle stress/density ratios, σ_1/γ* While specific measurements are highly relevant to shear stress analysis, a very interesting aspect is how this ratio changes with variations in the compacting stress. Coarse powders can generally withstand high static stresses with little change in density and are not usually poorly flowing by virtue of cohesive forces. Those powders which do alter in density significantly with increasing loading invariably exhibit poor flow characteristics, due to the progressively closer proximity of the particles and the increasing number of contact points allowing surface effects to increase in both frequency and magnitude. A load/compaction graph is one means of categorising powders for poor flow characteristics. Although this will not identify potential problems due to other factors, such as mechanical interlocking, thermal and chemical effects, it is a useful measurement related to basic properties of a powder.

7.5.2.6 *Internal angle of friction, δ* This measurement is derived by conducting a series of tests on shear cell equipment. For accurate results and detailed interpretation, as for hopper design, a few days work is required for one important result, and this task is usually the domain of the expert. However, useful results can be secured from more limited shear testing with relatively simple equipment, where check or comparison tests are sufficient, and it is not essential to undertake an extensive series of tests to derive a family of yield loci.

7.5.2.7 *Permeability,* ψ The rate at which air will flow through a packed bed has interest for applications where a bed of powder has to act as a seal against a pressure differential or to contain ambient differences. The measurement is also a guide to how easily the bulk will expand and compact in given circumstances and hence indicate problems of 'void air demand' resistance to initial flow, 'air retarded' flow from hoppers or sustained fluidity when dilate.

7.5.2.8 *Shear strength,* f This value provides a measure of solids 'flow' potential by controlled deformation of the bulk. Flow requires the material to shear; hence, the strength of the bulk material to resist shear is a key feature of interest in bulk storage and handling. Particulate solids differ from normal materials in that their masses have variable strength properties. The main variable is the forces acting within the mass, as these ultimately control the ratio between the forces acting to shear the plane and those acting at ninety degrees to the shearing surface during flow. In sustained shear, the material attains a unique relationship between these values, and for given conditions, the material will attain a specific condition of compaction of the bulk solids as characterised by its density. Such a condition is called the 'critical state' and is represented by the end-point of a yield locus for a bulk material in a particular condition of compaction.

In any consideration of bulk strength, one should consider a possible influence of the void gas. A particulate solid is a two-phase material. Any change in density means there is a volume change, and the voids must either expand or contract. If the material is coarse, the bulk is porous, and air can enter or leave the bulk without much difficulty. Fine powders offer greater resistance to permeation, and, until the void pressure reaches ambient, there is a positive or negative external force applied to the material as a result of the existing pressure differential. Gas trapped in the voids supports part of the external applied load, and the bulk strength is reduced because of the looser dilation of the mass. It is also easier for the material to expand, as no inflow of gas is required to suit the expanding volume.

In extreme cases of compaction such as with roll presses and tableting processes, unless steps are taken to reduce excess air, the resulting compact can break down by the internal pressure of the compacted gas, to flake or explode the compact. By contrast, a densely compacted material will resist expansion by the reduced void pressure exerting an air demand and creating counter-current flow, which creates an effective resisting force to the expansion.

Therefore, unless relevant to the application, shear testing of powders should be conducted at low strain rates to avoid complications arising due to pressure differentials caused by void gas volume changes.

For products that include free surface moisture, the system has three phases. Damp and wet products such as filter cakes and centrifuged materials therefore can be especially difficult to test because changes of density cause disproportionate changes in void condition, as the liquor is incompressible. If sufficient liquor is present to fill the voids at dense compaction, the liquor will then sustain any increase in external load without any change in shear strength, as hydrostatic pressure absorbs any additional externally applied pressure. However, on relaxation of the applied load, the material will retain its fully saturated state and have irreversibly changed from its initial more open textured condition. It is commonly found that a damp friable bulk material is irreversibly converted to a wet amorphous paste by 'working' and that its handling condition is seriously deteriorated.

However, if there is insufficient liquor present to fully saturate the void volume under compaction, the product will gain strength with compaction, and the influence of the significantly reduced unsaturated void volume has the effect of radically reducing the permeability of the mass. Its resistance to expansion on the release of the compacting stresses is then re-enforced by the ambient pressure compared with a near vacuum in the expanding void space.

The behaviour of damp and wet bulk materials is therefore strongly influenced by the proportion of loose surface moisture contained. Each moisture condition effectively creates a different bulk material in different degrees of consolidation that do not allow inferences to be drawn between one value and another. Changes due to drying out or absorption from relative humidity differences also lead to serious behavioural changes according to how the moisture changes are effected. The effect of temperature in these circumstances may also be relevant. Likewise, the effect on shear strength of ambient gasses other than air, and the effect of vacuum or elevated pressures, also has to be taken into account when considering such plant operation.

The starting conditions for flow of dry products under particular conditions of applied stress offer a different strength to its sustained shear strength, particularly if the bulk was formed in a different manner to how it is sheared for flow. Material filled into a hopper, for example, may suffer impact loading and then settle under the pressure of a deep bed of product. When discharge takes place, the movement of material alters the stress situation. Resistance to slip from the walls or the boundary of the flow channel within the bulk supports more of the dead weight of product, while a converging flow channel applies higher transverse stresses. As shown by the yield locus, material that is loosely compacted will shear initially at a low strength but then gain strength with continued deformation up to the critical state value. The impact of this on testing is that if a material is compacted and then sheared with the original compacting load still applied, it will change to a higher density because the addition of shear stress to the original compacting stress causes a larger compound stress to be applied.

However, if the compacting force is removed and a much lower normal load applied, the initial resistance to shear will fall off as the shear plane expands to the yield strength corresponding to the combination of shear and normal loading. The Jenike shear cell technique is initially devoted to finding the relationship between the compacting load and the normal load which allows the bulk to shear without change in volume, as this reflects flowing conditions in that particular state of density. With linear shear cells of the Jenike type, the critical shear strength at densities appropriate to the stored material must be found by trial, using samples prepared under differing loads. This condition is achieved in the annular shear cell in a single run, because the sample is sheared continuously until a steady state is reached.

A different approach to shear testing is adopted by Johanson, who measures the initial failure of a sample prepared at a value of compaction reflecting the storage bin conditions. These 'Indiciser' values have no direct relationship to Jenike values but are used with applied factors to reflect Johanson's experience in the field to secure safe values for bin outlet sizes that guarantee the commencement of flow.

The principles of the Johanson test are essentially based upon a vertical shear test where a sample is prepared under a consolidation load and then sheared in the same direction of the initial load after the load has been removed. Apart from potential use for bin design, the vertical shear test is a simple measure of a compact's strength and has value for reference and comparative values of material strength. A simple version of the vertical shear test is a useful basic tool. The essential feature is to undertake the shear tests on materials that have been prepared in conditions of compaction which reflect the plant conditions of interest.

Some care is required to evaluate the effect of isotropic materials and products that mechanically interact as an assembly. Shear is essentially directional and requires particles to move relative to each other. Overlapping rods, flakes, fibres, entwine or particles that are large in relation to the shear plane require special consideration. Mechanical, physical and chemical circumstances that influence the particle-to-particle interaction forces involve an extremely broad spectrum of surface contact and bulk mechanics. Nevertheless, a basic understanding of simple concepts can be usefully applied to give a general grasp of how bulk material may be expected to flow in given circumstances. Shear measurements are a valuable tool in this assessment.

Phenomenological test devices, such as recently introduced rheometers and the avalanche tester, measure the forces resisting shear in rotating vanes or in a dynamic failure mode. The stress conditions are most uncertain under these circumstances, but the devices can be used to compare the reaction of samples to a given dynamic condition and allow comparative judgements to be made. Experience gained in correlating the measurements may also be useful for process assessment, but the values cannot be related to fundamental properties of bulk materials.

7.6 Bulk characteristics other than flow

7.6.1 *Segregation*

The preferential accumulation of particles during flow causes many industrial problems. The root cause of this phenomenon is that particles that have different physical characteristics tend to respond differently under the action of commonly applied forces. Characterising the tendency for a product to segregate is therefore a compound process because there are many different mechanisms by which particles are induced to take different paths.

A common cause of segregation occurs in the formation of a pile during the loading of a hopper or container. The dynamic surface stream is subjected to preferential percolation of fines on the repose slope, leading to the accumulation of the smaller fractions in the centre of the stored contents and larger components on the periphery of the pile. However, this is not always the case. Dust can become airborne to congregate on the product surface or the container walls. Another form of segregation can occur if the mass has a predominately fine composition as coarse members penetrate the bed on impact from the falling flow stream and are captured, while the fines content is expressed to the periphery. These mechanisms can be replicated by a rotating drum test in which the axis of the test device is inclined to collect the material in the lower portion and the rotational speed adjusted to initially form a dynamic repose condition and then a trajectory stream as the angular velocity approaches the critical speed at which centrifugal force holds the material to the container wall. Tendencies to dust, segregate by dynamic repose and penetration segregation can be quantified by controlled methodology and sampling.

7.6.2 *Flushing*

The ability of fine powders to dilate and entrain excess air to the extent that the particles are prevented from interacting with sufficient friction to inhibit their relative motion gives rise to many solids handling and control problems. In extreme cases, such as the collapse of an arch or rathole, a fluid mass will flush through the outlet of a hopper and cause an uncontrolled spillage of large quantities of the solid as a hazard to operatives or a gross over-feed to subsequent equipment. The phenomenon can prove unwelcome at a small scale, such as filling sacks or big bags. There are capacity problems owing to the low density of the product, which can result in difficulty in achieving the required mass in the available volume. The resulting instability of the container can lead to difficulties with stacking, bag stitching and sealing. Pills and tablets will explode on release of the compacting stresses, and roll presses are unable to capture the product in the nip zone of the rollers.

These characteristics may be unexpected because the physical nature of the material that favours this behaviour causes the bulk to be in a cohesive, poor flow condition when it is fully settled or compacted. This conduct of the material is exacerbated when the product or air is warm or hot. This is because the viscosity of air increases with temperature and the escape of excess air in the material voids between the particles is slowed by the reduced permeability of the interstitial passages through the mass.

This nature of the material can be readily identified in a sample by vigorously shaking a quantity of the material in a closed, transparent vessel. This 'agitated bottle' test will show whether the product settles too quickly to a stable condition that will remain firm to hold an angle of repose when the vessel is tilted. To quantify the propensity of the material to flush, a sample should be placed in a tall fluidising cylinder and air passed through to fluidise the material. A graph of the rate at which the bulk material settles when the air is cut off will effectively characterise the material for this tendency. The bulk density at which the material attains a condition of stability can be established by tilting the container at different stages in the settling process and finding the state at which the surface supports a repose angle.

7.6.3 *Caking*

Products that develop strong particle-to-particle bonds cause many handling and flow difficulties. Rigid, brittle masses are formed when crystal bridges or chemical reactions join particles into masses of such strength that gravitational forces are totally inadequate to deform the bulk. In mild cases, the bulk material may agglomerate to form lumps of indeterminate size and strength that cause flow or blockage problems in flow channels. Adherence to hopper or chute walls, contact parts of equipment and setting in packets, drums, sacks, big bags or containers, gives rise to equipment performance and discharge problems. These effects are such that products susceptible to caking must be identified and steps taken to avoid or accommodate the behaviour as a prerequisite to proceeding with a project design scheme. Recognition of this characterisation feature is therefore of primary importance during any project evaluation involving the storage or movement of a bulk solid.

The effect can be established by way of exposing a sample to ambient variations similar to those that can apply to the product in service. The value of bulk strength may be measured by a modified tensile test device, if relatively weak, or by an unconfined failure test if high values of bulk strength are involved. The strength of the bulk subject to caking can well exceed the forces available to initiate gravity flow. These tests will show whether steps must be taken to eliminate the prospects of caking by environmental control or the use of anti-caking agents, or by equipment design to avoid or counter the problem.

7.7 Summary

Myths and false trails bedevil the history of powder testing. Currently, the technology is well established at the research level but not well structured or well presented at the scale of industrial application. The importance of quantifying powder properties with respect to flow has never been more important. An approach based upon the principles of powder technology allows a good assessment to be made of the degree and form of testing, which may be appropriate in any particular situation. Any form of powder testing which is found to give useful results in an economical manner may be used, but only if it is in line with these fundamental principles will they have universal application, free from anomalies of practice and interpretation.

The subject is far from exhausted. Means now exist for the industrialist to adopt more standard procedures and techniques for testing and for mitigating or avoiding unwanted phenomena. Useful support would be provided by the publication of reference data relating to common and well-defined bulk material operating problems and techniques related to proven applications. Such materials must be of an invariable or well-defined nature and the application details most clearly described.

Emphasis should be directed to achieving quick and inexpensive simple means of testing that yield fundamental values, the interpretation of which is also simple with their limitations well understood, rather than attempting to register 'once only' values for multiple applications. Wall friction is such a 'simple' feature of major importance that facilities should be available for routine measurement in virtually all solids-handling applications. Even this is sensitive to many factors, and great care must prevail in representing the actual methods and material of construction, as well as the range of material conditions that may be experienced. Above all, the need for a basic appreciation of variegates of bulk material properties is paramount to an understanding of the significance of their bulk characteristics to a solids-handling project.

References

Guide to the Specification of Bulk Materials for Storage and Handling Applications (1994) I. Mech. E., London.
Code of Practice for Hopper Design. BSI/British Materials Handling Board, London.
The Design of Chutes and Conveyor Transfer points (1989) MHEA Stockton on Tees.
Glossary of Terms in Powder and Bulk Technology (2004) British Materials Handling Board, London.
User Guide to Segregation (1997) British Materials Handling Board, Cheshire.
Vertical shear tester, wall friction and tensile tester: www.ajax.co.uk

8 A view from industry on the current industrial best practice use of particle property data and what will be needed in the future

TOM TAYLOR

8.1 Particle property characterisation data

Industry views particle characterisation as the quantitative description of the qualities of a single particle or group of particles. This characterisation includes the particle size, shape, density, surface, hardness and solubility. Quantification is important because 'what can be measured can be controlled'. Knowledge of particle characteristics is essential for the successful design and operation of solids-handling operations. For the design of bulk-solids-handling systems, it is essential to know, as an absolute minimum: the particle-size distribution, flow properties, segregation potential, cohesivity, friability, abrasivity and moisture content (IMechE 1994). If the industry is involved with developing new particulate products, then additional particle characteristics such as shape, density, dustiness, solubility and caking potential will be key in product specification. A typical modern particulate product specification is shown in Table 8.1.

Table 8.1 shows the important specified particle characteristics of mean size, size distribution, bulk density, moisture content and solubility in addition to the chemical properties of moisture content, impurities, trace metals, etc. A specification is important because it is central to *legally enforceable* commercial contracts. A safety data sheet is appended to such a specification giving information on hazards, toxicological, microbiological, handling and emergency procedures.

Both suppliers and users of solids-handling equipment use particle characterisation. Particulate product manufacturing industries and associated R&D organisations should possess a good knowledge of particle characterisation technology and methods in order to improve existing and develop novel products. There is a wealth of literature in scientific and technical journals on the methods for determining the characteristics of particles. Descriptions of new methods and modifications of old ones are constantly appearing. Numerous symposia on this subject have been held. Because of this plethora of information, the scientist or engineer faced with the task of determining the characteristics of particles in some material may be confused by the multitude of methods. Without considerable experience in making such determinations, there is a danger of choosing an inappropriate technique and of failing to realise its limitations. This chapter has been written primarily as a guide to the subject for these scientists, engineers and students.

One of the most important specified particle characteristics is the particle-size distribution. The industrial use of particle-size data is therefore covered first in the following sections.

Table 8.1 Typical modern particulate product specification.

Note	Property	Requirement	Test method
1	Appearance	Colourless and odourless powder or granule free from visible contamination	Visual inspection vs. standard
2	Reflectance at 460 nm %	95 minimum	Vs. standard
3	Active content anhydrous basis, %	99 minimum	Chemical method
4	Water content, %	1 max.	Loss on heating
5	Water insolubles 10 g/l at 20C %	0.2 max.	Filtration
6	Impurity X, mg/kg	100 max.	Chemical method
7	Sulphate, mg/kg	200 max.	Chemical method
8	Chloride, mg/kg	50 max.	Chemical method
9	Bulk density, kg/m^3		Weight per unit volume
10	Granulometry	Granular	
	Mean size, μm	400–800; 600 target	Rosin–Rammler or verified method
	Cumulative %		
	>1250 μm		
	>1000 μm	5 maximum	
	>710 μm		
	>500 μm	90 minimum	
	>355 μm		
	>250 μm		
	>150 μm		
11	Trace metals, mg/kg		
	Cr, Ni, Co, Pb, Hg	1 each maximum	Atomic absorption
12	pH	7.5–9	1% solution

8.2 Industrial best practice use of bulk-solids particle-size data

8.2.1 *Sieve data*

Sieving is one of the simplest methods of determining particle-size distributions, and is probably used in industrial laboratories more than any other method. It is a technique generally overlooked when carrying out fundamental research, which is unfortunate, since it can be very accurate. Sieving has been classed along with optical and electron microscopy as a method which classifies particles according to geometric similarity, regardless of density or optical properties.

Sieving methods for determining the size distributions of materials consisting of particles over 50 μm in diameter have great advantages over many methods. Sieves with openings smaller than about 50 μm are seldom used for determining the size of dry powders because they are prone to become blocked ('sieve blinding'). Above this limit, the results are capable of a high precision, although they are somewhat empirical, since they depend on the method and time of shaking.

Sieve analyses generally require little time. A complete sieve analysis often can be made in half an hour or less, in contrast with determination of size distribution using the optical or electron microscopes, which is much more time-consuming. All that is required is the ability to use a simple balance and to follow a brief set of directions. The equipment is relatively inexpensive. The method is therefore very useful in global applications where high-technology equipment and support may not be available. Another advantage of sieve tests is that the samples are physically separated into fractions according to size, and these fractions can be studied individually.

The use of sieves to separate particulate material into fractions of various sizes has occurred throughout the period of recorded history. Consequently, sieve analyses have a good 'track record' in many industries and continue in use even though more accurate or less laborious methods are available. The method is applied very widely over many disparate industries. These days, sieve analyses, quoted as mesh '% retained' or mesh '% passing' are essential to the specification of an enormous range of products ranging from agricultural fertiliser to zinc ores. Plant-side sieve analysis is also widely used for process control in industry, adjustments being made according to planned systems by operators, e.g. Statistical Process Control.

In industry, sieves are available with sieve openings from 5 μm to over 100 000 μm. For most practical particle-size analysis purposes, the range 100–1680 μm is normally used. Below 100 μm, sieves are prone to 'blinding' (blocking) owing to increased particle cohesivity and considerable error results. With the special technique of 'Air jet sieving', it is possible with some powders to measure below 100 μm, commonly 50 μm but occasionally as low as 10 μm. However, this requires specialist equipment and is very time-consuming. Above 1680 μm, particulates are not considered powders for bulk-solids-handling purposes. They may better be described as granules, pellets, prills, flakes, cubes or other shapes for which specialist techniques can be used.

There are two main procedures for sampling powders for sieve tests: obtaining a representative gross sample and then reducing its size to that suitable for sieve testing (usually 50–100 g). Both manual and automated methods are possible, but the sampling method used is crucial to the success of the analysis. It is useless having an accurate and repeatable sieve test method if it is carried out using non-representative samples.

Both manual and automated determinations of particle-size distributions using sieves are possible. Manual methods are still important because the equipment is robust and can be used by trained operators in the manufacturing plant as well as laboratory technicians. Automated methods require more technical support, which tends to inhibit their global application. The International Standard (ISO R565) can be used for a progression of sieve meshes based upon a 100 μm opening size. The maximum mesh opening is 1000 μm, this being considered the upper limit for powders.

Sieve analysis can be automated to provide detailed information about the particle-size distribution of a sample in a clear, concise manner. A dedicated 'sieve test analyser and recorder' is available which incorporates a precision balance linked to a microprocessor and printer (http://www.tyler.com). This device weighs and memorises the net weight of the sieves and receiver before sieving. It then compares these with the gross weight after sieving and makes the necessary calculations to provide a detailed analysis of the particle-size distribution. By doing all the necessary weighing and calculations and then storing the data and results, it saves a great deal of time particularly in volume work (Figure 8.1).

Wet sieving techniques are of value for substances that tend to form aggregates during shaking, or which are already suspended in a liquid. Water or some other liquid is mixed with the material to form a dilute suspension, which is run through one or more sieves. Additional liquid is poured over the residue, as required, to aid the separation.

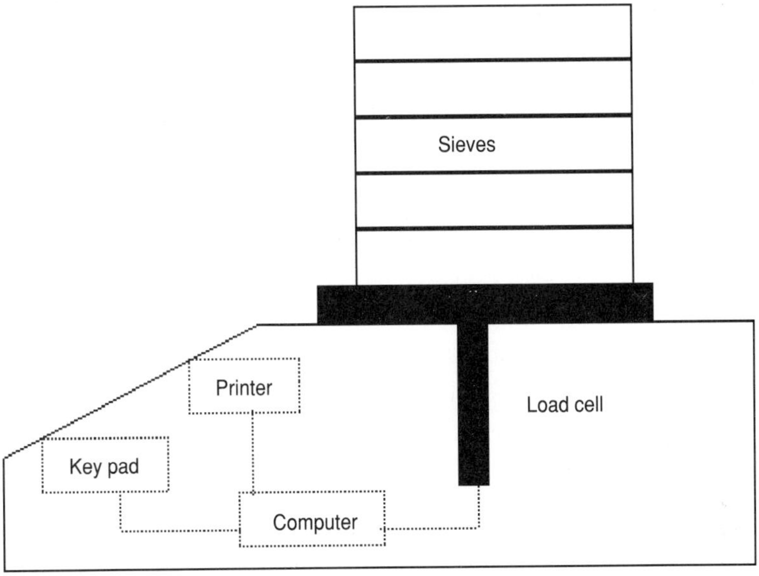

Figure 8.1 Automated sieve test apparatus.

The results of sieve analysis are often reported in tabular form, as shown in Table 8.2. Cumulative percentages, in addition to percentages and weights retained on each screen, may be reported. When large numbers of sieve analyses are being made, standard forms for recording the data are a great convenience. An example is shown in Table 8.2.

The results obtained by sieving can be considered to be classified data in which the class intervals, as usually expressed, are not of the same size, but differ by a constant ratio (in

Table 8.2 Example of a record of a sieve analysis.

Sieve opening (μm)	US sieve mesh number	Weights on or between sieves		Cumulative percentage	
		g	%	On	Passing
Retained on 1680	12	4.8	4.8	4.8	95.2
Retained on 1410	14	5.2	5.2	10.0	90.0
Retained on 1000	18	8.3	8.3	18.3	81.7
Retained on 707	25	11.7	11.7	30.0	70.0
Retained on 500	30	15.0	15.0	45.0	55.0
Retained on 354	45	35.0	35.0	80.0	20.0
Retained on 250	60	10.0	10.0	90.0	10.0
Retained on 177	80	5.1	5.1	95.1	4.9
Retained on 125	120	3.1	3.1	98.2	1.8
Base 0		1.8	1.8	100.0	0.0

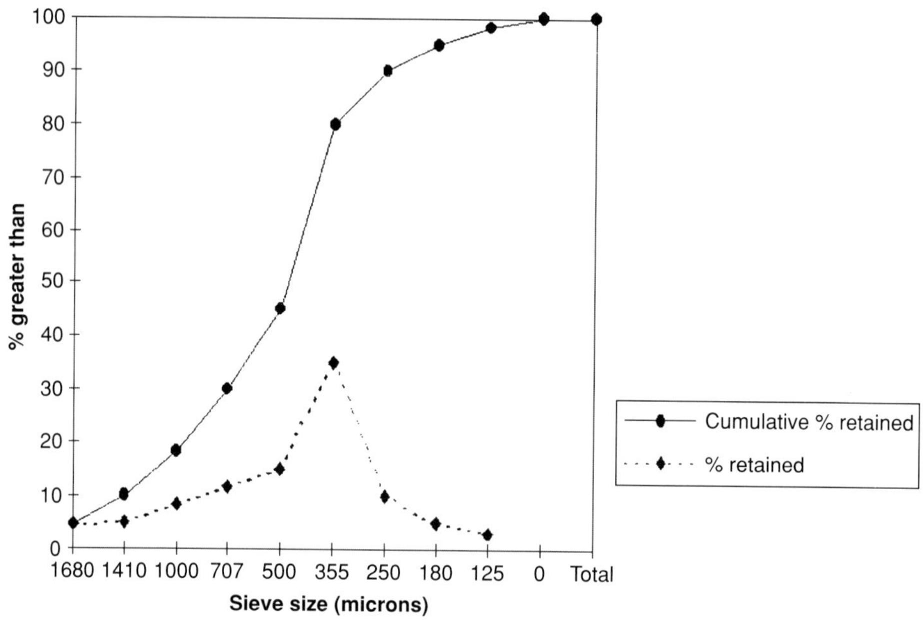

Figure 8.2 Frequency and cumulative particle size distribution.

Table 8.2). However, if the class intervals are expressed as differences in the logarithms of the screen openings, they usually do have the same size. A convenient way to express the information is by way of histograms, which use a logarithmic scale for the abscissa. The results of sieve tests are usually plotted as frequency or cumulative curves, as shown in Figure 8.2.

The ordinates may be linear or probability scales, while the abscissas may be linear or logarithmic scales. Graph paper prepared especially for recording the results of sieve tests is commercially available. The linear graph paper has the ordinate ruled into 100 divisions representing percent, and the abscissa is divided by lines placed so that the distances of any two successive lines from the origin are related by the factor $\sqrt{2}$. Thus, the rulings on the abscissa can be made to represent any set of screens chosen from the British Standard, US Standard or Tyler series. Similar paper is available for the closer sizing ratio $\sqrt[4]{2}$. The semilog paper has the same type of ordinate as the linear paper, but the abscissa is a logarithmic scale, which, in addition to the usual markings, has markings representing screen openings. This scale has the advantage over the linear scale that it compresses the curve for the coarse sizes and extends it for the small sizes.

Whenever a straight line on log probability paper can represent the data, plots of weight distribution can be transposed to plots of surface and number distribution. The mean diameters of these distributions can be calculated from:

$$\log d_g = \log d_{wg} - 6.908 \log^2 s, \tag{8.1}$$

where d_g is the geometric mean diameter, d_{wg} is the geometric mean of the weight distribution, and s is the standard deviation, or

$$\log d_{vs} = \log d_{wg} - 1.151 \log^2 s_g, \tag{8.2}$$

where d_{vs} is the volume surface mean diameter, and s_g is the geometric standard deviation.

The application of the Rosin–Rammler relationship to determining particle-size distributions is also possible:

$$R_w = 100 \exp -(d/d')^n, \tag{8.3}$$

where R_w is the weight of material less than d in diameter, d' and n are constants. d' is termed the 'Rosin–Rammler Mean Diameter', and n is a measure of the width of the distribution. The higher the value of n, the narrower is the width of the distribution. When d' is equal to d, then:

$$R_w = 100 \exp -(d'/d')^n = 100/e = 100/2.718 = 36.8. \tag{8.4}$$

Thus, the 'Rosin–Rammler Mean Diameter' is the particle size at which 36.8% of the particles cumulatively are retained on a sieve deck.

8.2.2 *Electrical sensing zone technique*

Wallace H. Coulter, co-founder, former chairman and president of the Coulter Corporation, is best known as the discoverer of the Coulter Principle, which is the most widely used method for counting and sizing microscopic particles suspended in a fluid. This method is also termed the 'electrical sensing zone' (or 'electrozone sensing') technique for particle-size analysis. This method exploits the change in electrical resistance occurring when a particle in an ionic solution passes through an aperture between two electrodes. The technology revolutionised medical research and diagnostics by eliminating the labour-intensive and often inaccurate method of manually counting blood cells.

The electrozone sensing technique is a versatile particle-size measurement method for a multitude of size range materials. It provides size and volume distributions in terms of number, volume, surface area and mass. More than 6000 documented references have used the Coulter Principle. There are nine approved ASTM methods, and the technology is defined by the International Standard ISO 13319. It has been used as the reference counter for cell biology, as defined in 1987 by the Expert Panel on Cytometry in the International Committee for Standardisation in Haematology. The method has a high resolution for particle counting and size distribution and provides a direct measurement of particle volume. It is capable of counting and sizing particles at concentration levels not detected by other technologies. The colour or refractive index does not affect results.

The electrozone sensing technique is used for biological or industrial applications, in quality control or research. The technique measures particle volume, a direct measurement of a physical property of the particle. The electrozone sensing method provides size and volume distributions in number, volume, surface area and mass, with an overall range of 0.4–1200 μm. A higher resolution is possible giving more detailed size information, resolution being the ability to differentiate between different particle sizes. It can discriminate between two particles close in size better than any other method. The instrument is easy to operate

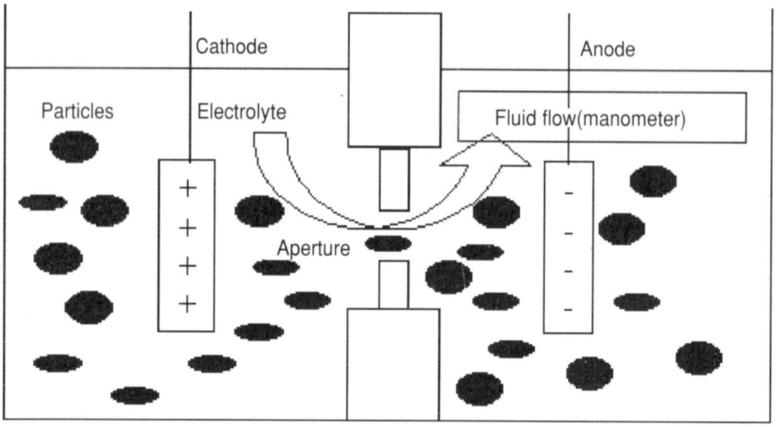

Figure 8.3 Modern electrozone sensing particle size analyser.

manually, but modern equipment is fully controlled through a computer with user-friendly software. A useful feature is that the resolution, range and other parameters can be changed at any time after the analysis without re-analysing the sample (Figure 8.3).

On modern instruments, the resolution can be selected from four to 300 channels at any selected range. It is therefore possible to detect any change in the sample over the length of the analysis, e.g. if the sample dissolves, agglomerates, flocculates, etc. The technique is suitable for both aqueous and organic electrolytes. Variable volumetric metering is possible; typical volumes can be selected from 50 to 2000 μl.

There are few problems involved in sampling suspensions in liquids when the particles are sufficiently small that they remain in reasonably uniform suspension during the period required for taking the sample. In such cases, the liquid is thoroughly mixed and a sample withdrawn by siphoning or pipetting. When the particles settle rapidly or when adequate mixing is practically impossible (for example when sampling muddy stream water), it is essential to obtain samples at various depths. This can be accomplished by means of a special thief or by moving the intake tube of a siphon down through the suspension. Obtaining representative samples of flowing suspensions is sometimes quite difficult. If the flow is horizontal, streamlined or only slightly turbulent, larger particles tend to concentrate near the bottom. This effect is largely eliminated when the flow is vertical. For this reason, whenever possible, suspensions flowing through pipes should be sampled from vertical sections of the pipe.

8.2.3 *Laser diffraction and laser scanning*

The laser diffraction method of particle-size analysis has been used increasingly in industrial and academic applications since its development in the 1970s. This is primarily because of its ease of use and flexibility in handling a wide range of samples using one system, e.g. for the measurement of sprays, emulsions, suspensions, pastes and dry powders. High-speed data acquisition is possible, at the rate of 500 Hz (one measurement every 2 ms). It is therefore used also for on-line measurement of sprays and dry powder inhalers. Transient effects, such as dissolution or agglomeration, can also be monitored. The laser diffraction method offers the maximum measurement versatility. It can measure spray-droplet size as well as wet and dry samples. With a wide range of sample dispersion units, it is the instrument of choice for many R&D departments. It allows robust, repeatable, representative sampling and measurement.

It is used in a wide range of industries and has become industry-standard technology. The equipment can be engineered into a rugged and robust on-line measurement platform. Sampling interfaces for dry and wet systems can be tailored to meet the needs of different processes and will accommodate mass flows from 1 to 500 000 kg/h. The most widely used instrument incorporates an air venturi to aspirate from the process line and disperse samples for measurement. Dynamic samplers are available for high tonnage, abrasive, solid streams. In-line flow cells are used for very dilute processes, and a new wet cell can be applied either in-line or on-line, according to the concentration of the liquid suspension.

A broad measurement range is possible from 0.05 to 3500 μm. Conventional laser light-scattering technology loses its applicability in the sub-micron particle-size region. However, polarisation intensity differential scattering extends the size measurement down to 0.05 μm. Equipment is available for off-line laboratory bench use and industrial on-line sizing. The off-line method is a proven and robust particle measurement technique, pioneered by Malvern Instruments, for more than 20 years and is the standard in most particulate-processing industries (Figure 8.4).

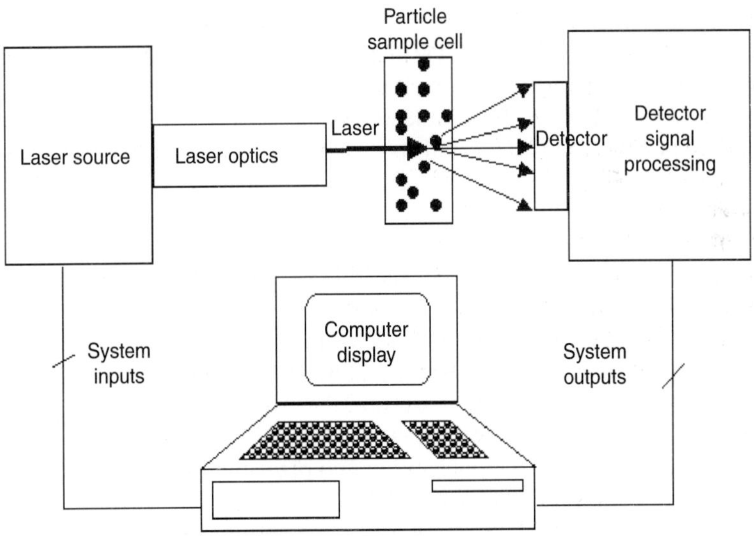

Figure 8.4 Typical bench-scale laser diffraction particle size analyser.

The on-line instrument has many benefits in withstanding the demands of the process environment. The rigorous Mie light scattering theory allows determination of the complete particle-size distribution. No on-site calibration is required, and there are no long data acquisition intervals that can mask process behaviour (Figure 8.5).

For sampling, many instrument suppliers provide a choice of dry powder feeders or large- and small-volume wet sample dispersion units, which allows the system to be configured to meet the application needs. Spray measurement capabilities are also available. In wet

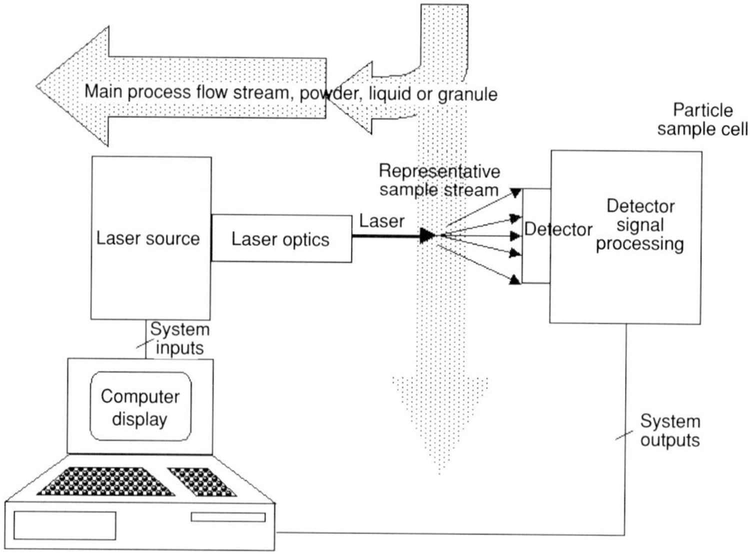

Figure 8.5 On-line laser-scattering particle size analyser.

dispersion, the 'obscuration' is important. This is a measure of the amount of sample added to the dispersant. It is usual to use an obscuration in the region of 10–20%, although this is dependent upon the scattering properties of the particles.

With on-line sizing, a range of instruments (sensors) can be applied to meet the requirements of different industrial processes. This includes a broad range of dry powder applications and processes. The instrument, process interface, analysis software, automation and data reporting can all be customised. Good manufacturing practice (GMP) compliance is possible with intrinsic safety for environments where gas or dust constitutes an explosion hazard. Tri-clamping and the use of 316 stainless steel parts facilitate GMP compliance. Continuous (24/7) on-line particle sizing is possible in bulk powders (including abrasive materials) with mass flows from 1 to 500 000 kg/h with measurement of particles in the range of 0.5–1000 μm. Changeable lenses in the sensor allow the instrument resolution to be tuned to individual process needs. High-quality, real-time data are possible with up to four measurements every second. A range of process interfaces allow sampling directly in-line or on-line using eductor full automation, and integration into plant control systems is possible using appropriate software links for an industrial SCADA package. This should meet industrial-grade electromagnetic compatibility (EMC) requirements for safe, reliable operation.

For high-concentration measurement, a special algorithm is needed to accommodate the multiple scattering that takes place at the high particle concentrations encountered in a process line. Where particle concentrations are low enough, turbidity is linear with concentration. However, this is not the situation at higher concentrations, where particles are so close together that the scattered radiation is re-scattered by other particles. This is not an issue in laboratory instruments because the user has control over the amount of sample measured. However, in a process instrument using continuous sampling techniques, measurement must be accurate, even when higher loadings are present in the process stream. This is especially significant during plant start-up, shutdown and process changes.

Rugged systems are needed from the outset for industrial use. Units must be compact, IP65-certified (dust and waterproof) and vibration-proof. Construction and robust technology ensure that they withstand the stresses of continuous use in the process environment. Since laser-scattering systems are able to take scattering patterns every second and deliver particle-size distribution in real time, a single instrument can make more than a million measurements every year. In practical use, alignment of the laser source and scattered light sensor is important. It may take an extremely long time for alignment if there is dirt on the cell windows or the instrument lenses. This diffuses the laser beam and makes it difficult for the alignment detectors to work properly. Therefore, cleaning the cell windows and instrument lenses will often cure this problem.

Results are often quoted as follows:

(1) $D(v,0.1)$: This is called the 10% cutoff point, as 10% of the distribution is below this size.
(2) $D(v,0.9)$: This is called the 90% cutoff point, as 90% of the distribution is below this size.
(3) $D(v,0.5)$ mean: This is the volume median diameter. Fifty percent of the distribution is above, and 50% is below this size; it divides the distribution exactly in half.
(4) $D(3,2)$: This is the surface area mean diameter or Sauter mean diameter; it is the mean of the diameters of the spheres having the same surface area as our real particles.
(5) $D(4,3)$: This is the volume mean diameter; it is the mean of the diameters of the spheres having the same volume as our real particles.

The distribution obtained from laser diffraction will be different to that obtained from sieving, unless the particles are perfectly spherical. The degree of difference will depend on the shape of the particle, as diffraction is based upon the volume of the equivalent sphere. Laser technology can provide meaningful data by ‘correlating’ the raw laser data to traditional test-sieving weight-distribution results. The information can be presented in size-analysis fractions or shape calculations.

8.2.4 *Optical microscopy*

Methods based on optical and electron microscopy for determining particle sizes and size distributions are often considered to be the most direct and fundamental methods for making such determinations. Exclusive use of such methods is often not justified because they are usually quite tedious and may be of necessity based on a definition of particle or particle sizing that is not appropriate. However, even a casual examination of particles with a microscope is often a useful supplement to other methods of particle-size determination. A good industrial microscope is shown in Figure 8.6.

The size range of greatest usefulness of the optical microscope is about 0.2–100 μm. Size distributions for particulate material larger than 100 μm in diameter are more easily obtained by sieving than by microscope techniques, whereas 0.2 μm is the lower limit of resolution of the microscope using visible light. The lower limit decreases to about 0.1 μm by using ultraviolet illumination and focusing mirrors or quartz lenses so that the ultraviolet light is not completely absorbed. The ultraviolet microscope can be used to obtain average sizes for particles as small as about 0.01 μm. However, both the ultraviolet microscope and the

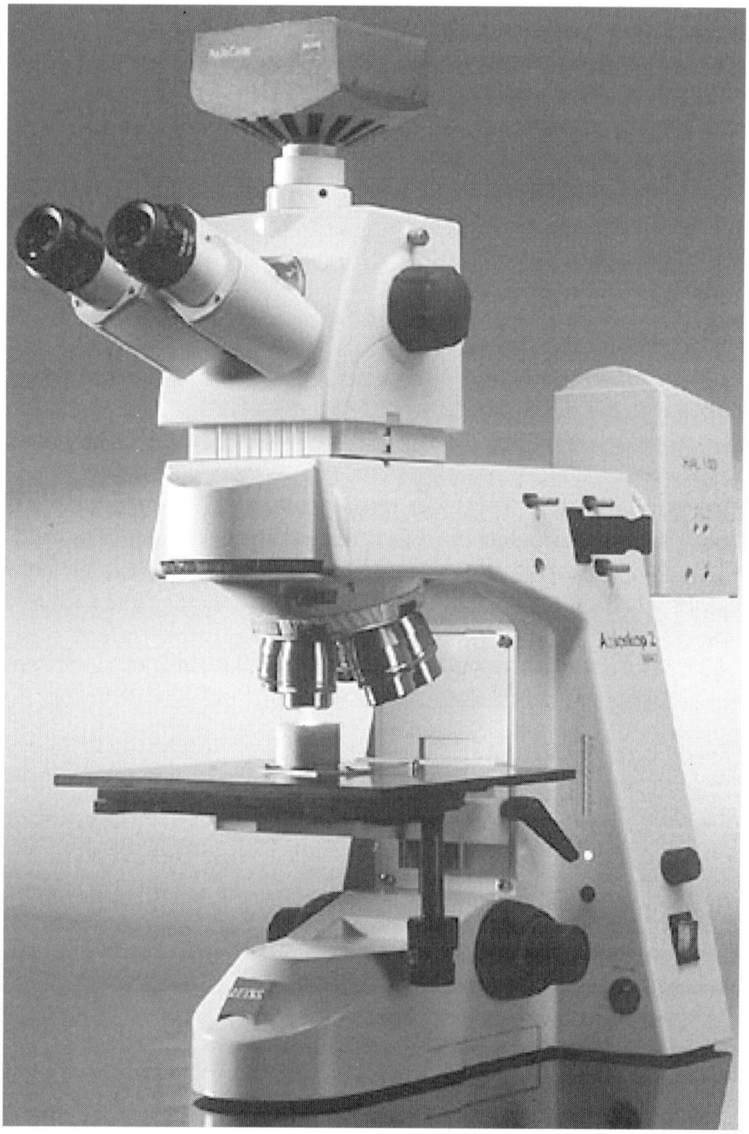

Figure 8.6 Modern optical microscope.

ultramicroscope have generally been superseded by the electron microscope for determining the sizes of particles less than 0.2 μm in diameter. A modern microscope technique, which can be used with some particles, is that of 'confocal microscopy'. Although conventional light and fluorescence microscopy allow the examination of particles and the observation of dynamic processes as they actually occur, certain problems do exist with these techniques. One of the main difficulties faced is out-of-focus blur degrading the image. Defocused information often obscures important structures of interest, particularly in larger particles.

In a conventional microscope set-up, not only is the plane of focus illuminated, but much of the specimen above and below this point is also illuminated at the same time. This results in out-of-focus blur from these areas above and below the plane of interest. This out-of-focus light leads to a reduction in image contrast and a decrease in resolution. The illumination in a confocal microscope system however, is sequential in nature. The specimen is not uniformly illuminated throughout its depth, the light being focused on a spot on one volume element of the specimen at a time. The dimensions of this spot will vary from one system to another and are also dependent on the specific illumination wavelength and the way in which the confocal microscope is set up. It is possible, however, to obtain illumination spots as small as 0.25 mm in diameter and 0.5 mm deep. The design of the confocal system is such that as the beam of illuminating light diverges above and below the plane of focus, volume elements receive less light as one moves away from the focal plane. This results in a reduction in some of the out-of-focus information.

Strictly speaking, spheres are the only particles which have diameters. However, the term diameter is, by convention, also applied to various linear properties of irregularly shaped particles. Such diameters are commonly called statistical diameters because the diameters of a large number of particles are usually measured. The results of optical microscopy particle-size analysis are often quoted as follows:

(1) Martin's diameter: the distance between sides of the particle, measured crosswise of the particle and on a line bisecting the projected area
(2) diameter of the circle whose area is the same as the projected area of the particle.
(3) diameter of a cube of equal volume
(4) diameter of a sphere of equal volume
(5) the shorter of the two dimensions exhibited
(6) the average of the two dimensions exhibited
(7) the average of the three dimensions of the particle
(8) the distance between two tangents to the particle measured crosswise of the field and the perpendicular to the tangents

The tedious nature of making particle-size determinations with a microscope has already been mentioned. However, such determinations can be made with great accuracy. In principle, the accuracy is limited only by the resolving power of the objective. In actual practice, sampling errors are probably the greatest source of uncertainty, except when the particles are close to the limit of resolution of the microscope objective.

More skill is required to make satisfactory particle-size determinations using a microscope than when using any of the other techniques. For routine work, however, many of the steps can be standardised, and technicians can be trained to carry them out. The fundamental equipment required is relatively inexpensive. Most laboratories are equipped with microscopes as standard items. However, some of the equipment, such as special microscopes, cameras and image analysis packages, costs thousands of dollars.

Unfortunately, it is rather difficult to obtain a representative small sample directly for microscope studies. When large amounts of powder in the undispersed state are being studied, the preparation of the slide may involve three sampling processes. The first two of these, collection of the gross sample and reduction of the sample to a convenient size to transport to the laboratory, are discussed elsewhere. The third sampling process involves taking a portion of the reduced gross sample for preparation of the slide. The reduced sample is thoroughly mixed, and a sample weighing a few grains is taken from the sample, preferably in several portions chosen at random. This entire sample is suspended in some dispersant, the nature of which depends on the mounting technique which is used. The resulting slurry is thoroughly mixed, and a small portion is transferred to the microscope with a small spatula or the end of a stirring rod. When large dense particles are included in the sample, the dispersing liquid should be sufficiently viscous to prevent appreciable sedimentation. An alternative sampling procedure is to transfer a very small sample of the powder (a few milligrams) directly to the microscope and disperse it in the liquid on the slide.

8.2.5 *Electron microscopy*

Electron microscopy is analogous to light microscopy in terms of the physical principles but uses a beam of electrons to probe the sample instead of a light beam. It was invented by Knoll and Ruska in 1930 and can these days magnify up to 1 000 000 × and resolve down to 0.3 nm. The electron microscope is used in particle characterisation when it is necessary to know details of fine particle size, shape, surface and internal structure. It is particularly suitable in the range of less than 20 μm down to 0.3 nm and it is used by a wide variety of industries involved in bulk-solids manufacture and handling from agricultural fertiliser to zinc-ore processing. Latest models allow for easy observation and analysis of insulating samples such as ceramics, polymers and semiconductors without coating.

Electron microscopes are available from a number of suppliers (e.g. Hitachi, Zeiss). There are many software and hardware solutions for the collection and distribution of digital information. These range from the digital capture of data from older analogue instrumentation (optical microscopes, analogue-electron microscopes, energy-dispersive analysers, etc.) to the displaying of high-resolution digital micrographs or live video through Internet/Intranet web-based technology.

Practically all size measurements made with an electron microscope involve measuring images on photographic plates, films or prints. Measurements are seldom made directly on the fluorescent screen because of the low intensity of the illumination and the possibility of damage to the specimen or specimen mount as a result of prolonged exposure to bombardment with electrons. The images on the plates or films can be projected for measurement on a screen. The image-analysis techniques used then for measuring the particles and classifying the data and the definitions of particle size are identical to those of optical microscopy.

One of the most common errors in the determination of particle size using an electron microscope is examining too few fields. It is good sampling practice to make measurements of several fields in each of several specimen screens. When the particle-size distribution is very wide, both optical and electron microscopy may be required to generate useful data. This involves fitting together the results obtained with the two instruments.

Determining particle-size distributions using an electron microscope as with the optical microscope is usually very time-consuming. The equipment is expensive, costing tens of thousands of dollars. A high degree of skill is required to prepare the samples, adjust and operate the microscope, and interpret the electron micrographs. However, the electron microscope is usually the most satisfactory method for determining the sizes of particles less than 0.2 μm in diameter. Methods based on the use of the electron microscope can produce results of great accuracy, but sampling errors can also be relatively large. The largest sampling errors are often those produced as a result of the choice of representative fields to be photographed.

8.2.6 *Image analysis*

The need to evaluate the total distribution of particles in a sample, typically under a microscope, can be classically addressed by image analysis. Computers equipped for particle counting and analysis can handle large amounts of data. The range of application is determined by the resolution of the image-gathering equipment, usually optical or electron microscopes. For particles, the range is therefore very wide from several nanometres to several centimetres.

The latest particle-counting software has the flexibility to accept a variety of camera inputs and microscope automation attachments, such as stage, focus and light control. It guides the user from set-up to the final reported data, which is produced in printed and electronic format. Methodologies are included for accuracy and repeatability 'sample to sample'. Stage, focus and light control integration is possible. Reports can be generated and data exported to popular applications with optional data encryption

Systems can be adapted to regulatory requirements, e.g. 21CFR part 11 compliance. Configuration is possible to individual site requirements and particle type. An audit is also available as part of the methodology. The best systems provide the ability to produce repeatable results and reports in a secure and traceable manner. For example, these systems ensure that the microscope is set up correctly and that the lighting levels do not influence the analysis. Automatic separation of touching objects is also possible.

Image capture, analysis and archiving is the key to many fields of science. Reliable, application-specific imaging systems with accessible technical support are essential to efficient, productive laboratory operation. In solids-handling industries, digital image analysis is used in a wide variety of ways. These measurements include:

(1) particle size
  - measurement of any object: pore, impurities, shrink voids, etc.
  - entire image or selected area
  - determination of particle fractions and size distribution

(2) grain size
  - determination of grain sizes, e.g. single-phased, ferritic and dual-phased, ferritic-pearlitic structures according to ASTM E112 or DIN 50601
  - calculation of anisotropy
  - determination of secondary phases in multi-phased structures

(3) coating thickness
  - interactive measurement of single- and multi-layered coating
  - determination of the mean coating thickness and the histogram of the coating thickness distribution

(4) dendrite arm spacing
  - determination of the mean dendrite arm spacing in cast aluminium

(5) cast-iron G
  - examination of globular cast iron material according to standard specifications EN ISO 945 and ASTM A247-67 (1998)

- assessment of graphite particle size
- determination of the proportions of graphite, ferrite and pearlite
- determination of nodularity

(6) cast iron L
- automatic classification of the configuration of flake graphite material according to standard specifications EN 945 and ASTM A247–67 (1998)

(7) phase fraction
- determination of the phase fractions in multi-phased structures

One method of sampling from bulk is the use of a spinning riffler. This technique provides a reliable representative sample for analysis by low-angle laser light scattering. However, for image analysis, a smaller representative sample is required, and a tumbling sampler better achieves this. Slides may be prepared for the microscope by drawing a glass slide through a flowing stream of powder sample. This is carried out until random juxtaposition of the particles is at a minimum (Figure 8.7).

Granule images, received by a charged-coupled device (CCD) camera, are continuously digitised by an A/D converter to yield an image. The image data are statistically treated in

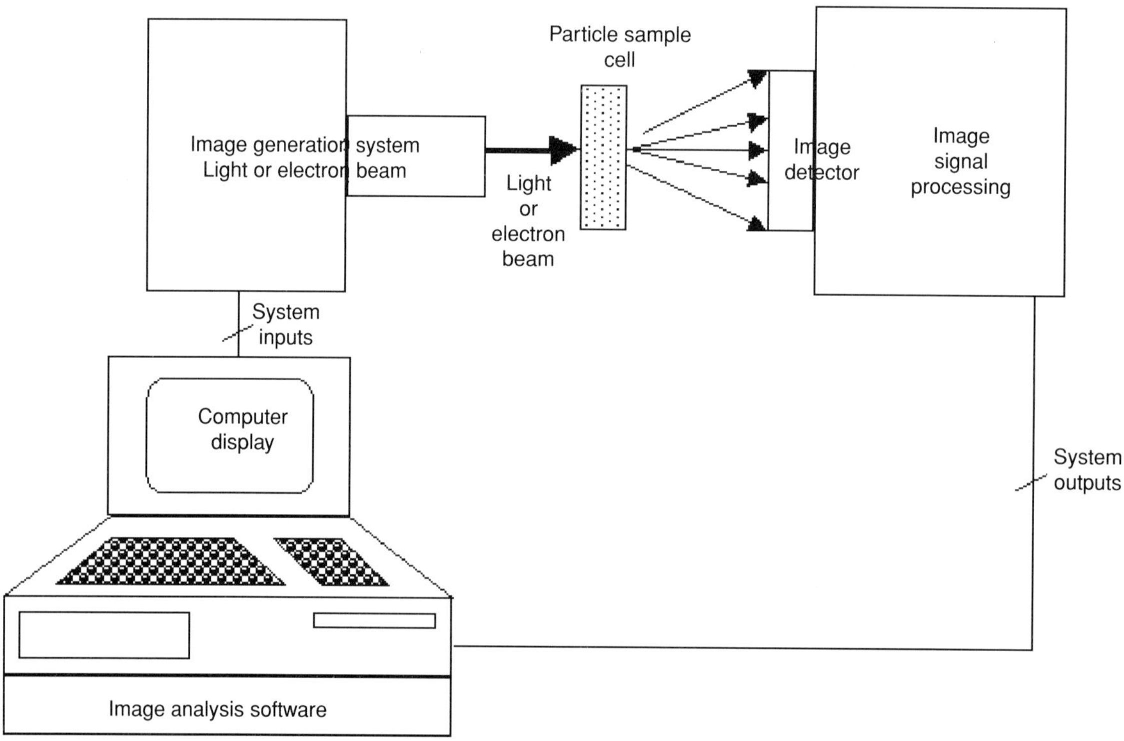

Figure 8.7 Schematic flow of image processing system.

a host computer to yield the particle-size distribution, median diameter and shape factor of granules. An image processing system can process 4000 granules within a few seconds. The three common methods for observing and recording the image are:

(1) Fluorescent screen: When electrons hit the screen, visible light is produced. This feature is useful for aligning the microscope.
(2) Film: Currently, the most commonly used method for recording data, as film is extremely sensitive and can record high-resolution data. Once developed, the film is scanned into a computer for analysis. The resolution is then limited by available computer memory and scanning resolution, which allows a best resolution of about 2 Å per pixel.
(3) CCD chips eliminate the need for film by transferring the image directly into a computer. However, these chips are very expensive, and their resolution is limited, but technology is constantly improving in this area and costs decreasing steadily.

Essentially, a sample is illuminated and presented so that the image analyser can resolve the particles. Appropriate pre-processing is performed on the image to ensure that all the particles are detected. Each of the particles is measured and counted; sometimes particle area and sometimes an appropriately defined length related to particle size. It is usual to exclude particles below a defined size. This is because there will be a lower limit to the size of particles resolved by the optics, and there are generally many very small particles. If a lower size limit is not defined, results will accrue which are different to an arbitrary degree for slightly different systems and very sensitive to slight changes in procedure. The results are usually put into classes and plotted as a histogram as well as being used to calculate a total number of particles and a mean size.

Particle-size analysis by laser diffraction has become a significant analytical tool for sizing up to about 2 mm and has made major inroads into sieve analysis methods over the last 10 years. For sizes of 100 μm and above, however, sieving is still very widely employed. This position is also becoming eroded by the introduction over the last few years of automated image-analysis methods. A comparison of laser light analysis and image analysis (Evans *et al.* 1998) is given in Figure 8.8.

8.2.7 *Particle-size analysis: sedimentation*

Probably the most widely used methods for determining the size distributions of particles in the subsieve range are based on sedimentation. Unlike optical or electron microscopy, which compare particles on the basis of geometric similarity, sedimentation techniques compare particles on the basis of their velocities of settling in some liquid. It is convenient and customary to use spherical particles as a basis for comparison. Thus, the percentage of any sample which is in a given size range using this technique is the percentage which settles at the same velocity as spheres having the same density as the particles in the sample. Actually, particle-size distributions obtained by sedimentation techniques are usually very similar to those obtained by microscopy. Exceptions are the distributions for particles that differ markedly in shape from spheres, particularly those that approach the form of needles

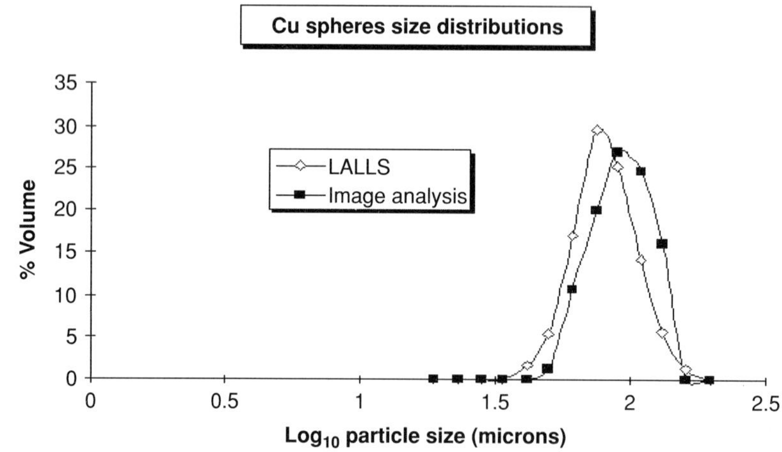

Figure 8.8 Comparison of particle size analysis by laser scattering and image analysis.

or plates. The literature concerning such techniques is extensive. This is because most of these techniques are simple to perform, give accurate results, and require inexpensive equipment. Many of the methods can be applied to materials that are dispersed in a liquid, without having to remove them from the liquid.

Sedimentation techniques are those in which the fluid is relatively motionless, and the particles are allowed to settle through the fluid either as a result of gravity or by centrifugal action. The fluid is usually water or an aqueous solution, but other liquids and even gases, especially air, are sometimes used. The gravitational sedimentation methods cover the range from equivalent spherical diameter 2–50 μm. An X-ray sedimentation instrument covers the range from equivalent spherical diameter 0.1–300 μm.

The X-ray sedimentation instrument is highly accurate and reproducible. The original Micromeritics SediGraph (Micromeritics 2004) was the first commercially available particle-sizing instrument utilising the principle of measuring the gravity-induced settling rates of different size particles in a liquid of known properties. The SediGraph 5100 measures particles with diameters in the range of 0.1–300 μm. The sedimenting sample is scanned by a narrow X-ray beam – less than 0.2% of the total distance scanned – permitting high resolution (Figure 8.9).

The Andreason pipette method is one of the most popular of all methods for determining particle-size distributions. The apparatus consists essentially of a cylindrical vessel, which contains the suspensions, and a pipette, which extends either down through the surface of the suspension or through the wall. Samples of the suspension are removed periodically by means of the pipette, and the concentrations of solids in the samples are determined by evaporating the water and weighing the residue. The preparation of the suspension is one of the most important parts of the procedure for carrying out sedimentation or elutriation tests. Most sedimentation, and some elutriation, techniques use a liquid as the suspending medium. Some of the desirable characteristics of the suspending liquid have already been described. The liquid should be sufficiently viscous that the Reynolds number is less than 1. However, it should not be so viscous that the sedimentation will be unnecessarily

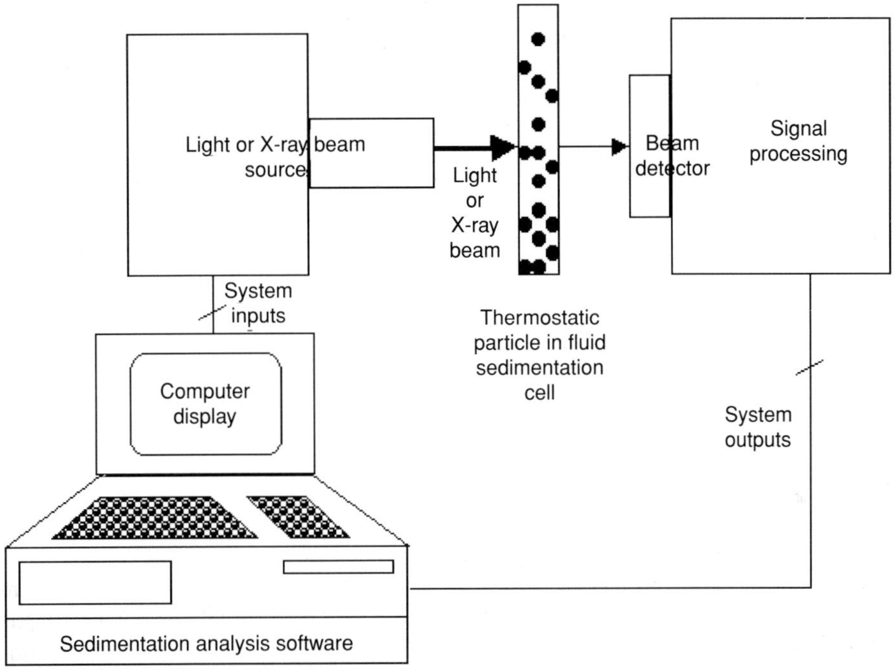

Figure 8.9 Principle of sedimentometer (SediGraph).

prolonged. It is of course essential that the liquid must not react chemically with the particles.

Acceptance of an alternative method to the pipette method depends on convenience, availability of qualified operators, efficiency, reproducibility, and the comparability of results with the pipette method. The SediGraph and the pipette methods of particle-size analysis have been compared using the results from replicate runs of reference sediment samples prepared by the US Geological Survey. This study showed that the SediGraph method fulfilled all criteria adequately and that the SediGraph method could be adopted by the US Geological Survey as one of the standard methods for particle-size analysis of sediment. Tests were designed to provide information on the operational characteristics of the SediGraph method including operator training needed, time required to make a complete analysis and reproducibility of results under varying conditions. The results indicated that the SediGraph method takes 50% less time to complete an analysis than does the pipette method. The training of qualified operators for the SediGraph method is less demanding and is not as critical as for the pipette method. Results of comparable accuracy also were obtained using varying sediment concentrations and different instrument cells. Agreement between the SediGraph and pipette results is good for samples having large percentages of fine material (50–60% less than 1 μm) and is acceptable for samples having relatively uniform distribution of particle size. For samples having a large percentage of coarse material (65–75% coarser than 10 μm), the SediGraph method results indicate a finer particle-size distribution than do the pipette-method results.

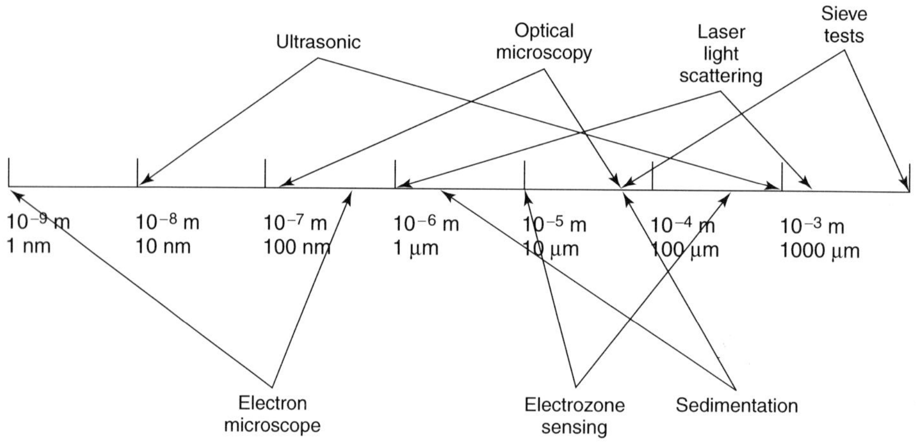

Figure 8.10 Range of sizes covered by particle size analysis instruments.

8.2.8 *Particle-size analysis technique selection*

We have considered above the eight basic size measurement techniques used most commonly today in industry. The methods cover a range of sizes, as shown in Figure 8.10. Not all available methods have been covered, but the most common. Of the eight described, sieve analyses and the sedimentation methods are the most widely used but laser light scattering and electron microscopy are also commonly employed for routine testing. Indeed, a laboratory equipped with a scanning electron microscope, optical microscope and sieve testing equipment is well placed to measure the size distribution and particle structure of most bulk solids handled today. Methods based upon optical and electron microscopy are often used as standard methods for comparison with other techniques, because these are the only methods which involve the examination of single particles. They are also very useful for rapidly obtaining an approximate breakdown of particle-size distribution. However, the quantitative determination of a size distribution can involve the very tedious manual task of counting thousands of particles. The advent of image-analysis software has enormously reduced the labour involved, but this approach involves incorporation of the software model assumptions.

Sieving methods, like optical or electron microscopy, are dependent upon particle geometry. However, they are simple and can be accurate. Generally, they are preferred for particles over 50 μm in size. It is good practice to supplement sieve tests with optical microscopy whenever possible as a 'sanity check'. In particular, the use of sieve tests for cohesive particles is highly suspect owing to the phenomenon of 'sieve blinding', where particles agglomerate and block the sieve apertures during the tests to give false results.

Sedimentation and elutriation methods depend upon the effect of gravitational or centrifugal forces on the motion of particle relative to a fluid, either liquid or gas (air). Particle sizes are calculated from Stokes law or a modification of this, assuming spherical particles. For industries involved in bulk-solids handling, the most common techniques for size analysis are:

(1) sieve testing
(2) laser light scattering
(3) sedimentation

Note that all the mean particle sizes measured by the above techniques are different, and there are different equivalent diameters depending on which 'equivalent' we wish to use. Industry resolves this paradox by using the technique that is most appropriate to the end use of the data, e.g. projected area for a paint pigment, a settling diameter for a gravity settling design, etc. Bulk-solids handling industries tend to develop an 'experience profile' relating the particle-size measurement technique to the user requirements and use this profile for specification, development, design and operation. An experience profile takes years to refine, but often this is accepted as standard by related industrial companies. For example, granular citric acid is specified in size according to sieve testing, and this is accepted as the most relevant standard by suppliers and users of the material based upon years of experience.

8.3 Industrial best practice use of bulk-solids particle shape data

Particle shape is important in solids handling because it can significantly influence the flow, porosity and friability properties of the particulate material. This in turn can significantly change the selection, design and operation of equipment and processes. As examples, the sedimentation rate of needle crystals in suspension is different from cubic crystals, and hence the size, capital and operating costs of settling tanks will differ. The sieving rate of dry needle crystals is also different from cubic crystals, which in turn influences the size, cost and operating parameters of the sieve separation equipment.

Strictly speaking, particle-size measurements can hardly be considered to be quantitatively significant unless some indication of particle shape has also been obtained. No matter how precisely a given definition of particle size is formulated or how carefully measurements are made, particle-size measurements cannot have the same significance for a needle-shaped particle as for a sphere. Heywood (1947) has discussed particle shape and the effect of shape on values of particle size determined by various methods. He has proposed the following method of indicating particle shape. The particle is assumed to be resting on a plane in the position of greatest stability. The breadth (B) is defined as the distance between two parallel lines tangent to the projection of the particle on the plane and oriented so that the distance between them is as small as possible. The length (L) is the distance between parallel lines tangent to the projection and perpendicular to the lines defining the breadth. The thickness (T) is the distance between two planes parallel to the plane of greatest stability and tangent to the surface of the particle. Flakiness is defined as B/T and elongation as L/B.The values of L, B, and T for individual particles can be obtained by means of a microscope equipped with a calibrated fine focusing adjustment.

There are numerous other ways of expressing particle shape. For example, s/d^2 and v/d^3 are called surface and volume shape factors, respectively, where s is the surface area, v the volume, and d the diameter of the particle. An important empirical finding is that in the case of irregularly or non-uniformly shaped particles, the ratio $\bar{v}/\bar{d}^3$ remains essentially the same for various grades of particle size of the same material, where $\bar{v}$ is the mean particle volume and $\bar{d}$ the mean particle size (Rubin 1952).

8.3.1 *Sphericity*

The Sphericity, S, of a particle is the surface area of a sphere of equivalent volume divided by the true surface area of the particles. This is a fraction for every shape other than a sphere:

$$S = \frac{\pi d_e^2}{a_s}, \tag{8.5}$$

where d_e is the equivalent diameter of sphere, and a_s is the true surface area of particles. A corollary of this is that for non-spherical particles, dividing the calculated specific surface (for spheres) by the value of the sphericity (a fractional value) gives the true specific surface.

8.3.2 *Fractal number*

A fractal is a geometric shape, which is self-similar and has a fractional (fractal) dimension. Fractal geometry was a concept introduced by Benoit Mandelbrot (1983) to describe hierarchies with the property that successive levels are geometrically similar. Many shapes of particles in nature manifest the same level of irregularity over a progressively smaller scale as they are magnified. Fractal geometry and chaos theory provide a new perspective to view the world. For centuries, a definitive line has been used as a boundary or basic building block to understand the objects around us. Chaos science uses a different geometry called fractal geometry. Fractal geometry employs a new language used to describe, model and analyse complex forms that are found in nature. A few things that fractals can model are: cells, agglomerates particles, plants, weather, fluid flow, geologic activity, human-body rhythms, animal group behaviour, socio-economic patterns, music.

The concept of fractal dimension has evolved to be a powerful tool. Now, mathematicians are able to measure forms that were previously immeasurable such as complex agglomerates. The fractal dimension indicates the degree of detail in the agglomerate and how much space it occupies between the Euclidean dimensions. Fractal geometry provides a valid method and justification for representing these relatively complicated particle attrition distributions with a simple linear relationship. Kaye (1989) describes the use of a size-frequency plot to describe a distribution of broken particles and later develops the term fragmentation fractal (Kaye 1993). The fragmentation fractal has since been used to describe several mineral breakage systems (Hasson and Rodgers 1995; Brown *et al.* 1996; Brown 1997).

8.4 Industrial best practice use of bulk-solids particle density data

Particle density is of prime importance to industries which market products based upon weight per unit volume or bulk density. This is determined by the particle density and particle packing, as shown in the equation:

$$\mathrm{BD} = \rho_\mathrm{P}(1 - \varepsilon_\mathrm{B}) \tag{8.6}$$

where BD is bulk density (kg/m^3), ρ_P is particle density (kg/m^3), and ε_B is the particle bed porosity (dimensionless).

Control of bulk density and its prime determinant, the particle density, is therefore of huge importance in many solids-handling industries. The failure to control particle and bulk density can result in excessive 'give away' in packing operations. Product losses can be very considerable, in some cases as high as 10% of product sales. Controlling the density of finished goods is often desirable for product performance as well as for economic considerations. For this reason, many manufacturers strive to develop materials that are partially porous in order best to fulfil their function, while others seek an end product of maximum density. For example, the quality of a thermal insulating material is related to low density (maximum closed cell void space), while the quality of a laminated material is indicated by maximised density (minimised unfilled voids). Precise, accurate and highly sensitive density determinations therefore play a key role in these applications.

Since particle density is a key determinant of particle 'structure', along with particle size, shape, colour, odour and solubility, it is a key variable used to design particle in-use properties. Particle density is varied to suit the product properties required and, in particular, the bulk density. If a lighter particulate product is required, then it is necessary to reduce the particle density and vice versa. A lighter, more porous particle can sometimes be easier to dissolve or more prone to float on a liquid depending upon which attribute is required at point of sale. Where density influences the overall quality of manufactured products, density measurements guide or monitor the manufacturing process. For this reason, particle-density measurements are used widely to characterise raw materials as well as to determine the density of finished products.

The range of particle densities varies from the very low, e.g. flour at a particle density of 700 kg/m^3 to phosphate rock at 4700 kg/m^3. Many food and chemical particle densities are in the range of 800–1800 kg/m^3. Both manual and automated methods are used for measurement of particle density.

8.4.1 *Manual method*

A cheap manual method for the measurement of particle densities is to use Archimedes' principle with a specific-gravity bottle, termed a pyknometer, and a liquid in which the particles are insoluble. The volume of the pyknometer to be used in the test is established first by weighing both when empty and full of water. Next, the density of the liquid to be used is confirmed by weighing the pyknometer full of the liquid. Once these calibrations have been completed, the particle density of a test specimen of particles can be determined. Three masses are measured: the mass of an empty pyknometer, the mass of the pyknometer containing a test specimen of the particles and the mass of the pyknometer containing the test specimen and topped up with a liquid. Pyknometers of nominal capacity 50 ml are used, and the test specimen mass should be about 10 g. A vacuum desiccator is used to remove air prior to weighing a pyknometer containing the test specimen and topped up with liquid. Before a pyknometer is to be weighed containing the liquid, it is placed in a water bath at 25.0 ± 0.1 °C for at least 60 min. All weights are determined to an accuracy of 0.01 g.

8.4.2 *Automated method*

Gas pyknometers are used for this purpose. They are specifically designed to measure the true volume of solid materials by employing the same Archimedes' principle of fluid (gas) displacement and the technique of gas expansion. True densities are measured using helium gas, since it will penetrate every surface flaw down to about 1 Å, thereby enabling the measurement of powder volumes with great accuracy. The measurement of density by helium displacement often can often reveal the presence of impurities and occluded pores that cannot be determined by any other method.

Some gas pyknometers offer three sizes of interchangeable sample cells: 135 cc, 20 cc, and 4.5 cc. In addition, there are three different calibrated reference volumes, which provide peak performance for each cell size. The operating sequence pressurises the reference volume first, then expands the gas into the sample cell. Some analysers provide high-speed, high-precision volume and density measurements on both powders and slurries. They are also capable of measuring open- and closed-cell foam materials. Most analyses are complete in 2–3 min. A temperature-control version allows analysis at user-selected temperatures. Glovebox models allow analysis in controlled environments.

8.5 Industrial best practice use of bulk-solids particle-hardness data

Particle hardness is of interest in bulk-solids handling because of:

(1) the effect of the handling system on the particles, e.g. attrition
(2) the effect of the particles on the bulk system, e.g. abrasive wear
(3) the need for particle grinding systems, e.g. for product manufacture

Particle attrition is important for many reasons but especially its effect on end-use properties such as bulk density, flow, dustiness and solubility. For example, catalyst carrier beads used in the oil industry experience breakage in fluidised reactors. Abrasive attrition is characterised *in situ* by collection of elutriated fines as they are generated during the fluidised-bed operation. Particle hardness is particularly important for the suppliers and users of grinding equipment, e.g. flour milling, animal feeds and coal. It is highly relevant to the industries marketing household goods, e.g. detergents and pharmaceuticals. All industries supplying and using solids-handling equipment will want to know the effects of particle attrition and equipment wear.

Abrasive wear is that due to hard particles forced against and moving along a solid surface. Two-body abrasive wear occurs when one surface (usually harder than the second) cuts material away from the second, although this mechanism very often changes to three-body abrasion as the wear debris then acts as an abrasive between the two surfaces. Adhesive wear is also possible due to localised bonding between contacting surfaces. This leads to material transfer between the two surfaces or loss from either surface. Erosion can also be important due to mechanical interaction between a surface and a fluid. This can be a multi-component fluid, or impinging liquid or solid particles on a surface.

In many industries, grinding is used as a comminution process after crushing, through which it is possible to obtain a very small particle size. In most cases, material grinding is due to the action of milling elements positioned inside a machine. This operation can be dry or wet, based on the material and the required reduction degree, and can be carried out through one or more grinding stages. The most commonly used machines are:

(1) rod mills
(2) pebble mills
(3) pendular mills
(4) tower mills
(5) pin mills

Grinding can be realised in open, closed or mixed circuits or through a complex combination of operative stages. Closed circuits are realised by the insertion of classification equipment whose task is to separate the particles in the ground product according to size and to recycle those particles not yet reduced to the desired size back through the mill.

Particle hardness is of relevance in a number of situations. First, the knowledge is needed when designing particles to withstand handling or in the development of novel products. An example might be when developing a new selective catalyst for fluid-bed reactors. In this situation, it is helpful to rank the particulates, among other parameters, according to hardness. Second, particles can be used to change the wear properties of materials. For example, thermally spray-coated Cr steel shows equal or superior wear resistance to uncoated steel. It should be understood that the real area of contact between two solid surfaces compared with the apparent area of contact is invariably very small, being limited to irregular points of contact between surface asperities. The load applied to the surfaces is transferred through these points of contact, and the localised forces can be extremely high. The intrinsic surface properties of a material such as hardness, strength, ductility, work hardening, etc. are very important factors for wear resistance, but other factors like surface finish, lubrication, load, speed, corrosion, temperature and properties of the opposing surface, etc. are equally important.

Investigations of grinding in stirred-media mills reveal that both fracture and attrition play significant roles in the breakage process. The fracture mechanism generally leads to first-order breakage, and its variation with operating variables such as media size and density follows a similar pattern to that seen in tumbling ball mills. The attrition mechanism is found to be important in the breakdown of relatively large particles and has been investigated by monitoring weight loss in individual coarse particles. Evaluation of the derived attrition rate laws indicates that the specific rates decrease with the extent of attrition. Rotor impact mills are utilised for the comminution of semi-hard materials, i.e. materials with a Mohs hardness smaller than 4. The fineness of the grinding products depends to a great extent on impact velocity. However, other parameters, such as the geometry of the grinding zone, have to be considered as well. According to Rumpf, comminution of particles can be caused by four types of stress: stress applied between two solid bodies, at a single solid body, by a surrounding medium and non-mechanically. The second type, i.e. comminution at one solid body, corresponds to impact comminution. This is used for semi-hard and brittle materials, without abrasive contamination such as small amounts of quartz particles. Usual impact grinding machines are universal mills, pin mills or, especially for product sizes smaller than 10 μm, mills with integrated air classifiers and fluid-energy mills. The particle impact velocities for the first group are in the order of 40–120 m/s, whereas, in fluid-energy mills, 300–600 m/s can be achieved. The product size distribution depends to a high degree on the impact velocity, i.e. the relative velocity between the impact partners. Increasing velocity yields higher amounts of fine particles and causes the fracture of smaller particles. In rotor impact mills, therefore, high circumferential velocities have to be used.

Grindability is particularly important if a coal is to be burnt in the pulverised state. In this case, significant work must be done to reduce the coal down to particles of sufficient size for combustion. The Hardgrove grindability index is calculated by applying a standard amount of work on a sample of coal and determining the increase in surface area. The value, G, is based on the fraction of coal initially sieved with a size 16 mesh, passing through mesh size 30 after a standard mill. The value ranges between 20 and 100 for most coals. The higher the value of G, the easier it is to grind the coal. If the grindability index (work index) of a material is known, software is available for the sizing of the mill required.

Young's modulus, hardness and fracture toughness of larger particles can be measured using a nanocrusher manufactured by Micromaterials Ltd. These properties reflect the particle resistance to elastic deformation, plastic deformation and crack propagation, respectively. Young's modulus can be measured by the method, based on Hertz analysis of elastic contacts, previously used by Ouwerkerk (1991). The dependence of the applied load, P, on the displacement, s, is given by:

$$P = 0.47R^{0.5}Es^{3/2}, \tag{8.7}$$

where R and E are the particle radius and Young's modulus, respectively. Each particle tested is loaded and unloaded twice, e.g. the range 2.0–3.5 N. The second unloading is used to calculate E from the slope of the load–displacement curve in the above equation.

Particle hardness is measured by quasi-static Vickers indentation using an Instron machine. This method has previously been used to determine the Vickers hardness of spherical porous silica particles (Ghadiri *et al.* 1995).

$$H = 18\,453\frac{P}{l_{\mathrm{d}}^2}. \tag{8.8}$$

Hardness, H, is calculated from the load, P, and diagonal length, l_{d}, of the impression area produced by the indenter. Fracture toughness is measured by quasi-static indentation using an Instron machine and an indentation fracture method. Most brittle materials form radial cracks when indented with a sharp Vickers indenter. The length of the surface traces of these cracks, C, varies with the load, P, and the fracture toughness, K_{c}, can be derived using:

$$K_{\mathrm{c}} = Xv(E/H)^{1/2}\frac{P}{C^{3/2}}, \tag{8.9}$$

where Xv is a calibration constant.

For surface hardness testing, e.g. pipes and surfaces, there are a number of options. The Brinell hardness test can be used where a tungsten sphere is forced into the material for 15 s. The ball is 10 mm in diameter, and the force is 3000 kg. In the Vickers hardness test, the indentation formed is usually so small that a microscope has to be used to measure it. Therefore, a square diamond is forced into a material with a force of 10 kg. The Rockwell hardness test is less accurate but is useful for rapid checks. The Rockwell method measures the permanent depth of the indentation produced by force acting on an indenter. First, a preliminary test force (A), also called pre-load or minor load, is applied to a sample using a diamond indenter. This is the zero or reference position that breaks through the surface to reduce the effects of surface finish. Then, an additional test force or major load (B) is applied to reach the total required test force. This force is held for a predetermined amount of time to allow for elastic recovery of the metal (C). The additional test force is then released, and the final position (D) is measured against the preliminary position and converted to a hardness number. The Brinell hardness test results and Vickers hardness test results are listed in Tables 8.3 and 8.4, respectively.

Microindentation hardness testing, also known as microhardness testing, is a common test method for determining the hardness of a variety of materials. The microindentation test is a

Table 8.3 Brinell hardness test results.

Material	Brinell hardness
Brass	60–100
Steel	120–140
Cast iron	150–240
Stainless steel	170

Table 8.4 Vickers hardness test results.

Material	Vickers hardness
Talc	32
Silver	60
Copper	163
Quartz	820
Diamond	7000

reliable, proven means of determining the hardness of small precision parts, thin material or wire, coatings, and performing case depth determinations. Wilson® offers microhardness testers built around precision force sensors and electromechanical drive systems to produce repeatable, error-free and accurate test results. The MicroRockwell hardness tester is a hybrid instrument that combines typical test loads and a quality depth measurement system to perform high-speed direct-reading microhardness tests.

The surface hardness test methods, e.g. the Rockwell Tester, can be calibrated as defined in ASTM E-18. This is the most commonly used hardness tester operation method, since it is generally easier to perform and more accurate than other types of hardness testing. Rockwell hardness testers can be used on all metals except in conditions where the test metal structure or surface conditions would introduce too much variation, where the indentations would be too large for the application or where the sample size or shape prohibits its use. Particle hardness testing equipment cannot be calibrated because of the lack of absolute particle standards. However, equipment can be verified by reference to commercially available standards, e.g. glass or quartz. These methods are essentially limited to larger size particles and agglomerates. The hardness of small particles must be determined by crushing, shear or impact tests.

Conversion comparison between the various hardness test methods is approximate and becomes even less reliable for materials other than low alloy steels or where the properties have been dramatically altered by, say, cold working or as a thermal spray coating. This can be further complicated by whether the hardness test is measuring microhardness or macrohardness (particle hardness or bulk hardness) and whether the material is homogeneous or heterogeneous in structure. Defining what is micro or macro is a grey area, but generally the larger the load and indentation, the better the result for macrohardness and bulk property. Wear resistance (abrasive) has a relationship with hardness. Again, the correlation can be good between some very similar materials like carbon steels, but not necessarily when comparing different materials.

8.6 Future needs for particle characterisation in bulk-solids handling

The bulk-solids-handling industry is rapidly changing in response to global manufacturing and trade needs. This in turn requires new and improved equipment, methods and research into the fundamentals of handling and storage. The changing demands and their implications for particle characterisation are grouped below.

8.6.1 *Environment*

As we recognise the fragile nature of the world we live in, companies are becoming more responsible about protecting our environment for future generations. In solids handling, improved dust emission, spillage and waste will require research into:

(1) predicting dustiness from measurements of the particle-size distribution,
(2) modelling of elutriation effects and subsequent segregated deposition of fine particulates,
(3) predicting particle attrition from measurements of particle strength,
(4) software to simulate handling regimes using the particle characteristics identified above,
(5) designing equipment that is breaking new ground and setting the standards for the future by meeting or exceeding the current environmental requirements.

8.6.2 *Storage*

Storage fulfils valuable functions: it provides a break in the supply and transportation chain, and/or the transit and consumption chain and is essential to participants operating in specific markets in a number of ways. 'Caking' or 'lumping' of materials in storage can be a serious problem in terms of not only product-quality issues but also operation of equipment. Many types of product can be prone to 'caking' if handled or stored inappropriately, and as such it is an ongoing problem that costs the industry substantial sums in terms of downtime and reject material.

Novel forms of storage are being increasingly used, for example 25 years ago, only a few concrete domes were being used for bulk storage. Now, they often seem to be the storage structure of choice, especially for applications of large capacity. Domes have been constructed for storing bulk cement, clinker, fertiliser, flyash, gypsum, petroleum coke, and grains.

A rapid assessment of solid material-handling properties is required for:

(1) testing handleability of solids,
(2) pre-test of coal hopper loads,
(3) pre-test of grain silo loads,
(4) aid to hopper/silo design,
(5) measurement of uniaxial unconfined compression strength,
(6) modelling effects of consolidation pressure,
(7) assessing reliability of solid material flow characteristics.

While it is not currently practical to assess these values from a knowledge of the primary particle properties, owing to their many interacting characteristics, one goal is to be able to determine potential bulk behaviour from measurements derived from individual particles.

In tackling future storage needs, research in particle characterisation is needed in:

(1) prediction of caking from measurements of particle properties,
(2) prediction of the time dependence of caking,
(3) software to simulate caking behaviour in bulk arrays of any size and shape,
(4) lower cost and more flexible storage equipment with soundly based scale-up rules.

8.6.3 *Conveying*

Conveying research needs include:

(1) analysis of erosion in pneumatic conveying pipelines,
(2) modelling of flows in the vicinity of local exhaust ventilation,
(3) segregation of free-flowing particulate materials,
(4) pressure loss in pneumatic conveying pipelines,
(5) particle degradation in pneumatic conveying systems,
(6) design of silos and hoppers for reliable gravity flow,
(7) analysis and control of pulverised fuel in pneumatic conveying pipelines,
(8) development of a method for predicting wear in hoppers and chutes,
(9) the influence of temperature and work hardening on the solid particle erosion of steels,
(10) predictive model for determining the life of pneumatic conveyor bends subject to erosive wear options for a sustainable energy future.

8.6.4 *Energy*

Bulk-solids handling is widely used in energy generation, for example in the mining, storage and conveying of coal. The future needs for solids handling in this area are likely to increase and diversify. Four broad avenues are potentially open to us to meet the world's future energy needs in a sustained and affordable manner:

(1) Renewable energies (in particular, hydro, wind, solar, wave, geothermal, modern biomass and hydrogen from non-fossil fuel sources) generally have low or no emissions and some are potentially well suited for meeting rural energy needs.
(2) Conservation and energy efficiency represent the substantial potential for the same task to be achieved with either less energy or to produce the needed energy with less fuel.
(3) Cleaner fossil fuels systems enable us to use fossil fuels with less environmental impact. Fossil fuels are cheap and convenient to use, and their reserves will last for a long time to come. If we can use them cleanly, they can play an immensely valuable role.
(4) Nuclear energy is one of the few options currently available because ample uranium resources worldwide support electricity supply without greenhouse gases.

Particle characterisation research will be required for any of the above thrusts, for example:

(1) handling solid and suspended biomass particulates,
(2) increased energy efficiency in particle grinding,
(3) improved dust particle separation systems,
(4) lower cost and safer handling of radioactive particulate fuels.

8.6.5 *Long-term needs*

The following research issues need to be tackled.

(1) Is it feasible to develop a generic unifying software framework for modelling processes involving particulates?
(2) Can we couple continuum methods (such as computational fluid dynamics) with discrete methods (such as discrete element method)?
(3) How do we include the underlying particle characteristic measurements?
(4) What capabilities might be offered by molecular modelling, for example: prediction of properties and design of solids with particular characteristics?
(5) The lower cost and generic measurement, monitoring and control of processes involving particulate materials.

References

Brown, G.J. (1997) A fractal description of the progeny of double impact single particle breakage. *Miner. Eng.*, **10**, 229–35.

Brown, G.J., Miles, N.J. & Jones, T.F. (1996) A fractal description of the progeny of single impact single particle breakage. *Miner. Eng.*, **9**, 715–26.

Evans, A., Ansell, R.O., Jones, M.G. & Knight, E.A. (1998) The use of low angle laser light scattering (lalls) and image analysis techniques for the dry characterisation of particulate solids, Paper 43. In: *Proceedings of World Congress in Particle Technology*. Brighton, UK, 6–9 July 1998.

Ghadiri, M., Papadopoulos, D.G., Bassam, F. & Ning, Z. (1995) *Attrition of Particulate Solids*. IFPRI Final Report, FRR 16-06, University of Surrey, UK.

Hassan, M.K. & Rodgers, G.J. (1995) Models of fragmentation and stochastic fractals. *Phys. Rev. Lett. A*, **208**, 95–98.

Heywood, H. (1947) *Symposium on Particle Size Analysis*. IChemE, London, 4 February.

http://www.tyler.com

http://www.micromeritics co.uk

IMechE (1994) *Guide to the Specification of Bulk Solids For Storage Applications*.

Kaye, B.H. (1989) Ch. 9, *A Random Walk Through Fractal Dimensions*. VCH Verlagsgesellschaft, Weinheim.

Kaye, B.H. (1993) Fractal dimensions in data space; new descriptors for fineparticle systems. *Part. Part. Syst. Charact.*, **10**, 191–200.

Mandelbrot, B.B. (1983) *The Fractal Geometry of Nature*. W.H. Freeman & Co., San Francisco.

Ouwerkerk, C.E.D. (1991) A micro-mechanical connection between the single-particle strength and the bulk strength of random packing of spherical particles. *Powder Technol.*, **65**, 125–38.

Rubin, S. (1952) *J. Atmos. Terr. Phys.*, **2**, 130.

Index